NANO MATERIALS

NANO MATERIALS

SECOND EDITION

A K BANDYOPADHYAY
Professor
Government College of Engineering and Ceramic Technology
Kolkata

Publishing Globally

NEW AGE INTERNATIONAL (P) LIMITED, PUBLISHERS
LONDON • NEW DELHI • NAIROBI
Bangalore • Chennai • Cochin • Guwahati • Hyderabad • Kolkata • Lucknow • Mumbai
Visit us at www.newagepublishers.com

Published by New Age International (P) Ltd., Publishers
First Edition: 2007
Second Edition: 2010
Reprint: 2020

GLOBAL OFFICES

- **New Delhi** **NEW AGE INTERNATIONAL (P) LIMITED, PUBLISHERS**
7/30 A, Daryaganj, New Delhi-110002, (INDIA)
Tel.: (011) 23253771, 23253472, **Telefax:** 23267437, 43551305
E-mail: contactus@newagepublishers.com • Visit us at www.newagepublishers.com
- **London** **NEW AGE INTERNATIONAL (UK) LTD.**
27 Old Gloucester Street, London, WC1N 3AX, UK
E-mail: info@newacademicscience.co.uk • Visit us at www.newacademicscience.co.uk
- **Nairobi** **NEW AGE GOLDEN (EAST AFRICA) LTD.**
Ground Floor, Westlands Arcade, Chiromo Road (Next to Naivas Supermarket)
Westlands, Nairobi, KENYA, **Tel.:** 00-254-713848772, 00-254-725700286
E-mail: kenya@newagepublishers.com

BRANCHES

- **Bangalore** 37/10, 8th Cross (Near Hanuman Temple), Azad Nagar, Chamarajpet, Bangalore- 560 018
Tel.: (080) 26756823, **Telefax:** 26756820, **E-mail: bangalore@newagepublishers.com**
- **Chennai** 26, Damodaran Street, T. Nagar, Chennai-600 017, **Tel.:** (044) 24353401, **Telefax:** 24351463
E-mail: chennai@newagepublishers.com
- **Cochin** CC-39/1016, Carrier Station Road, Ernakulam South, Cochin-682 016, **Tel.:** (0484) 2377303, **Telefax:** 405130
E-mail: cochin@newagepublishers.com
- **Guwahati** Hemsen Complex, Mohd. Shah Road, Paltan Bazar, Near Starline Hotel, Guwahati-781 008,
Tel.: (0361) 2513881, **Telefax:** 2543669, **E-mail: guwahati@newagepublishers.com**
- **Hyderabad** 105, 1st Floor, Madhiray Kaveri Tower, 3-2-19, Azam Jahi Road, Near Kumar Theater, Nimboliadda Kachiguda, Hyderabad-500 027, **Tel.:** (040) 24652456, **Telefax:** 24652457
E-mail: hyderabad@newagepublishers.com
- **Kolkata** RDB Chambers (Formerly Lotus Cinema) 106A, 1st Floor, S N Banerjee Road, Kolkata-700 014
Tel.: (033) 22273773, **Telefax:** 22275247, **E-mail: kolkata@newagepublishers.com**
- **Lucknow** 16-A, Jopling Road, Lucknow-226 001, **Tel.:** (0522) 2209578, 4045297, **Telefax:** 2204098
E-mail: lucknow@newagepublishers.com
- **Mumbai** 142C, Victor House, Ground Floor, N.M. Joshi Marg, Lower Parel, Mumbai-400 013
Tel.: (022) 24927869, **Telefax:** 24915415, **E-mail: mumbai@newagepublishers.com**
- **New Delhi** 22, Golden House, Daryaganj, New Delhi-110 002, **Tel.:** (011) 23262368, 23262370, **Telefax:** 43551305
E-mail: sales@newagepublishers.com

ISBN: 978-81-224-2781-3
C-19-05-12103

Printed in India at Ajit Printing Press, Delhi.
Typeset at In-house, Delhi.

NEW AGE INTERNATIONAL (P) LIMITED, PUBLISHERS
7/30 A, Daryaganj, New Delhi-110002
Visit us at **www.newagepublishers.com**
(CIN: U74899DL1966PTC004618)

DEDICATIONS

I would like to thank my wife Soma, son Anik and my sister Sujata for giving me a constant support during the preparation of this book

Preface to the Second Edition

The publication of "Nano Materials" a couple of years ago created a lot of interest among a wide range of readers starting from undergraduate to post-graduate students, academicians in colleges and universities as well as many research scientists both in India and abroad.

As it was one of the first books on the subject in India, the importance of this book was also discussed with a variety of concerned academicians both inside and outside this field. As it is an emerging field, it is but natural that we either add a few new topics or we incorporate some new sub-chapters within the concerned chapters. However, a majority of those academicians suggested to keep the book without making any changes, since this might be documented as an *'early development'* after many years in this particular field that is ever expanding in its coverage and contents.

With due apology to the readers, therefore, the previous edition has been kept unchanged except some mistakes and/or omissions, as pointed out by our kind readers. First of all, 20 references have been added in the last chapter (Chapter-8) that was inadvertently omitted. A whole lot of mistakes in the formulations and equations were duly corrected. Finally, a 'subject index' was also created to help the readers identify a particular topic of interest within the book that was not included in the first edition.

It is the earnest hope of the author that many new readers in the field as well as the old ones will still find this book interesting and useful for their purpose of teaching and research work in future. It is also expected that the first chapter on basics of quantum mechanics with an inclination towards 'application' will make the undergraduate students interested in different disciplines to arouse confidence in the subject of *Nano Technology.*

Author

Preface

About 20–25 years ago, when we used to work on materials with small particles even in the range 4–5 nm, particularly in a magnetic or electronic material, we were not aware that we were actually dealing with nano materials. These materials showed very interesting magnetic or electronic properties, which are the main properties of great concern in the field of modern nano materials or nano composites.

Recently, during the last ten years or so, there has been a surge of scientific activities on the nano materials, or even on commercial products in the market place, that are called nano products. Any material containing particles with size ranging from 1 to 100 nm is called nano material, and in this particle size range, these materials show peculiar properties, which cannot be adequately explained with our present-day knowledge. So, the surge on the research activities and the consequent enthusiasm are on the rise by the day.

In the world of materials, like ceramics, glasses, polymers and metals, there has been a considerable activity in finding and devising newer materials. All these new materials in question have extraordinary properties for the specific applications, and most of these materials have been fabricated by newer techniques of preparation. Moreover, they have been mostly characterized by some novel techniques in order to have an edge on the interpretation of the experimental data to be able to elucidate the observed interesting properties.

For example, for oxide and non-oxide ceramics and many metal-composites in powder metallurgy, the sintering of the material is of utmost importance in making hi-tech materials for high performance applications, *e.g.* in space, aeronautics and in automobiles as ceramic engine parts. The sintering of these materials to a high density, almost near to the theoretical density, has been possible by using the 'preparation techniques', which allow the creation of nano sized particles. Our knowledge on solid state physics and chemistry tells us that these are the materials, or rather the 'preparation techniques' to make them, that are fundamentally important to achieve our goal of creating high strength and high performance materials of today's necessity.

Some of these techniques of preparation and characterization of nano materials are elucidated in this book. The subject of 'nano' is quite a nascent field and consequently the literature on this emerging subject is not so extensive. Hence, such an attempt, even at the cost of restricting ourselves to a fewer techniques of materials preparation and characterization, is worth in the context of dissemination of knowledge, since this knowledge could also be useful for other nano materials for many other applications.

This book is concerned with the technique of attrition milling for the preparation of nano particles like two important ceramic materials for hi-tech applications, *e.g.* silicon carbide and alumina. Some more techniques, particularly the recent interest on sol-gel method, will also be elaborated in the case of zirconia. This is described in the chapters 2 and 3, after a quite exhaustive discussion on the relevant theoretical aspects in chapter 1. Since one of our goals is to make high strength materials, the chapter 4 is devoted on the mechanical properties of nano materials together with an adequate dose of fracture mechanics, which is important to understand the behaviour of fracture of brittle materials or 'materials failure' during their use.

The small sized nano particles of magnetite embedded in a glass-ceramic matrix showing interesting magnetic properties would also be highlighted. This is done together with some novel techniques, like Mössbauer Spectroscopy for super-paramagnetic behaviour of nano-sized magnetite and Small Angle Neutron Scattering (SANS) for the determination of nucleation and crystallization behaviour of such nano particles in the chapter 5. The electronic and optical properties of nano particles, which are created within a glassy matrix, would also be elaborated in the chapters 6 and 7, with some mention of the latest developments in these interesting fields of research.

The recent subject like nano-optics, nano-magnetics and nano-electronics, and some such newer materials in the horizon, are also briefly included in this book in the chapter - 8 in order to highlight many important issues involved in the preparation and application of these useful materials.

There is a slight inclination to the theoretical front for most of the subjects, including the mechanical part, discussed in this book. This cannot be avoided by considering the immensity of the problem. The whole attempt in this book is devoted to the interests of the materials scientists and technologists working in diverse fields of nano materials. If it raises some form of interest and encouragement to the newer brand of engineers and scientists, then the purpose of the book will be well served.

A. K. Bandyopadhyay

ACKNOWLEDGEMENTS

First of all, I would like to thank Professor D. Chakravorty, Ex-Director, Centre of Materials Science, I.I.T. Kanpur, and Ex-Director of Indian Association for the Cultivation of Science, Kolkata, for introducing me to the field of nano materials many years ago.

I would like to thank Professor J. Zarzycki, Ex-Director of Saint Gobain Research, Paris, and Ex-Director, CNRS Glass Lab at Montpellier (France) for giving me a lot of insights on sol-gel materials for preparing nano particles. Of course, I would like to thank Professor J. Phalippou of CNRS Glass Lab at Montpellier, for his help to make me understand the subject of sol-gel processing.

I would like to thank both Dr. J. Chappert and Dr. P. Auric of Dept. of Fundamental Research, Centre of Nuclear Studies at Grenoble (France), for inducting me to the world of magnetic materials and Mössbauer spectroscopy.

I would also like to thank Dr. A.F. Wright of Institute Laue Langevin at Grenoble (France) for introducing me to the subject of Small Angle Neutron Scattering for the study of nano particles.

I would also like to thank M.S. Datta for doing experiments painstakingly on attrition milling and sintering of silicon carbides creating a possibility to prepare a large amount of nano particle-sized materials for many applications like ceramic engines, and many more hi-tech applications for the future.

I would like to thank some of my colleagues in my college, who have shown a lot of interest on nano materials and doing some useful work, and to many staff members, particularly to Mr. S.M. Hossain, to help me prepare the manuscript.

Finally, I would like to give special thanks to Dr. P.C. Ray, Dept. of Mathematics, Govt. College of Engineering and Leather Technology, Kolkata, for his constant help on many scientific issues involving nano physics and continuous encouragements, and special thanks are also due to Dr. V. Gopalan of Materials Science Deptt., Penn. State University (USA), for giving me a lot of insights on Photonics based on ferroelectric materials, and whose work in this field has been a great inspiration to me.

Author

Contents

Chapter 1

General Introduction

PREAMBLE

In 1959, the great physicist of our time Professor Richard Feynman gave the first illuminating talk on nano technology, which was entitles as : **There's Plenty of Room at the Bottom.** He consciously explored the possibility of "direct manipulation" of the individual atoms to be effective as a more powerful form of 'synthetic chemistry'.

Feynman talked about a number of interesting ramifications of a 'general ability' to manipulate matter on an atomic scale. He was particularly interested in the possibility of denser computer circuitry and microscopes that could see things much smaller than is possible with 'scanning electron microscope'. The IBM research scientists created today's 'atomic force microscope' and 'scanning tunneling microscope', and there are other important examples.

Feynman proposed that it could be possible to develop a 'general ability' to manipulate things on an atomic scale with a 'top → down' approach. He advocated using ordinary machine shop tools to develop and operate a set of one-fourth-scale machine shop tools, and then further down to one-sixteenth-scale machine tools, including miniaturized *hands* to operate them. We can continue with this particular trend of down-scaling until the tools are able to directly manipulate atoms, which will require redesign of the tools periodically, as different forces and effects come into play. Thus, the effect of gravity will decrease, and the effects of surface tension and Van der Waals attraction will be enhanced. He concluded his talk with challenges to build a tiny motor and to write the information from a book page on a surface 1/25,000 smaller in linear scale.

Although Feynman's talk did not explain the full concept of nano technology, it was K. E. Drexler who envisioned self-replicating 'nanobots', *i.e.,* self-replicating robots at the molecular scale, in **Engines of Creation: The Coming Era of Nano Technology in 1986,** which was a seminal 'molecular nano technology' book.

That brings us to the end of the brief history on how the concept of nano technology emerged.

1.1. INTRODUCTION

In the usual and standard language, when we talk about 'materials science' and 'materials technology', we normally mean ceramics or crystalline materials, glasses or non-crystalline materials, polymers or heavy chain molecular materials and metals or cohesively-bonded materials. All these materials have a wide variety of applications in the diverse fields towards the service for the betterment of human life.

The world of materials is rapidly progressing with new and trendiest technologies, and obviously novel applications. Nano technology is among these modern and sophisticated technologies → which is creating waves in the modern times. Actually, nano technology includes the concept of physics and chemistry of materials. It beckons a new field coming to the limelight. So, nano technology is an interesting but emerging field of study, which is under constant evolution offering a very wide scope of research activity.

1.1.1. What is Nano Technology ?

Nano technology is an advanced technology, which deals with the synthesis of nano-particles, processing of the nano materials and their applications. Normally, if the particle sizes are in the **1-100 nm** ranges, they are generally called nano particles or materials. In order to give an idea on this size range, let us look at some dimensions : **1 nm** = 10 Å = 10^{-9} meter and 1 μm (*i.e.*, 1 micron) = 10^{-4} cm = **1000 nm.** For oxide materials, the diameter of one oxygen ion is about 1.4 Å. So, seven oxygen ions will make about 10 Å or **1 nm**, *i.e.*, the 'lower' side of the nano range. On the higher side, about 700 oxygen ions in a spatial dimension will make the so-called 'limit' of the nano range of materials.

1.1.2. Why Nano Technology ?

In the materials world, particularly in ceramics, the trend is always to prepare finer powder for the ultimate processing and better sintering to achieve dense materials with dense fine-grained microstructure of the particulates with better and useful properties for various applications. The fineness can reach up to a molecular level **(1 nm – 100 nm),** by special processing techniques. More is the fineness, more is the surface area, which increases the 'reactivity' of the material. So, the densification or consolidation occurs very well at lower temperature than that of conventional ceramic systems, which is finally 'cost-effective' and also improves the properties of materials like abrasion resistance, corrosion resistance, mechanical properties, electrical properties, optical properties, magnetic properties, and a host of other properties for various useful applications in diverse fields.

1.1.3. Scope of Applications

The deviations from the bulk phase diagram may be exploited to form certain compositions of alloys that are otherwise unstable in the bulk form. In addition, the thermal stability of interfacial regions is typically less than that of the bulk material : thus the nano phase materials are often sintered or undergo phase transformation at temperatures below those of the bulk material. This is a characteristic which has numerous applications to material processing.

By improving material properties, we are able to find the applications as varied as semiconductor electronics, sensors, special polymers, magnetics, advanced ceramics, and membranes. We need to improve our current understanding of particle size control and methodologies for several classes of nano-phase materials and address the issues of their characterization. We should also explore the fields in which there are foreseeable application of nano-phase materials to long standing materials problems, since these 'issues' have to be tackled by us.

As said earlier, there is a scope of wider applications in different fields such as : (*a*) Electronics in terms of Thin Films, Electronic Devices like MOSFET, JFET and in Electrical Ceramics, (*b*) Bionics, (*c*) Photonics, (*d*) Bioceramics, (*e*) Biotechnology, (*f*) Medical Instrumentation, etc.

1.2. BASICS OF QUANTUM MECHANICS

It was mentioned above that about 7 oxygen ions make the lowest side of nano particles. Below this level or even at this level, the concept of 'quantum mechanics' is useful. If we do not understand the atoms themselves, then how we can aspire to know more about the behaviour of the "nano particles", which are either embedded within a particular matrix or just remain as a mixture in a 'particulate assembly'.

In order to talk about quantum mechanics, we must clarify different aspects of mechanics—which is a pillar in science since the era of 'Newtonian Mechanics'. Actually, there are four realms of mechanics, which will put quantum mechanics in proper perspectives. The following diagram simply illustrates this point :

Speed ↑

QUANTUM FIELD THEORY (Pauli, Dirac, Schwinger, Feynman, et al.)	**RELATIVISTIC MECHANICS (Einstein)**
QUANTUM MECHANICS (Planck, Bohr, Schrödinger, de Broglie, Heisenberg, et al)	**CLASSICAL MECHANICS (Newton)**

Distance→

Some people say that the subject of quantum mechanics is all about 'waves' and that's why sometimes we call it 'wave mechanics' in common parlance, yet many textbooks on this subject do not explicitly clarify how the 'waves' are created through the mathematical route. When this part is made clear, it has been observed that many readers find quantum mechanics quite interesting. Hence, a simple attempt is made here towards this objective.

1.2.1 Differential Equations of Wave Mechanics

There are so many problems in wave mechanics, which can be described as the 'solutions' of a differential equation of the following type :

$$\frac{d^2y}{dx^2} + f(x)\,y = 0 \qquad \text{...(1.1)}$$

The readers studying this subject must thoroughly understand this equation. Here, $f(x)$ is a function of the independent variable x. With this equation, we can plot y *vs.* x, when the values of y and $\frac{d^2y}{dx^2}$ are provided for an arbitrary value of x.

We can also make an equivalent statement : Two independent solutions of y_1 and y_2 exist and that $(\text{A}y_1 + \text{B}y_2)$ is the 'general solution'; this is also possible to be shown graphically.

The simplest case of equation (1.1) is that where $f(x)$ is constant. Two cases are possible for this situation as

1. If $f(x)$ is a positive constant, *i.e.* $f(x) = k^2$, we can write the solution as:

$$y = A \cos kx + B \sin kx$$

or,
$$y = a \cos (kx + \varepsilon)$$

where, A, B, a and ε are all arbitrary constants. This particular solution is clearly shown in **Fig. 1.1(*a*).**

2. If $f(x)$ is constant, but negative, *i.e.,* by setting $f(x) = -\gamma^2$, we get the solutions as $e^{-\gamma x}$ and $e^{\gamma x}$, with the general solution as :

$$y = A\,e^{-\gamma x} + B\,e^{\gamma x}$$

These solutions are depicted in **Fig. 1.1(*b*).**

In the general case, where $f(x)$ is not a constant, it is easy to show that if $f(x)$ is positive → y is an oscillating function. If $f(x)$ is negative, y is of exponential form.

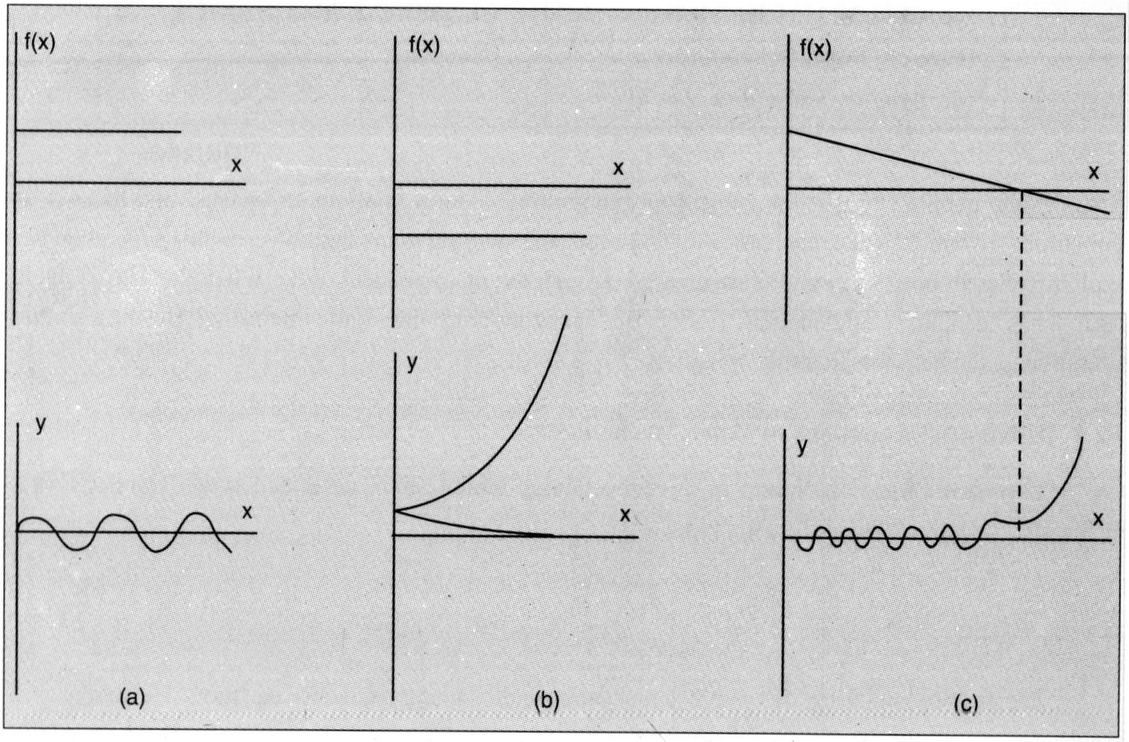

Figure 1.1 : Solutions of the differential equations $y'' + f(x)y = 0$. (*a*) for $f(x) = k^2$, (*b*) for $f(x) = -\gamma^2$, (*c*) for an arbitrary form of $f(x)$ that changes sign.

This is due to the fact that if $f(x)$ is positive, both y and $\frac{d^2y}{dx^2}$ have the opposite sign, as shown in **Fig. 1.2(*a*).**

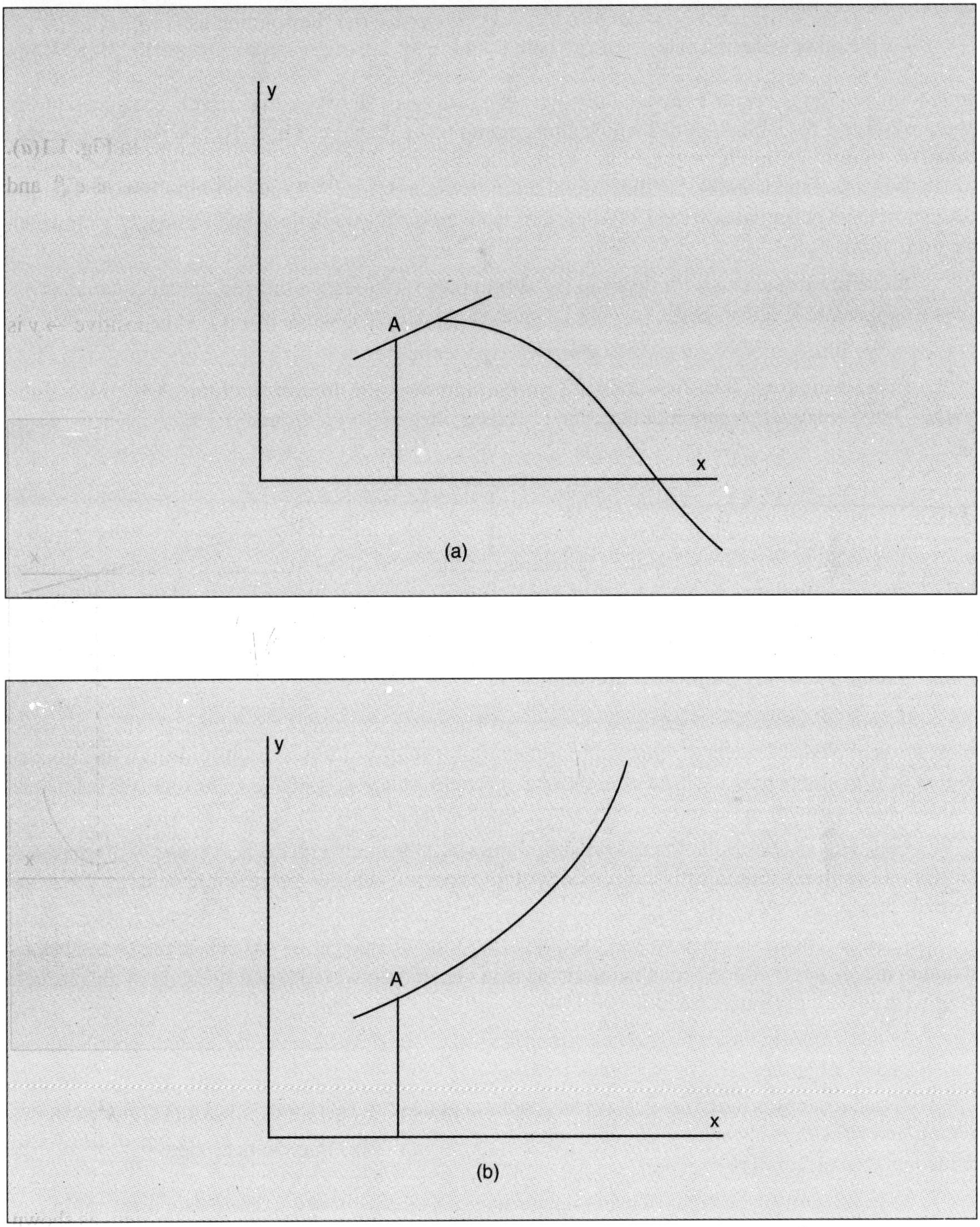

Figure 1.2 : *y vs. x* plot, (*a*) for a decreasing function, (*b*) for an increasing function.

On the other hand, if $f(x)$ is negative, both y and $\frac{d^2y}{dx^2}$ are of the same sign, and the slope at any point will increase giving an exponentially increasing curve, as shown in **Fig. 1.2(*b*)**. The general form of the solution y for a function $f(x)$ which changes sign is as shown in **Fig. 1.1(*c*)**. When $f(x)$ becomes negative, y goes over to the 'exponential' form. Generally speaking, there will always be one solution which decreases exponentially, but the general solution will increase. When we consider that the solution consists of an exponential decrease, this determines the phase of oscillations in the range of x for which the oscillations occur.

A useful method exists for determining approximate solutions of the differential equation (1.1), known as Wentzel-Kramers-Brillouin (WKB) method, which is written as :

$$y = \alpha\, e^{i\beta} \qquad ...(1.2)$$

Here, a and b are both functions of x and α represents the amplitude of the oscillations, β the phase. The approximate solution of equation (1.2) is written as :

$$y = \text{Const.}\, f^{-1/4} \exp\left[i \int_{x_0}^{x} f^{1/2}\, dx\right]$$

Immediately, it follows that the 'amplitude' of the oscillations increases, as f becomes smaller and the wavelength increases (as shown in **Fig. 1.1*c***). So that sums up the basics of waves through a simple mathematical route, which should clarify the point mentioned above.

1.2.2. Background of Quantum Mechanics

First of all, it could be stated that a knowledge of quantum mechanics is indispensable to understand many areas of physical sciences. Quantum mechanics is a branch of science, which deals with 'atomic' and 'molecular' properties and behaviour on a microscopic scale, *i.e.*, useful to understand the behaviour of the "nano" particles in the microscopic level. Some salient points can be mentioned as :

It is known that while 'thermodynamics' may be concerned with the heat capacity of a gaseous sample → quantum mechanics is concerned with the specific changes in 'rotational energy states' of the molecules.

While 'chemical kinetics' may deal with the 'rate of change' of one substance to another → quantum mechanics is concerned with the changes in the vibrational states and structure of the reactant molecules as they get transformed.

Quantum mechanics is also concerned with the 'spins' of atomic nuclei and 'population of excited states' of atoms.

Spectroscopy is based on changes of various quantized energy levels. Thus quantum mechanics seem to merge with many other areas of modern science from nuclear physics to organic chemistry to semiconductor electronics.

The modern applications of quantum mechanics have their roots in the development of physics around the turn of the 20th century. Some of the classic experiments date back to 100 years, which provides a solid physical basis for interpretation of quantum mechanics.

The names attached to much of the early times are due to the work of Planck, Einstein, Bohr, de Broglie and Heisenberg, who are legendary in the realm of physics. A brief review of their work is necessary before going to the details of quantum mechanics.

1.2.3 Origin of the Problem : Quantization of Energy

So, a little bit of history needs to be told. Before 1900, with proper statistical considerations, the physicists assumed that the laws governing the 'macroworld' were valid in the 'microworld' → this posed problems in terms of inadequate 'theory of black-body radiation' due to Wien's and Rayleigh-Jeans' radiation laws.

So, the quantum theory was developed, which had its origin in the lapses of 'Classical Mechanics', *i.e.* mechanics, electromagnetism, thermodynamics and optics, to explain experimentally observed energy (E) *vs.* wavelength or frequency (ν) curves, *i.e.*, distribution, in the 'continuous spectrum' of black-body radiation. Actually, we need to explain the colour of light emitted by an object when heated to a certain temperature. Here, the extraordinary efforts made by Planck needs to be little bit elaborated on how the concept of 'quantization' came into existence.

A correct theory of black-body radiation was developed by Max Planck (1857–1947) in 1900, by assuming that the absorption and emission of radiation still arose from some sort of oscillators, which requires that the radiation be 'quantized'. The fundamental assumption of Planck was that only certain frequencies were possible/permissible for the oscillators instead of the whole range of frequencies, which are normally predicted by classical mechanics. These frequencies were presumed to be some multiple of a fundamental frequency of the oscillators, ν.

Furthermore, Planck assumed that the energy needed to be absorbed to make the oscillator move from one allowed frequency to the next higher one, and that the energy was emitted at the frequency lowered by ν.

Planck also assumed that the change in energy is proportional to the fundamental frequency, ν. By introducing a constant of proportionality, h, *i.e.*, $E = h\nu$ (h = Planck's Constant = 6.63×10^{-27} erg sec = 6.63×10^{-34} Joule sec). This famous equation predicted the observed relationship between the frequency of radiation emitted by a blackbody and the intensity.

In 1905, Albert Einstein (1879–1955) further developed the 'concept' of energy quantization by assuming that : this phenomenon was a property of the radiation itself and this process applied to both absorption and emission of radiation. By using the above quantization concept, Einstein developed a correct theory of the 'photoelectric' effect.

In 1913, Neils Bohr (1885-1962) combining classical physics and quantization concept postulated theory for the observed spectrum of hydrogen atom as follows :

A. The electron in the hydrogen atom moves around the nucleus, *i.e.* proton, in certain circular orbits (*i.e.*, stationary states) without radiating energy.

B. The allowed 'stationary states' are such that $L = m V r = n \hbar$ (where, L = angular momentum of the electron, r = radius of the orbit, m = mass of the electron, $\hbar = \frac{h}{2\pi}$, V = velocity of the electron with n = principal quantum number).

C. When the electron makes a transition from a state of energy E_1 to E_2 ($E_1 > E_2$), electromagnetic radiation (*i.e.*, photon) is emitted from the hydrogen atom. The frequency of this emission process is given by : $\nu = \frac{(E_1 - E_2)}{h}$.

Bohr's theory was applied to other atoms with some success in a generalized form by Wilson and Sommerfeld. By about 1924, it was clear that all we needed is a 'new theory' to interpret the basic properties of atoms and molecules in a proper manner [1-4]

1.2.4 Development of New Quantum Theory

In 1925, Heisenberg (1901) developed a system of mechanics where the knowledge of classical concepts of mechanics was revised. The essence of his theory is : Heisenberg assumed that the atomic theory should talk about the 'observable' quantities' rather than the shapes of electronic orbits (*i.e.,* Bohr's theory), which was later developed into matrix mechanics by matrix algebra. Then came the theory of wave mechanics, which was inspired by de Broglie's (1892-) wave theory of matter :

$\lambda = \dfrac{h}{p}$ (p = momentum of a particle and λ = wavelength).

Almost parallel to the advancement of matrix mechanics, in 1926 Schrödinger (1887-1961) introduced an 'equation of motion' based on 'partial differential equation' for matter waves, which proved that wave mechanics was mathematically equivalent to matrix mechanics, although its physical meaning was not very clear at first.

But, Why it is so ?

Schrödinger first considered the 'de Broglie wave' as a physical entity, *i.e.,* the particle, electron, is actually a wave. But this explanation has some difficulty, since a wave may be partially reflected and partially transmitted at a 'boundary' → but an electron cannot be split into two component parts, one for transmission and the other for reflection.

This difficulty was removed by the statistical interpretation of de Broglie wave by Max Born (1882-1970), which is now widely accepted → known as 'Born Interpretation'. The entire subject was very rapidly developed into a cohesive system of mechanics → called Quantum Mechanics. Since it deals with the waves, we may sometimes call it wave mechanics.

Incidentally, it may be mentioned that the famous German Mathematician David Hilbert suggested to Heisenberg to try the route of 'partial differential equations'. If Heisenberg listened to Hilbert, then the famous 'partial differential wave equation' would be to his credit, but Schrödinger got the Nobel Prize for this most important discovery of the past century in 1933 with Paul Dirac. So, this is the short story of quantum mechanics.

How Schrödinger Advanced His Ideas ?

For the 'Wave Equation for Particles', Schrödinger assumed a 'Wave Packet' and used Hamiltonian's 'Principle of Least Action'. During the development of wave mechanics, it was known to Schrödinger that :

A. Hamilton had established an analogy between the Newtonian Mechanics of a particle and geometrical or ray optics called Hamiltonian Mechanics, and

B. Equations of wave optics reduced to those of geometrical optics, if the wavelength in the former is equal to zero.

Hence, Schrödinger postulated that the classical Newtonian Mechanics was the limiting case of a more general 'wave mechanics' and then derived a 2nd order wave equation of particles.

In order to make a complete description of the 'motion of the particle' by the 'motion of a wave', we must do the following:

(*A*) To find a suitable 'wave representation' of a single particle, and

(*B*) To establish the 'kinematical equivalence' of a ray and a particle trajectory

A localized wave whose amplitude is zero everywhere, except in a small region, is called the 'wave packet', which will satisfy the condition (*A*), but we have to also satisfy the condition (*B*).

To Prove the 'Kinematic Equivalence' → How to Start ?

A monochromatic 'plane wave' in one dimension can be represented by :

$$\varphi_k(x, t) = \varphi(k) \exp[i(kx - \omega t)] \quad ...(1.3)$$

where, k is the x-component of the propagation vector, denoted by k or $|k| = \frac{2\pi}{\lambda}$, and $\omega = \omega(k) = 2\pi\nu$. It has to be noted that a superposition of a group of plane waves of nearly the same wavelength and frequency that interfere destructively everywhere except in a small region gives rise to a 'wave packet'. In the one-dimensional case, such a 'wave packet' can be represented by 'Fourier Analysis', by taking an 'wave packet' centered at k which extends to $\pm \Delta k$ so that the 'Fourier Integral' can be used between $k - \Delta k/2$ and $k + \Delta k/2$ as :

$$\psi(x, t) = \frac{1}{\sqrt{2\pi}} \int_{-\infty}^{\infty} F(k) \exp\,[i(kx - wt)]\, dk \quad ...(1.4)$$

A maximum, *i.e.*, constructive interference, will occur when $(kx - \omega t) = 0$, since 'the sum' over the 'oscillating exponential function' for different values of k would result on an average in a flat pattern, *i.e.*, a destructive interference.

Let us assume that the form of $\psi(0, 0)$ is

$$\psi(0, 0) = \frac{1}{\sqrt{2\pi}} \int_{k_0 - \Delta k/2}^{k_0 + \Delta k/2} F(k)\, dk \quad ...(1.5)$$

At a later time Δt and a further distance Δx, the form of ψ is

$$\psi(\Delta x, \Delta t) = \frac{1}{\sqrt{2\pi}} \int_{k_0 - \Delta k/2}^{k_0 + \Delta k/2} F(k) \exp[i(k\Delta x - \omega\Delta t)]dk$$

$$= \frac{1}{\sqrt{2\pi}} \exp[i(k_0\Delta x - \omega_0\Delta t)] \int_{k_0 - \Delta k/2}^{k_0 + \Delta k/2} F(k) \exp\{i[(k - k_0)\Delta x - (\omega - \omega_0)\Delta t]\}dk \quad ...(1.6)$$

In order to get a 'maximum', we require that

$$i[(k - k_0)\Delta x - (\omega - \omega_0)\Delta t] = 0$$

or,

$$\frac{\Delta x}{\Delta t} = \frac{(\omega - \omega_0)}{(k - k_0)} \quad ...(1.7)$$

By expanding $\omega(k)$ in a Taylor series about k_0, we get

$$\omega = \omega_0 + (k - k_0)\left(\frac{d\omega}{dk}\right)_{\omega = \omega_0} + (k - k_0)^2 \left(\frac{d^2\omega}{dk^2}\right)_{\omega = \omega_0} + ... \quad ...(1.8)$$

By neglecting the 2nd derivative of ω and the higher order terms in the above expansion, ultimately we find that the equation (1.7) becomes

$$\frac{\Delta x}{\Delta t} = \left(\frac{d\omega}{dk}\right)_{\omega = \omega_0}$$

or,

$$V_{group\ velocity} = \lim_{\Delta t \to 0}\left(\frac{\Delta x}{\Delta t}\right) = \frac{d\omega}{dk} \qquad \text{...(1.9)}$$

Now, it is clear how we establish the 'kinematical equivalence' of a 'ray' and a 'particle' trajectory, *i.e.,* the condition (*B*) as explained above, by requiring that the 'group velocity' of the 'wave packet' equals the velocity of the particle → which means that

$$V_g = V_p \qquad \text{...(1.10)}$$

The velocity of the particle, V_p, is given by

$$V_p = \frac{dE}{dp} = \frac{d(E/H)}{d(1/\lambda)} \qquad \text{...(1.11)}$$

Since, the energy is written as

$$E = \frac{p^2}{2m} + V$$

and the momentum is written as

$$p^2 = 2m[E - V] = \frac{H}{\lambda}$$

the group velocity of the wave can be written as

$$V_g = \frac{d\omega}{dk} = \frac{d\omega}{d(2\pi/\lambda)} = \frac{d\left(\frac{\omega}{2\pi}\right)}{d\left(\frac{1}{\lambda}\right)} \qquad \text{...(1.12)}$$

Since we require that $V_g = V_p$, we must have the following relation

$$\frac{E}{H} = \frac{\omega}{2\pi} = \nu$$

or,

$$E = H\nu = h\nu = \text{Planck's Constant} \times \nu \qquad \text{...(1.13)}$$

Now, we have finally established the fact that it is reasonable to consider → describing the motion of a particle by the use of a 'localised wave', *i.e.,* wave packet → if we require that E = Constant × ν. Surprisingly, this is exactly the 'Planck's Quantization of Energy Condition', where H = *h* (*i.e.* the Planck's Constant), as described earlier.

1.2.5 Quantum Mechanical Way: The Wave Equations

In order to familiarize with procedures and terminology, we can start by stating the 'postulates' of quantum mechanics and showing some of their uses.

The Postulate 1 → For any possible 'state of a system', there is a function ψ, of the coordinates of the parts of the system and time that completely describe the 'system'.

For a single particle described by the Cartesian coordinates, we can write it as

$$\psi = \psi(x, y, z, t) \quad ...(1.14)$$

For two particles, the coordinates of each particle must be specified so that

$$\psi = \psi(x_1, y_1, z_1, x_2, y_2, z_2, t) \quad ...(1.15)$$

For a general system, we can use generalized coordinates q_i , and it is written as

$$\psi = \psi(q_i, t) \quad ...(1.16)$$

Since the model is that of a wave, the function is called a 'wave function'. The state of the system which is described by this function is called the 'quantum state'.

The meaning of this wave function is that ψ^2 is proportional to the probability. Since ψ may be complex, we are interested in $\psi\psi^*$, where ψ^* is the complex conjugate of ψ. The complex conjugate is the same function with i replaced by $-i$, where $i = \sqrt{-1}$.

For example → If we square the function $(x + ib)$, we obtain : $(x + ib)(x + ib) = x^2 + 2ibx + i^2 b^2 = x^2 + 2ibx - b^2$ and the resulting function is still complex. Now, if we multiply $(x + ib)$ by its complex conjugate $(x - ib)$, we obtain : $(x + ib)(x - ib) = x^2 - i^2 b^2 = x^2 + b^2$, which is real. Hence, for the calculation of probability, it is always done by multiplying a function with its complex conjugate.

The quantity $\psi\psi^* dV$ is proportional to the probability of finding the particle of the system in the volume element, $dV = dx\,dy\,dz$. We require that the total probability be unity so that the particle must be somewhere, *i.e.,* it can be expressed as

$$\int_V \psi\psi^* \, dV = 1 \quad ...(1.17)$$

If this condition is met, then ψ is normalized. In addition, ψ must be 'Finite', 'Single Valued' and 'Continuous'. These conditions describe a "well behaved" wave function. The reasons for these requirements are as follows:

1. **Finite.** A probability of unity denotes a 'sure thing'. A probability of zero means that a particular event cannot happen. Hence, the probability varies from zero to unity. If ψ were infinite, the probability could be greater than unity.
2. **Single valued.** In a given area of space, there is only one probability of finding a particle. For example, there is a single probability of finding an electron at some specified distance from the nucleus in a hydrogen atom. There cannot be two different probabilities of finding the electron at some given distance.
3. **Continuous.** If there is a certain probability of finding an electron at a given distance from the nucleus in a hydrogen atom, there will be a slightly different probability if the distance is changed slightly. The probability function does not have 'discontinuities' so the wave function must be continuous.

If two functions ψ_1 and ψ_2 have the following property

$$\int \psi_1^* \psi_2 \, dV = 0$$

or,
$$\int \psi_1 \psi_2 ^* dV = 0 \quad ...(1.18)$$

They are said to be orthogonal. Whether the integral vanishes or not may depend on the 'limits of integration', and hence we always speak of the "orthogonality" within a certain interval.

Therefore, the 'limits of integration' must be clear. In the above case, the integration is carried out over the possible range of coordinates used in dV. If the coordinates are x, y and z, the limits are from $-\infty$ to $+\infty$ for each variable. If the coordinates are r, θ and ϕ, the limits of integration are 0 to ∞, 0 to π, and 0 to 2π, respectively.

Postulate 2. For every 'dynamical variable' (classical observable), there is a corresponding "operator".

This postulate provides the 'connection' between the quantities which are classical observables and the quantum mechanical techniques for doing things.

But what are the dynamic variables ?

These are such quantities as energy, momentum, angular momentum and position coordinates.

The operators are symbols which indicate that some mathematical operations have to be performed. Such symbols include $(\ )^2$, $\frac{d}{dx}$ and $\int$. The coordinates are the same in operator and classical forms, *e.g.*, the coordinate x is simply used in operator form as x. Some operators can be combined, *e.g.*, since the kinetic energy is $\frac{mV^2}{2}$, it can be written in terms of the momentum p, as $\frac{p^2}{2m}$.

The operators that are important in quantum mechanics have two important characteristics :

1. First, the operators are linear, which means that :

$$\alpha(\phi_1 + \phi_2) = \alpha\phi_1 + \alpha\phi_2 \quad ...(1.19)$$

where α is the operator and ϕ_1 and ϕ_2 are the 'functions' being operated on. Also, if C is a constant, we get

$$\alpha(C\phi) = C\ (\alpha\ \phi) \quad ...(1.20)$$

The linear character of the operator is related to the superposition of 'states' and waves reinforcing each other in the process.

2. Secondly, the operators that we encounter in quantum mechanics are Hermitian. If we consider two functions ϕ_1 and ϕ_2, the operator α is Hermitian if we have the following relation :

$$\int \phi_1 ^* \alpha\phi_2\, dV = \int \phi_2\, \alpha^* \phi_1^*\, dV \quad ...(1.21)$$

This requirement is necessary to ensure that the calculated quantities are real. We will come across these types of behaviour in the operators that we use in quantum mechanics.

The Eigenvalues

Postulate 3. The permissible values that a dynamical variable may have are those given by $\alpha\phi = a\phi$, where ϕ is the eigenfunction of the operator α that corresponds to the observable, whose permissible values are "a".

The postulate can be stated in terms of an equation as

$$\underset{\text{operator}}{\alpha} \quad \underset{\text{wave function}}{\phi} = \underset{\text{constant (eigenvalue)}}{a} \quad \underset{\text{wave function}}{\phi} \qquad ...(1.22)$$

If we are performing a particular operation on the 'wave function', which yields the 'original function' multiplied by a 'constant', then ϕ is an 'eigenfunction' of the operator α. This can be illustrated by letting the value of $\phi = e^{2x}$ and taking the operator as $\frac{d}{dx}$. Then, by operating on this function with the operator, we get

$$\frac{d\phi}{dx} = 2\, e^{2x} = \text{constant} \,.\, e^{2x} \qquad ...(1.23)$$

Therefore, e^{2x} is an 'eigenfunction' of the operator α with an 'eigenvalue' of 2. For example, if we let $\phi = e^{2x}$ and the operator be $(\ )^2$, we get $(e^{2x})^2 = e^{4x}$, which is not a constant times the original function. Hence, e^{2x} is not an eigenfunction of the operator $(\ )^2$. If we use the operator for the z component of angular momentum,

$$L_z = \left(\frac{\hbar}{i}\right) \frac{\partial}{\partial\phi} \qquad ...(1.24)$$

Operating on the function $e^{in\phi}$ (where n is a constant), we get

$$\left(\frac{\hbar}{i}\right) \frac{\partial}{\partial\phi} (e^{in\phi}) = in \left(\frac{\hbar}{i}\right) e^{in\phi} = n\hbar \,.\, e^{in\phi} \qquad ...(1.25)$$

which is a constant ($n\hbar$) times the original (eigen)function. Hence, the 'eigenvalue' is $n\hbar$.

The Expectation Value

For a given system, there may be various possible values of a 'parameter' we wish to calculate. Since most properties (such as the 'distance' from the nucleus to an electron) may vary, we desire to determine an average or 'expectation' value. By using the operator equation $\alpha\phi = a\phi$ where ϕ is some function, we multiply both sides of this equation by ϕ^*

$$\phi^* \, \alpha \, \phi = \phi^* \, a \, \phi \qquad ...(1.26)$$

However, it has to be noted that $\phi^* a\phi$ is not necessarily the same as $\phi a\phi^*$. In order to obtain the sum of the probability over all space, we write this in the form of the integral equation as

$$\int_V \phi^* \alpha \, \phi \, dV = \int_V \phi^* \, a \, \phi \, dV \qquad ...(1.27)$$

But 'a' is a constant and is not affected by the order of operations. By removing it from the integral and solving for 'a' yields

$$a = \frac{\int_V \phi^* \alpha \, \phi \, dV}{\int_V \phi^* \phi \, dV} \qquad ...(1.28)$$

It has to be remembered that since α is an operator, $\phi^*\alpha\,\phi$ is not necessarily the same as $\alpha\phi^*\phi$, so that the order of $\phi^*\alpha\,\phi$ must be preserved and α cannot be removed from the integral.

Now, if ϕ is normalized, then by definition $\int \phi^*\alpha\,\phi\,dV = 1$, and we get

$$\bar{a} = \langle a \rangle = \int \phi^*\alpha\,\phi\,dV \qquad ...(1.29)$$

Where, $\bar{a}$ and $\langle a \rangle$ are the usual ways of expressing the average or expectation value. If the wave function is known, then theoretically an expectation or average value can be calculated for a given parameter by using its operator.

A Concrete Example → The Hydrogen Atom

Let us consider the following simple example, which illustrates the 'application' of these ideas.

Let us suppose that we want to calculate the 'radius' of the hydrogen atom in the **1*s*** state. The normalized wave function is written as

$$\psi_{1s} = \left(\frac{1}{\sqrt{\pi}}\right)\left(\frac{1}{a_0}\right)^{3/2} e^{-r/a_0} = \psi^*_{1s} \qquad ...(1.30)$$

Where a_0 is the Bohr radius. This equation becomes

$$\langle r \rangle = \int \psi^*_{1s}\ (\text{operator})\ \psi_{1s}\,dV \qquad ...(1.31)$$

Here the operator is just r, since the position coordinates have the same form in operator and classical forms. In polar coordinates, the volume element $dV = r^2 \sin\theta\,dr\,d\theta\,d\phi$. Hence, the problem becomes integration in three different coordinates with different limits as

$$\langle r \rangle = \int_0^{\infty}\int_0^{\pi}\int_0^{2\pi} \frac{1}{\sqrt{\pi}}\left(\frac{1}{a_0}\right)^{3/2} e^{-r/a_0}\,(r)\,\frac{1}{\sqrt{\pi}}\left(\frac{1}{a_0}\right)^{3/2} \times e^{-r/a_0}\,r^2 \sin\theta\,dr\,d\theta\,d\phi \qquad ...(1.32)$$

While this may look rather complicated, it simplifies greatly, since the operator r becomes a multiplier and the function r can be multiplied. Then the result is written as

$$\langle r \rangle = \int_0^{\infty}\int_0^{\pi}\int_0^{2\pi} \left(\frac{1}{\pi a_0^3}\right) r^3 e^{-2r/a_0} \sin\theta\,dr\,d\theta\,d\phi \qquad ...(1.33)$$

By using the technique from the calculus, which allows us to separate multiple integrals as :

$$\int f(x)\,g(y)\,dx\,dy = \int f(x)\,dx \int g(y)\,dy \qquad ...(1.34)$$

We can write equation (1.33) as

$$\langle r \rangle = \left(\frac{1}{\pi a_0^3}\right)\int_0^{\infty} r^3 e^{-2r/a_0}\,dr \int_0^{\pi}\int_0^{2\pi} \sin\theta\,d\theta\,d\phi \qquad ...(1.35)$$

It can be easily verified that

$$\int_0^{\pi}\int_0^{2\pi} \sin\theta\,d\theta\,d\phi = 4\pi \qquad ...(1.36)$$

and the exponential integral is a commonly occurring one in quantum mechanics. It can be easily evaluated by using the formula

$$\int_0^{\infty} x^n e^{-bx}\, dx = n!\,/b^{n+1} \qquad ...(1.37)$$

In this case, $n = 3$ and $b = 2a_0$. Therefore, we get

$$\int_0^{\infty} r^3 e^{-2r/a_0}\, dr = \frac{3!}{(2/a_0)^4} \qquad ...(1.38)$$

so that we can write the 'expectation value' as

$$< r > = \left(\frac{4\pi}{\pi a_0^{\,3}}\right) \frac{3!}{(2/a_0)^4} = \left(\frac{3}{2}\right) a_0 \qquad ...(1.39)$$

Thus, finally, we can get the 'expectation value' for the hydrogen 1*s* state as

$$< r >_{1s} = 1.5\, a_0 \quad (a_0 = 0.529 \text{ Å}) \qquad ...(1.40)$$

With a probability curve of the electron in 1*s* state as a function of the distance from the nucleus, this typical value of the 'expectation value' is shown in **Fig. 1.3,** where the meaning of the maximum of probability of finding an electron at certain distance and its expectation value are quite different.

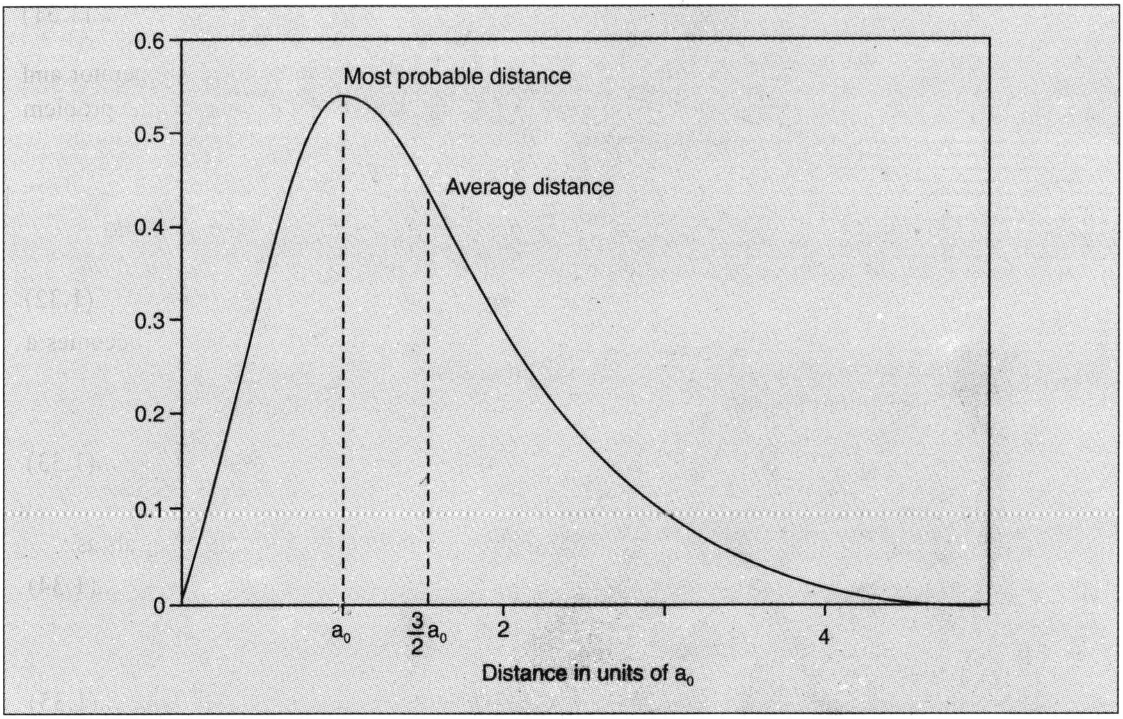

Figure 1.3: Probability of finding a 1s electron as a function of distance from the nucleus.

1.2.6 The Wave Function

Postulate 4. The 'state' function, ψ, is given as a solution of : Hψ = Eψ, where, H is the operator for total energy, the Hamiltonian Operator.

This postulate provides a starting point for formulating a problem in quantum mechanical terms, because we usually seek to determine a wave function to describe the system being studied. The Hamiltonian function in classical physics is the total energy, K + V, where K is the translational (kinetic) energy and V is the potential energy. In operator form

$$H = K + V \qquad ...(1.41)$$

Where K is the operator for kinetic energy and V is the operator for potential energy. If we write in the generalized coordinates, q_i , and time, the starting equation becomes

$$H\psi(q_i, t) = -\left(\frac{\hbar}{i}\right) \partial\psi(q_i, t)/\partial t \qquad ...(1.42)$$

The kinetic energy can be written in terms of the momentum as

$$K = \frac{mV^2}{2} = \frac{p^2}{2m} \qquad ...(1.43)$$

Now, we can write it in three dimensions as

$$K = \frac{p_x^2}{2m} + \frac{p_y^2}{2m} + \frac{p_z^2}{2m} \qquad ...(1.44)$$

By putting this in operator form, we make use of the momentum operators as

$$K = \left(\frac{1}{2m}\right)\left(\frac{\hbar}{i}\frac{\partial}{\partial x}\right)^2 + \left(\frac{1}{2m}\right)\left(\frac{\hbar}{i}\frac{\partial}{\partial y}\right)^2 + \left(\frac{1}{2m}\right)\left(\frac{\hbar}{i}\frac{\partial}{\partial z}\right)^2 \qquad ...(1.45)$$

However, we can write the square of each momentum operator as

$$\left(\frac{\hbar}{i}\frac{\partial}{\partial x}\right)^2 = \left(\frac{\hbar}{i}\frac{\partial}{\partial x}\right)\left(\frac{\hbar}{i}\frac{\partial}{\partial x}\right) = \frac{\hbar^2}{i^2}\frac{\partial^2}{\partial x^2} = -\hbar^2\frac{\partial^2}{\partial x^2} \qquad ...(1.46)$$

so that we get

$$K = -\frac{\hbar^2}{2m}\left(\frac{\partial^2}{\partial x^2} + \frac{\partial^2}{\partial y^2} + \frac{\partial^2}{\partial z^2}\right) = -\left(\frac{\hbar^2}{2m}\right)\nabla^2 \qquad ...(1.47)$$

where ∇^2 is Laplacian operator or simply Laplacian. The general form of the potential energy can be written as

$$V = V(q_i, t) \qquad ...(1.48)$$

So that the operator equation becomes

$$\left\{-\left(\frac{\hbar^2}{2m}\right)\nabla^2 + V(q_i, t)\right\} \psi(q_i, t) = -\left(\frac{\hbar}{i}\right)\frac{\partial\psi(q_i, t)}{\partial t} \qquad ...(1.49)$$

This is the famous Schrödinger time dependent equation or, Schrödinger second equation.

In many problems, the classical observables have values that do not change with time, or at least their average values do not change with time. Therefore, in most cases, it would be advantageous to simplify the problem by the removal of the dependence on the 'time'.

How to do it ?

The well known 'separation of variable technique' can now be applied to see if the time dependence can be separated from the 'joint function'. First of all, it is assumed that $\psi(q_i, t)$ is the product of two functions : one a function which contains only the q_i and another which contains only the 'time' (t). Then, we can easily write it as

$$\Psi(q_i, t) = \psi(q_i)\tau(t) \qquad ...(1.50)$$

It has to be noted that Ψ is used to denote the complete 'state' function and the lower case ψ is used to represent the 'state' function with the time dependence removed. The Hamiltonian can now be written in terms of the two functions ψ and τ as

$$\mathrm{H}\Psi(q_i, t) = \mathrm{H}\psi(q_i)\tau(t) \qquad ...(1.51)$$

Therefore, the equation (1.49) can be written as

$$\mathrm{H}\psi(q_i)\,\tau(t) = -\left(\frac{\hbar}{i}\right)\frac{\partial}{\partial t}\,[\psi(q_i)\,\tau(t)] = -\left(\frac{\hbar}{i}\right)\psi(q_i)\frac{\partial\tau(t)}{\partial t} \qquad ...(1.52)$$

By dividing equation (1.52) by the product $\psi(q_i)\tau(t)$, we get

$$\frac{\mathrm{H}\psi(q_i)\,\tau(t)}{\psi(q_i)\,\tau(t)} = -\left(\frac{\hbar}{i}\right)\psi(q_i)\,\frac{\left[\frac{\partial\tau(t)}{\partial t}\right]}{\psi(q_i)\tau(t)} \qquad ...(1.53)$$

Then, we get :

$$\left[\frac{1}{\psi(q_i)}\right]\mathrm{H}\psi(q_i) = -\left(\frac{\hbar}{i}\right)\left[\frac{1}{\tau(t)}\right]\frac{\partial\tau(t)}{\partial t} \qquad ...(1.54)$$

It has to be noted that $\psi(q_i)$ does not cancel, since $\mathrm{H}\psi(q_i)$ does not represent H times $\psi(q_i)$, but rather H operating on $\psi(q_i)$. The left-hand side is a function of q_i and the right-hand side is a function of 'time' (t), so each can be considered as a constant with respect to changes in the values of the other variable. Both sides can be set equal to some new parameter, X, so that

$$\left[\frac{1}{\psi(q_i)}\right]\mathrm{H}\psi(q_i) = \mathrm{X}$$

and,

$$-\left(\frac{\hbar}{i}\right)\left[\frac{1}{\tau(t)}\right]\frac{\partial\tau(t)}{\partial t} = \mathrm{X} \qquad ...(1.55)$$

From the first of these equations, we get

$$\mathrm{H}\psi(q_i) = \mathrm{X}\psi(q_i) \qquad ...(1.56)$$

and from the second one, we get

$$\left[\frac{1}{\tau(t)}\right]\frac{d\tau(t)}{dt} = -\left(\frac{i}{\hbar}\right)X\tau(t) \quad ...(1.57)$$

The differential equation involving the 'time' can be solved readily to give

$$\tau(t) = e^{-(i/\hbar)Xt} \quad ...(1.58)$$

By substituting this result into equation (1.50), we find that the total 'state' function, Ψ, is

$$\Psi(q_i, t) = \psi(q_i)\ e^{-(i/\hbar)Xt} \quad ...(1.59)$$

Therefore, the equation (1.52) can be written as :

$$e^{-(i/\hbar)Xt}\ H\psi(q_i) = +\left(\frac{\hbar}{i}\right)\left(\frac{i}{\hbar}\right)X\psi(q_i)\,e^{-(i/\hbar)Xt} \quad ...(1.60)$$

or,

$$e^{-(i/\hbar)Xt}\ H\psi(q_i) = X\psi(q_i)\ e^{-(i/\hbar)Wt} \quad ...(1.61)$$

The factor $e^{-(i/\hbar)Xt}$ can be dropped from both sides of equation (1.61), which results in

$$H\psi(q_i) = X\psi(q_i) \quad ...(1.62)$$

which clearly shows that the time dependence has been separated.

Here, neither the Hamiltonian operator nor the wave function is time dependent. It is this form of the equation that could be used to solve many problems. Hence, the time-independent wave function, ψ, will be normally indicated when we write $H\psi = E\psi$.

For the hydrogen atom, $V = -\frac{e^2}{r}$, which remains unchanged in the operator form. Hence, we can write it as :

$$H = -\left(\frac{\hbar^2}{2m}\right)\nabla^2 - \frac{e^2}{r} \quad ...(1.63)$$

which gives rise to the following equation as :

$$H\psi = E\psi = -\left(\frac{\hbar^2}{2m}\right)\nabla^2\psi - \left(\frac{e^2}{r}\right)\psi \quad ...(1.64)$$

or,

$$\nabla^2\psi + \left(\frac{2m}{\hbar^2}\right)[E - V]\psi = 0 \quad ...(1.65)$$

This is the Schrödinger wave equation for the hydrogen atom. Several relatively simple models are capable of being treated by the methods of quantum mechanics. In order to treat these models, we use the above four 'postulates' in a relatively straight forward manner. For any of these models, we always begin with

$$H\psi = E\psi \quad ...(1.66)$$

and use the approximate expression for the operators corresponding to the potential and kinetic energies. In practice, we will find that there is a rather limited number of potential functions, the most common being a Coulombic (electro-static) potential [1 – 4].

The quantum mechanical models need to be presented, because they can be applied to several systems which are of considerable interest. For example :

(*a*) The 'rigid rotor' and 'harmonic oscillator' models are useful as models in rotational and vibrational spectroscopy, and obviously for understanding the thermal properties of materials.

(*b*) The 'barrier penetration phenomenon' has application as a model for nuclear decay and transition state theory (not discussed here).

(*c*) The particle in a box model has some utility in treating electrons in metals or conjugated molecules (also not discussed here due to limited applicability).

Out of the above utilities or applications of quantum mechanics, only (*a*) or Harmonic Oscillator problem has direct relevance to explain many thermal behaviour of materials, since we need heat to produce a wide range of materials including the "nano materials".

1.3 THE HARMONIC OSCILLATOR

1.3.1 The Vibrating Object

The vibrations in molecular systems constitute one of the most important properties, which provide the basis for studying molecular structure by various spectroscopic methods (I. R./FTIR, Raman Spectroscopy). Let us start with a **vibrating object →** For an object attached to a spring, Hook's law describes the system in terms of the force (F) on the object and the displacement (x) from the equilibrium position as

$$F = -kx$$

where k = Spring Constant or Force Constant (Newton m or Dynes/cm)

The negative sign means that the restoring force or spring tension is in the direction opposite to the displacement. The work or energy needed to cause this displacement (*i.e.* Potential Energy) is expressed by the "Force Law", which is integrated over the interval, 0 to x, that the spring is stretched

$$\int_0^x F(x)\,dx = \int_0^x -kx\,dx = \left(\frac{1}{2}\right)kx^2$$

If the mass (m) is displaced by a distance of x and released, the object vibrates in simple harmonic motion. The 'angular frequency' of this vibration (ω) is given by

$$\omega = \sqrt{\frac{k}{m}}$$

where the classical or vibrational frequency (ν) is given by

$$\nu = \left(\frac{1}{2\pi}\right)\sqrt{\frac{k}{m}}$$

It is now clear that $\omega = 2\pi\nu$. The maximum displacement from the equilibrium position is called the "amplitude" and the variation of the displacement with time is given by Newton's 2nd Law of

Motion, F = *m* . *a*. The velocity is the 1st derivative of distance with time $\left(\frac{dx}{dt}\right)$ and acceleration is the derivative of velocity with time $\left(\frac{d^2x}{dt^2}\right)$. Therefore, the force (F = *m* . *a*) can be written as:

$$m\,\frac{d^2x}{dt^2} = -\,kx$$

or,
$$\frac{d^2x}{dt^2} + \left(\frac{k}{m}\right)x = 0$$

This is a linear differential equation with constant coefficients, which can be solved by using the formalism presented above. It is thus seen that the problem of the classical vibrating object serves to introduce the terminology and techniques for the quantum mechanical oscillator, which is much more complex than the classical harmonic oscillator [5 – 6].

1.3.2 Quantum Mechanical Harmonic Oscillator

For studying molecular vibrations and the structure, the harmonic oscillator is a very useful model in quantum mechanics. It was shown in the above description that for a vibrating object, the potential energy (V) is given by

$$V = \frac{1}{2}kx^2$$

which can also be written as

$$V = \frac{1}{2}\,m\,x^2\,\omega^2$$

The total energy is the sum of the potential energy and kinetic energy. Now, we must start with the Schrödinger equation as

$$H\psi = E\psi$$

Before we write the full form of the Schrödinger equation, we have to find out →

What is the form of the Hamiltonian Operator?

Before we find the form of this Hamiltonian operator, the kinetic energy (K) must be known, which is written as

$$K = \left(-\frac{\hbar^2}{2m}\right)\frac{\partial^2}{\partial x^2}$$

The potential energy is $2\pi^2\nu^2 mx^2$, so that the Hamiltonian operator can now be written as

$$H = \left\{\left(-\frac{\hbar^2}{2m}\right)\frac{\partial^2}{\partial x^2} + 2\pi^2\nu^2\,mx^2\right\}$$

Therefore, the Schrödinger wave equation $(H\psi = E\psi)$ becomes

$$\left\{\left(-\frac{\hbar^2}{2m}\right)\frac{\partial^2}{\partial x^2} + 2\pi^2\nu^2 mx^2\right\}\psi = E\psi$$

By simplifying this equation by multiplying by $-2m$ and dividing by $\hbar^2$ gives

$$\left(\frac{\partial^2}{\partial x^2} - \frac{4\pi^2\nu^2 m^2x^2}{\hbar^2}\right)\psi = \left(-\frac{2mE}{\hbar^2}\right)\psi$$

which can be rearranged by putting $\alpha = 2m\frac{E}{\hbar}$, $\beta = 2\pi\nu\frac{m}{\hbar} = m\frac{\omega}{\hbar}$ as

$$\frac{\partial^2\psi}{\partial x^2} + \{\alpha - \beta^2 x^2\}\psi = 0 \qquad \text{...(1.67)}$$

This is the usual form of Schrödinger equation. In this equation, the potential varies as x^2, and since it is a non-linear function, this is much more complex than the classical harmonic oscillator or the particle in the one-dimensional box.

A close look of the above wave equation shows that the "solution" must be a function such that its second derivative contains both the original function and a factor of x^2. For very large x, we could assume that a function like $\exp(-\beta x^2)$ satisfies the requirement as

$$\psi = c[\exp(-bx^2)]$$

where, b $(= \beta/2)$ and c are constants.

The other solution is

$$\psi = c[\exp(+bx^2)]$$

which is not a viable solution, since this solution becomes infinity as $x \to \pm\infty$, which is in direct violation of one of the 'Born conditions'.

In order to check the first solution, we start by taking the required derivatives as

$$\frac{d\psi}{dx} = -2bxc[\exp(-bx^2)]$$

$$\frac{d^2\psi}{dx^2} = -2bc[\exp(-bx^2)] + 4b^2cx^2[\exp(-bx^2)] \qquad \text{...(1.68)}$$

Now, working with the second term of the equation (1.67), we notice that

$$-\left\{\frac{2mE}{\hbar^2} - \beta^2x^2\right\}\psi = -\left(\frac{2mE}{\hbar^2}\right)c[\exp(-bx^2)] + \left(\frac{m^2\omega^2}{\hbar^2}\right)x^2c[\exp(-bx^2)] \qquad \text{...(1.69)}$$

It should be noted that both the equations (1.68) and (1.69) contain terms in x^2 and terms that do not contain x except in the exponential. Hence, we can equate the terms that contain x^2 as

$$\left(\frac{m^2\omega^2}{\hbar^2}\right)x^2\, c[\exp(-bx^2)] = 4b^2\, cx^2[\exp(-bx^2)]$$

By canceling the common factors from both sides, we get

$$4b^2 = \left(\frac{m^2\omega^2}{\hbar^2}\right)$$

or,
$$b = \frac{m\omega}{2\hbar}$$

By working with the terms that do not contain x as a factor, we get

$$E = b\left(\frac{\hbar^2}{m}\right)$$

or,
$$E = \frac{1}{2}\omega\hbar \qquad \text{...(1.70)}$$

Therefore, when $b = \frac{m\omega}{2\hbar}$ and $E = \frac{1}{2}\omega\hbar$, the function: $\psi = c[\exp(-bx^2)]$ satisfies the Schrödinger equation. By using the value obtained for b, we can write the 'solution' as

$$\psi = c\left[\exp\left(-\frac{m\omega x^2}{2\hbar}\right)\right]$$

In fact, this is the solution for the harmonic oscillator in its lowest energy state.

The "solution" of the harmonic oscillator problem will now be addressed by starting with the wave equation as

$$\frac{d^2\psi}{dx^2} + \frac{2m}{\hbar^2}\left\{E - \frac{1}{2}kx^2\right\}\psi = 0$$

Now let us put $\alpha = \frac{2mE}{\hbar^2}$ and $\beta = \frac{(mk)^{1/2}}{\hbar}$, then the differential equation becomes

$$\frac{d^2\psi}{dx^2} + (\alpha - \beta^2x^2)\psi = 0$$

In order to solve this 'eigenvalue' problem, it is now necessary to find a set of wave equations (ψ) which satisfies this equation from $-\infty$ to $+\infty$. The function must also obey four 'Born Conditions'. In order to solve the above differential equation, a first solution is found in the limit that x becomes large. Once $x \to \infty$ solution is found, a "power series" is introduced to make the large x solution valid for all x. This is normally called the 'Polynomial Method', known since 1880. By the change of variables to z and after following 'Born Condition', we arrive at the famous Hermite's equation. Without going into a lengthy process, the general form of the Hermite polynomials is written as

$$H_n(z) = (-1)^n \exp(z^2)\frac{d^n}{dz^n}\exp\left(-\frac{z^2}{2}\right)$$

The first few Hermite polynomials can be written as

$$H_0(z) = 1,$$
$$H_1(z) = 2z,$$
$$H_2(z) = 4z^2 - 2,$$
$$H_3(z) = 8z^3 - 12z,$$
$$H_4(z) = 16z^4 - 48z^2 + 12$$

The wave functions for the 'harmonic oscillator' (ψ_i) have to be expressed as a normalization constant (N_i) times $H_i(z)$ to be able to ultimately give a set of normalized wave functions as

$$\psi_0 = N_0 \exp\left(-\frac{z^2}{2}\right)$$

$$\psi_1 = N_1 (2z) \exp\left(-\frac{z^2}{2}\right)$$

$$\psi_2 = N_2 (4z^2 - 2) \exp\left(-\frac{z^2}{2}\right)$$

$$\psi_3 = N_3 (8z^3 - 12z) \exp\left(-\frac{z^2}{2}\right)$$

The above is just an outline of some of the necessary steps in the full solution of the 'harmonic oscillator' model using quantum mechanics [5 – 6]

The above description is useful for vibrational spectroscopy for the determination of various 'vibration states' between 'two bonding atoms' in a molecule of nano-size or higher. This is obviously useful for some thermal properties like thermal expansion containing various phonon branches, specific heat involving optical and acoustic phonon vibrations in both the longitudinal and transverse modes, and finally on the thermal conductivity that involves the motion of both phonons and electrons. However, in order to have a better knowledge on nano materials, there are many useful properties, like magnetic, electronic and optical, which have to be properly understood in the context of quantum mechanics [7]. These properties are discussed in some details in order get a basic idea about these topics in the following subsections.

1.4 MAGNETIC PHENOMENA

Preamble

In classical physics, a classical charge distribution with angular momentum, which is a 'rotating' or 'spinning' charge distribution, gives rise to magnetic moment. Similarly, in quantum mechanics, the 'angular momentum' is referred to as 'electron spin'. In the mathematical treatment of hydrogen by Schrödinger, we get only "three" quantum numbers : n, l and m_l. The subscript l on the magnetic quantum number m_l shows that this m_l value is associated with the orbital angular momentum, which gives rise to s, p, d and f orbitals, but this treatment does not include the intrinsic angular momentum of the electron. Paul Dirac solved the hydrogen atom problem both from the point of non-relativistic quantum mechanics and the theory of relativity, and showed an additional quantum number m_s which is associated with the 'intrinsic angular momentum' of the electron, called s. The m_s is actually the projection of this momentum on the z-axis, and due to this momentum, the electron has a 'permanent magnetic dipole'. This is how we can see the relation between the 'electron spin' and the 'magnetic moment'.

Stern and Gerlach made a 'dramatic demonstration' of the existence of 'electron spin' by heating silver atom, and making the vapour pass through a baffle and then through an inhomogenous magnetic field (defined as the z-axis) onto a glass plate. This showed two distinct spots for the unpaired 5s electron of silver $\left(s = \frac{1}{2}\right)$ with two possible projections on the z-axis with $m_s = \pm\,\frac{1}{2}$. Obviously, it was already known that a magnetic dipole in an inhomogeneous magnetic field experiences a force, which deflects the 'dipole' in a direction that depends on the orientation of the dipole relative to the magnetic field.

1.4.1 Fundamentals of Magnetism

The "spin state" of electrons in an atom determine the magnetic property of the atom. Depending on this spin state of the electrons, the atoms may be classified into mainly : Diamagnetic and Paramagnetic.

(*a*) Diamagnetic atoms (or ions) are those in which there are no uncoupled or uncompensated electron spins.

(*b*) Paramagnetic atoms (or ions) are those with uncoupled or unpaired electron spins in the orbital giving rise a "net magnetic (spin) moment".

The magnetic moment of an uncoupled electron is given as Bohr magneton, μ_B. One Bohr magneton represents the magnetic moment of 'one uncoupled' electron. Therefore, for example, the net magnetic moment for the Fe^{3+} atom is $5\mu_B$, since it has '5 uncoupled' electrons. Different types of magnetism are shown below :

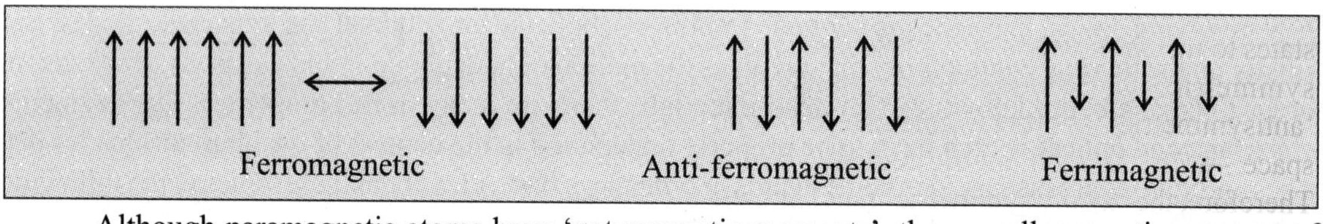

Although paramagnetic atoms have 'net magnetic moments', the overall magnetic moment of crystalline solid may be zero due to the interaction of the atoms in the crystalline lattice. Depending on the 'nature' of this interaction, the atoms may be further classified as 'ferromagnetic', 'anti-ferromagnetic', 'ferrimagnetic'.

A crystal is called a ferromagnetic, if the participating atoms are 'paramagnetic' and their 'directions' are aligned in one direction, which can be switched in the 'opposite direction'. Anti-ferromagnetic crystal is the one whose 'total magnetic moment' is zero, since the magnetic moment of the atoms aligned in one direction is compensated by other atoms, whose magnetic moments are aligned in the opposite direction.

In a ferrimagnetic crystal, the magnetic moments of the atoms are arranged in a similar way to anti-ferromagnetic crystal. However, if the magnetic moment in one direction is larger than the other direction, the result is a "non-zero" overall magnetic moment. Even in ferromagnetic crystal, both up and down magnetic moments coexist. The difference between ferromagnetic and ferrimagnetic is that up and down moments coexist 'intrinsically' within the crystal for ferrimagnetic → while they are formed by two distinct regions (called 'domains') of the crystal in ferromagnetic materials. Hence, only 'one direction' of magnetic moment exists in a domain of ferromagnetic crystals.

In this book, we are dealing with nano materials. In **chapter – 5**, the magnetic properties of the nano particles of magnetite (*i.e.,* a ferrite with a spinel structure) that are embedded in a glassy diamagnetic matrix are discussed in details. These small particles show a phenomenon of 'super-paramagnetism' and also within the nano domain for a slightly higher particle size, these nano particles show 'ferrimagnetism', as per the above description. In the continuation of this section, we would like to give some details on diamagnetism and paramagnetism along with symmetrization and antisymmetrization, which are very important concepts. Moreover, Pauli's principle on the 'electron spin' is discussed with a mathematical treatment in terms of 'determinants' so that a theoretical understanding is developed for the readers to appreciate some intricate details of the 'magnetic properties', *i.e.,* the 'spin properties', of nano materials. The necessary concept on the 'magnetic properties' of the nano materials can be obtained from the theoretical aspects of Mössbauer and ESR spectra, as detailed in the **sections – 1.6.1 and 1.6.2.**

The concept of antisymmetrization is important in the quantum level, which can have some consequences on the magnetic properties of solids containing nano particles. So, here is a brief discussion on this subject.

1.4.2 Antisymmetrization

It is known that the 'total wave function' of an electron in a solid consists of a 'spatial function' and a 'spin function', and for a 'symmetric spatial function', the spin function is 'antisymmetric', and vice versa. The 'spatial' part can be symmetric or antisymmetric. The 'spin' part can be symmetric or antisymmetric → but the 'total function' must be antisymmetric, since the totally symmetric wavefunction are not valid wave functions.

No additional terms in the Hamiltonian can mix symmetric and antisymmetric states. For the states to mix, a non-zero matrix element must exist. The Hamiltonian including any additional terms is symmetric. The product of a symmetric function and an antisymmetric function is always 'antisymmetric'. In order to calculate the value of a matrix element, an integration is performed over all space → *i.e.,* in a symmetric region. Since the product is 'antisymmetric', the integral will vanish. Therefore, the 'totally symmetric' functions and the 'totally antisymmetric' functions are two completely independent to each other, which never mix → *i.e.,* they can have no interaction with each other and hence only one of the two types of states can exist in nature. This point should be clearly understood.

The Question is → Which Type of Function Occurs in Nature?

This question has been answered by experiments. All experimental observations show that the states occurring in nature are totally 'antisymmetric'. Let us take an example of He atom, whose ground state is triply degenerate ($s = 1$) spin state, where the electron spins are 'unpaired', if the wavefunction is totally symmetric → whereas for a 'non-degenerate' ground state ($s = 0$), the spin state, where the electron spins are 'paired', if the wavefunction is totally antisymmetric.

The $s = 1$ state would be paramagnetic because of the combined magnetic dipole moments of the two electrons. By contrast, the $s = 0$ state is diamagnetic. The experiments show that the ground state of He is not paramagnetic. For example → a non-magnetic glass container with liquid helium cannot be picked up by a magnet. For the totally symmetric states, the $s = 0$ (singlet) states are lower in energy than the $s = 1$ (triplet) states for the states involving the same orbitals. The opposite is true for totally antisymmetric states.

Actually, spectroscopic experiments-which involves 'transitions' between two different energy states -- show that for the same orbitals, the triplet states are 'always' lower in energy than the singlet states.

This is due the 'qualitative' statement of the "Pauli Exclusion Principle" that →

All Total Many-Electron Wavefunctions Must be Antisymmetric

Since the 'permutation' symmetry of many-body wavefunctions can only be determined experimentally, it is necessary to bring the "Antisymmetrization Property" into quantum mechanical theory as an 'assumption'.

The Assumption →

The wavefunctions representing an actual state of a system containing two or more electrons must be completely 'antisymmetric' in the coordinates of the electrons → *i.e.,* "on interchanging the coordinates of any two electrons" → ***"the sign of the wavefunction must change".***

In fact, this assumption is the 'quantum mechanical statement' of the Pauli Exclusion Principle'. This can be better understood by two important properties of the 'determinants', which should be elaborated little further as follows:

First Property of the Determinant →

The antisymmetric wavefunctions can be expressed as determinants : Let us take A(1) as an "orbit × spin" function for one electron, such as $1s\alpha$, and B, C, ... N and others, then the total wave function $\psi(x)$ can be written as

$$\begin{vmatrix} A(1) & B(1) & \cdots & N(1) \\ A(2) & B(2) & \cdots & N(2) \\ \cdots & \cdots & \cdots & \cdots \\ A(N) & B(N) & \cdots & N(N) \end{vmatrix} \qquad ...(1.71)$$

which is completely antisymmetric in the N electrons, since any interchange of any two rows change the sign of the determinant, which is the 'first' property of the determinant. The number in each row is the 'electron label'. Each electron is in every orbital.

Now, let us take an actual example → He atom and write its ground state as a determinant, which will involve the $1s$ orbital, and α and β spin states. The 2 × 2 determinant is formed and expanded to obtain totally antisymmetric wavefunction

$$\begin{vmatrix} 1s(1)\,\alpha(1) & 1s(1)\,\beta(1) \\ 1s(2)\,\alpha(2) & 1s(2)\,\beta(2) \end{vmatrix} \qquad ...(1.72)$$

$$= 1s(1)\alpha(1).1s(2)\beta(2) - 1s(2)\alpha(2).1s(1)\beta(1)$$

$$= 1s(1)1s(2)[\alpha(1).\beta(2) - \beta(1).\alpha(2)] \qquad ...(1.73)$$

This equation (1.73) correctly represents antisymmetric ground state function. The spatial part is symmetric and the spin part is antisymmetric. The spin part here corresponds to $\boldsymbol{s = 0, m_s = 0}$ → *i.e.,* the electron spins are paired.

Second Property of the Determinant →

Another important property of the determinant is that if two columns are equal → it vanishes. In fact, this property of the determinant builds the 'Pauli Exclusion Principle' into mathematical formalism. For a given one-electron orbital, there are only two possible spatial × spin functions, *i.e.* those obtained by multiplying the spatial function by either of the two spin functions α and β. Thus, no more than two electrons can occupy the same orbital in an atom or molecule and those two must have their spins opposed → *i.e.,* no two electrons can have the same set of values of all four quantum numbers : n, l, m_l and m_s.

But, why it is so?

If all of the quantum numbers are the "same" → then two columns in the 'determinant form' of the wavefunction will be exactly equal → and the wavefunction (*i.e.,* the determinant) will vanish, which is not desirable. Hence, the requirement that all many-electron 'total wavefunctions' (spatial × spin) must be antisymmetric with respect to the interchange of all pairs of electrons → forces the wavefunctions to satisfy the 'Pauli Exclusion Principle'.

Again, let us take an example → for the ground state of the He atom, if we assume that both electrons are in the 1s orbital and both have α spin, then it is written as

$$\psi = \begin{vmatrix} 1s(1) & 1s(1) \\ 1s(2) & 1s(2) \end{vmatrix} \qquad ...(1.74)$$

If we expand this determinant, we find that

$$1s(1)\alpha(1).1s(2)\alpha(2) - 1s(2)\alpha(2).1s(1)\alpha(1) = 0 \qquad ...(1.75)$$

This is the price we pay → if we attempt to write a wavefunction, which "violates" the 'Pauli Exclusion Principle' → a wavefunction which is not totally antisymmetric → gives zero, which is not an 'acceptable' wavefunction. Hence, the concept of 'antisymmetrization' is clear through the 'Pauli Exclusion Principle'.

Since the 'spin state' decides the magnetic property of materials, it is important to discuss various 'spin states' that exist within a quantum system, like singlet state and triplet state.

1.4.3 Concept of Singlet and Triplet States

The Definition

A given 'state' can be obtained by multiplying a symmetric spatial function by the 'single' antisymmetric spin function. For each 'spatial configuration', there is only one such 'state' with $s = 0$ and $m_s = 0$. These states are called "singlet states".

A given 'state' can also be obtained by multiplying an antisymmetric spatial function by a symmetric spin function. Each antisymmetric spatial function can be actually multiplied by 'three' symmetric spin functions with $s = 0$ and $m_s = 1, 0$ and -1. Since each antisymmetric spatial function gives rise to three 'total' spatial × spin functions, these states are called "triplet states".

The three triplet states differ by the value of m_s associated with each one. They are not necessarily degenerate → for example, the 'spin-orbit' coupling can split the degeneracy of the triplet states giving rise to multiplets in the observed spectra. The symbols : ^{1}S, ^{3}S and ^{1}P are called 'term symbols'. ^{3}S means a triplet state derived from 's' states, while ^{1}P means a singlet state from 's' and 'p' states. They are useful in understanding the spectroscopic behaviour of solids.

The Energy Consideration

The triplet states have inherently lower energies than the singlet state, which arises from the same orbital configurations. For example → the He excited triplet states involving 1*s*, 2*s* orbitals are lower in energy than the 1*s*, 2*s* singlet state, in spite of the fact that the spatial wavefunctions contain the same orbitals. This difference in energy arises from the permutation symmetry of the spatial wavefunctions.

The triplet states have antisymmetric spatial functions, while the singlet states have symmetric spatial function. The antisymmetric spatial function has a node, *i.e.*, it vanishes if the two electrons are at the same spatial locations, while the symmetric spatial function does not vanish. Let us consider an antisymmetric spatial function

$$\frac{1}{\sqrt{2}}\,[1s(1)2s(2) - 2s(1)1s(2)] \qquad \text{...(1.76)}$$

Since the electron labels, 1 and 2, represent the coordinates of the two electrons, if the two electron are at the same location, *q*, the function is expressed as

$$\frac{1}{\sqrt{2}}\,[1s(q)2s(q) - 2s(q)1s(q)] = 0$$

The triplet electrons are anti-correlated. A plot of a function like (1.76) around the point *q* would show that the probability of finding the two electrons near each other is small and the probability vanishes at *q* for all *q*. By contrast, if the electrons are located at *q*, the symmetric spatial function is expressed as

$$\frac{1}{\sqrt{2}}\,[1s(q)2s(q) + 2s(q)1s(q)] \neq 0$$

Thus, the symmetric spatial function allows the 'two electrons' to reside at the same point in space, and the probability is not necessarily small for finding the two electrons at the same location or near each other.

The anti-correlation of the triplet electrons, which causes them to be further apart than the singlet electrons on average, reduces the 'triplet-state energy' by reducing the magnitude of 'electron-electron' repulsion. The electron-electron repulsion is a type of 'interaction', which increases the energy of a given state. Hence, the energy of the state is not solely determined by the orbitals, which make up the spatial wavefunction. The 'permutation symmetry' of the spatial wavefunction can play a significant role in determining the energy of a state.

1.4.4 Diamagnetism and Paramagnetism

The physical quantities like the magnetic field strength, H, and the magnetic induction, B, in vacuum, are related by the following relation

$$B = \mu_0 H \qquad \text{...(1.77)}$$

where, $\mu_0 = 4\pi \times 10^{-7}$ Vs/Amp is the permeability of free space. The magnetic state of a system (often called Sommerfeld System) will be specified by the magnetization M, which is related to B and H as

$$B = \mu_0 (H + M) \qquad \text{...(1.78)}$$

The magnetization M is equal to the density of magnetic dipole moments *m* as

$$M = m\,(N/V) \qquad \text{...(1.79)}$$

Generally, instead of the external field H, it is convenient to introduce an external 'induction' $B_0 = \mu_0 H$, and we can call the quantity B as simply the 'magnetic field strength'. In most cases in solids, there is a linear relation between the 'field' B_0 and the magnetization M as

$$\mu_0 M = \chi B_0 \quad ...(1.80)$$

where, χ is the 'magnetic susceptibility. The value of χ is important in determining whether the material is diamagnetic or paramagnetic.

If the value of χ is negative, then the induced magnetic polarization is opposite in sign to the applied field. Such behaviour is termed as "diamagnetic", while the reverse behaviour is termed as "paramagnetic", and is characterized by $\chi > 0$. In general, the susceptibility of atoms, and hence of solids, consist of a dia-magnetic and a para-magnetic component, which are denoted by χ_D and χ_P. The paramagnetic component is related to the 'intrinsic magnetic moments', which originate from the angular momentum and the spin of the electrons. For example, the magnetic dipole moment of an electron due to angular momentum can be written as

$$m = -\left(\frac{e}{2m}\right) \Sigma_i\, r_i \times p_i \; = -\mu_B\, L \quad ...(1.81)$$

with $\hbar\, \mathbf{L} = \mathbf{\Sigma_i}\, \boldsymbol{r_i} \times \boldsymbol{p_i}$ and the Bohr Magneton $\mu_B = \dfrac{e\hbar}{2m} = 5.7884 \times 10^{-5}$ eV/T $= 9.2742 \times 10^{-24}$ J/T ($1T = 1$ Tesla $= 1\ Vs/m^2 = 10^4$ Gauss).

The negative sign in (1.81) follows from the fact that the electric current has the opposite sense to that of the particle current due to the negative charge of the electron. Apart from the magnetic moment due to the angular momentum, the electrons also possess a magnetic moment due to 'spin', and the summation of this factor gives the spin moment of the whole atom as

$$m = \mu_B\, g_0\, \Sigma_i\, s_i = \mu_B\, g_0\, S \quad ...(1.82)$$

Here, g_0 is the 'electronic *g* factor' or 'Lande *g* factor' (= 2.0023 for free electron value) and s_i are the (negative) electron spins. As already shown in (1.81) and (1.82), L and S can be treated as 'operators'. The choice of sign of the 'spin operator' is best made so that the spin operator and the magnetic moment have the same sign. At this stage, it is better to talk about the spin-orbit coupling or L – S coupling.

Spin-Orbit Coupling

The importance of these operators needs to be understood in the light of 'spin-orbit' coupling, which lifts the 'degeneracy' of electronic states. In the Schrödinger treatment of hydrogen like atoms, the terms in the Hamiltonian arising from the electron spin have not been included. Thus, the six states of a single *p* electron in sodium, for example, would be degenerate, which is important for the understanding of the fluorescence of Na – D lines. However, the 'intrinsic angular momentum' of the electron gives rise to a magnetic dipole, which is moving in the electric field of the atom or molecule. The 'interaction' of a moving dipole with an electric field produces 'changes' in the electronic energy → and that's how the 'degeneracy' is lifted by the spin-orbit coupling or L – S coupling. For a hydrogen like atom, the spin-orbit coupling portion of the Hamiltonian is given by

$$H_{so} = \frac{Ze^2}{8\pi\varepsilon_0\, m^2 c}\left(\frac{1}{r^3}\right)[r \times p].S$$

$$= \frac{Ze^2}{8\pi\varepsilon_0\, m^2 c}\left(\frac{1}{r^3}\right) \text{L.S} \qquad ...(1.83)$$

since $(r \times p)$ is the orbital angular momentum (L). S operates on the electron spin wavefunction. L operates on the orbital angular momentum wavefunction (the spherical harmonics), and $\frac{1}{r^3}$ operates on the radial part of the wavefunction, and finally the Hamiltonian is expressed as

$$H_{so} = a(r) \text{ L.S}$$

For hydrogen like atoms, the value of the coefficient of L.S is expressed as

$$a(r) \propto Z^4$$

The Z^4 dependence gives rise to 'heavy atom effect' → *i.e.,* the 'spin-orbit' coupling increases very rapidly with nuclear charge (Z) of the atoms or molecules. Since atomic or molecular wavefunctions are spatially 'extended', there is some probability of finding a molecule's electrons on the 'heavy atoms' of the solvent. The electron delocalization is responsible for the 'external' heavy atom effect.

Finally, by evaluating the expectation values of the operators L and S for atoms, it can be seen that a non-vanishing expectation value results only for open shells. For closed shells, the sum of the angular momentum and spins is zero. In solids, we have 'open shells' for transition metals and rare earths. The paramagnetic behaviour is thus expected for solids containing both these elements with a finite expectation value (see **section 1.2.5** for expectation value).

Apart from paramagnetism due to the electrons, we must also consider diamagnetism. The latter results from the induction of 'eddy' currents by an external magnetic field. According to the Lenz's rule, the magnetic moment of these induced currents is opposed to the applied field. The susceptibility thereby acquires a negative diamagnetic contribution. The susceptibility relation is written as

$$\chi = -\left(\frac{e^2 n}{6m}\right)\mu_0\, \Sigma_i < \phi \mid r_i^2 \mid \phi > \qquad ...(1.84)$$

where n is the number of atoms per unit volume. In the sum over the matrix elements, the electrons in the outer shells are naturally of great importance, since their mean square distance from the nucleus is the largest. If the number of outer electrons is Z_a and we include the square of the atomic or ionic radii r_a in place of the mean of r_i^2, then we get

$$\chi = -\left(\frac{e^2}{6m}\right)\mu_0\, n\, Z_a\, r_a^2 \qquad ...(1.85)$$

The measured values of the diamagnetic susceptibility for atoms and ions with closed shells are indeed found to be in good agreement with $Z_a r_a^2$. A similar order of magnitude results for paramagnetic contributions. For a typical solid state density of 0.2 mol/cc, the susceptibility is about 10^{-4} (SI units), *i.e.,* small compared with 1. Therefore, apart from the solids showing ferromagnetism, the magnetic susceptibility of solids is small. By contrast, the electric susceptibility is of the order of 1 or larger.

Now, this should explain why in solid state spectroscopy with 'electromagnetic radiation', which is one of the most important experimental methods, we usually consider only electronic effects.

1.5 BAND STRUCTURE IN SOLIDS

Preamble

The semiconductivity is an important aspect in solids arising out of our knowledge in quantum mechanics with diverse applications in material components and devices in the field of electronics. In order to understand this most important semiconducting property of solid materials, it is of fundamental interest to know about the band structure in solids, *i.e.*, how 'allowed' and 'forbidden' energy gaps arise in solids. It is through a mathematical technique called Wronskian dealing with dependent and independent solutions of vector functions, the picture of the band gap will be made clear. But, in order to go to that step, an idea about the periodic Bloch function is necessary, *i.e.*, the potential of the electrons in a periodic crystal.

1.5.1 The Bloch Function

Before discussing the Bloch function, it is important to write a few lines on the 'Electron Kinetics' in solids :

Electron Kinetics. A convenient approach to the electronic properties of a solid consists of using the 'one-electron' motion, where core electrons are treated almost completely 'localised', and the energy eigenvalue problem for the valence electron system reduces to → solving the Schrödinger wave equation for each valence electron → as it moves in the spatially-dependent potential energy [V(r)].

As we deal with a large number of electrons in a solid, their interactions are important to describe by various electronic transport phenomena in solids. Although the 'nuclear motion' is not stationary, we still assume it to be at rest → so we need 'approximation', like Born-Oppenheimer approximation. In such an approximation, the total wave function for the system is given by a combination of wave functions. Each of these involves the 'coordinates' of only one electron → which gives rise to "one-electron approximation".

Within this framework → there are two different approaches

1. **Heitler – London or Valence Bond Scheme.** when the atoms are far apart from each other *i.e.,* the electrons are in 'localised' states.
2. **Bloch Approach.** where the electron is considered to be belonging to the three-dimensional crystal as a whole rather than to a particular atom.

The problem here involves an electron in a potential with the periodicity of the crystal lattice, which leads to a "natural distinction" between metals, insulators and semiconductors → in terms of band structure of these solids. However, first of all, we have to consider the 'analogy' between the electronic motion and elastic waves.

The Analogy

(*a*) The propagation of 'elastic waves' in a continuum or in a periodic structure, and

(*b*) Electronic motion in a constant or a periodic potential.

(*a*) **Elastic Waves**

For elastic waves in a continuous medium, the frequency (ν) is inversely proportional to the wavelength (λ) with a 'linear' relation between ν and the 'wave number' or the 'wave vector' → which implies that the "velocity of propagation" is independent of λ, with no upper limit for the 'frequency of vibrational modes' in a continuous medium. However, when we consider the 'modes of vibration' in a "lattice" of discrete points → which form a periodic structure → two characteristic features appear

There exists ALLOWED frequency bands, which are separated by FORBIDDEN regions of frequency or energy.

The frequency is no longer proportional to the "**wavenumber**", but a periodic function of the "**lattice**".

(*b*) **Electronic Motion**

In a constant potential, *i.e.,* the 'free' electron theory, the energy of the electron as a function of 'wave vector' $\boldsymbol{k}$ is given by : $E = \frac{\hbar^2 k^2}{2m}$, where $\boldsymbol{k} = \frac{2\pi}{\lambda} = \frac{p}{\hbar}$, λ = wavelength, p = momentum of the electron [$V(x) = 0$]. In this case, there is no upper limit to the energy → *i.e.,* the energy is 'quasi-continuous'.

However, if we consider the motion of an electron in a 'periodic potential' → we arrive at the following results :

There again exists 'allowed' energy bands, which are separated by 'forbidden' regions of energy, and the energy function $E(\boldsymbol{k})$ are periodic in $\boldsymbol{k}$.

In (*a*) above, we deal with the "elastic waves", and in (*b*) we deal with the 'waves' associated with the electrons → and hence this analogy is not surprising → both waves are moving in a periodic structure of a given lattice. It should be remembered that the 'discrete nature of solids' creating separate 'allowed' and 'forbidden' energy regions manifest in the observation of sharp resonance-like structures in the optical spectra of solids. Now, let us get into the most important theorem in formulating the 'band structure' in solids [7, 8].

1.5.2 The Bloch Theorem

In the 'free electron' theory → what do we assume?

Actually, we assume that an electron moves in a constant potential V_0 leading to a Schrödinger equation for a one-dimensional case

$$\frac{d^2\psi}{dx^2} + \left(\frac{2m}{\hbar}\right)[E - V_0]\psi = 0$$

The solution of this equation is given by the 'plane waves' of the type as

$$\psi(x) = e^{\pm ikx} \qquad ...(1.86)$$

Upon substitution, we get the 'kinetic energy' of the electron as

$$E_{kin} = E - V_0 = \frac{\hbar^2 \boldsymbol{k}^2}{2m} = \frac{p^2}{2m} \qquad ...(1.87)$$

The physical meaning of $\boldsymbol{k}$ is that it represents the momentum of electron divided by $\hbar$. In order to get a complete solution of the above wave function with time, we multiply $\psi(x)$ by $\exp(-i\omega t)$, where $\omega = \frac{E}{\hbar}$, so that the solutions of the type of equation (1.86) represents waves propagating along the x-axis.

The Periodic Potential

Now, let us consider the Schrödinger equation for an electron moving in a one-dimensional periodic potential. Therefore, the potential energy [V(x)] must satisfy the equation

$$V(x) = V(x + a) \quad ...(1.88)$$

where, a = period of the 'lattice'. The Schrödinger equation is then expressed as

$$\frac{d^2\psi}{dx^2} + \left(\frac{2m}{\hbar}\right)[E - V(x)]\psi = 0 \quad ...(1.89)$$

For a solution of this equation, there is an important 'theorem' → which states that there exists solutions of the form

$$\psi(x) = e^{\pm ikx}\, u_k(x) \quad \text{with } u_k(x) = u_k(x + a) \quad ...(1.90)$$

In other words, the solutions are 'plane waves', which are modulated by the function $u_k(x)$ → which has the same periodicity as the 'lattice'. This theorem is known as the "Bloch Theorem". In the theory of differential equations, it is called Floquet's theorem. The functions of this type (1.90) are called "Bloch Functions". The Bloch function has the property

$$\psi(x + a) = \exp[ik(x + a)]\, u_k(x + a) = \psi(x)\exp(ika)$$

since $u_k(x + a) = u_k(x)$. In other words, the Bloch functions have the property

$$\psi(x + a) = Q\,\psi(x), \quad \text{with} \quad Q = \exp(\pm\, ika) \quad ...(1.91)$$

Now, it is evident that → if we can show that the Schrödinger equation (1.89) has solutions with the property (1.91), the solutions can be written as Bloch functions, and the Bloch theorem is then proven.

The Proof

Let us suppose A(x) and B(x) are two 'real' independent solutions of the Schrödinger equation. We know that a differential equation of the 2nd order has only two independent solutions, and all other solutions can be expressed as a 'linear combination' of the independent ones. Hence, since A($x + a$) and B($x + a$) are also solutions of the Schrödinger equation, we get the following relations

$$A(x + a) = \alpha\, A(x) + \beta\, B(x)$$
$$B(x + a) = \gamma\, A(x) + \delta\, B(x) \quad ...(1.92)$$

where, α, β, γ and δ are real functions of E. The solution of the Schrödinger equation may be written in the form

$$\psi(x) = F\, A(x) + G\, B(x)$$

where, F and G are two arbitrary constants. According to (1.92), we get

$$\psi(x + a) = (F.\alpha + G.\gamma)\, A(x) + (F.\beta + G.\delta)\, B(x)$$

In view of the 'property' of the Bloch function in equation (1.91), let us now choose F and G in such a manner, so that we get

$$F.\alpha + G.\gamma = Q\, F$$
$$F.\beta + G.\delta = Q\, G \quad ...(1.93)$$

where, Q = Constant. In this way, we have obtained a function $\psi(x)$ with the following 'property'

$$\psi(x + a) = Q\,\psi(x) \quad ...(1.94)$$

i.e., equation (1.91). Since equation (1.93) have the 'non-vanishing' solutions for F and G, only if the "determinant" of their coefficients vanishes, so that we have the equation for Q as

$$\begin{vmatrix} \alpha - Q & \gamma \\ \beta & \delta - Q \end{vmatrix} = 0$$

or, $$Q^2 - (\alpha + \delta)Q + \alpha.\delta - \beta.\gamma = 0 \qquad ...(1.95)$$

Now, it is possible to show that $\alpha.\delta - \beta.\gamma = 1$ in the following **manner** : we can derive from equation (1.92)

$$\begin{vmatrix} A(x+a) & B(x+a) \\ A'(x+a) & B'(x+a) \end{vmatrix} \begin{vmatrix} A(x) & B(x) \\ A'(x) & B'(x) \end{vmatrix} \begin{vmatrix} \alpha & \beta \\ \gamma & \delta \end{vmatrix} \qquad ...(1.96)$$

where $A'(x) = \dfrac{dA(x)}{dx}$ and $B'(x) = \dfrac{dB(x)}{dx}$. Now, if we multiply the Schrödinger equation for B(*x*) by A(*x*), and the equation for A(*x*) by B(*x*), and we do the required subtraction, we get

$$0 = A\,B'' - B\,A'' = \frac{d}{dx}(A\,B' - B\,A')$$

Hence, in this case, the so-called "Wronskian" is a constant :

$$W(x) = \begin{vmatrix} A(x) & B(x) \\ A'(x) & B'(x) \end{vmatrix} = \text{Constant} \qquad ...(1.97)$$

This result together with equation (1.96) leads to the conclusion that $\alpha.\delta - \beta.\gamma = 1$. Therefore, instead of equation (1.95), we may write it as

$$Q^2 - (\alpha + \delta)Q + 1 = 0 \qquad ...(1.98)$$

Then, in general, there are two 'roots' : Q_1 and Q_2 , *i.e.*, there are two functions $\psi_1(x)$ and $\psi_2(x)$, which exhibit the property (1.93). It should be noted that the product $Q_1Q_2 = 1$.

For a certain range of energy E, *i.e.*, for those corresponding to $(\alpha + \delta)^2 < 4 \rightarrow$ The two roots Q_1 and Q_2 will be "complex", and since $Q_1Q_2 = 1$, they will be "conjugates". In those regions of energy, we may then write them as

$$Q_1 = \exp(\boldsymbol{ika})$$

and, $$Q_2 = \exp(-\boldsymbol{ika}) \qquad ...(1.99)$$

Then, the corresponding eigenfunctions $\psi_1(x)$ and $\psi_2(x)$ have the 'property' written as

$$\psi_1(x + a) = \exp(ika)\psi_1(x)$$

and, $$\psi_2(x + a) = \exp(-\,ika)\psi_2(x) \qquad ...(1.100)$$

Thus, following equation (1.90), they are the "Bloch Functions".

In other regions of energy E, where, if we put $(\alpha + \delta)^2 > 4$, the two 'roots' are real and reciprocals of each other. The roots corresponding to the solution of Schrödinger equation are

$$\psi_1(x) = \exp(\mu x)u(x)$$

and, $$\psi_2(x) = \exp(-\,\mu x)u(x)$$

where, $\mu = a$ real quantity. Although they are mathematically all right, they cannot be accepted as "wave functions" describing the electrons, since they are not bounded. Thus, there are no electronic states in the energy regions corresponding to the real 'roots' Q_1 and Q_2 .

Therefore, we get the "notion" that the energy spectrum of an electron in a periodic potential consists of → ALLOWED and FORBIDDEN energy regions or bands.

1.5.3 Band Structure in Three-Dimensions

From a general point of view, we can discuss the 'motion of electrons' in a 3-dimensional lattice. The most fundamental property of an 'infinite' crystal with "primitive" translation vectors : a, b, c, is that if we make a translation corresponding to any vector d as

$$d = n_1 a + n_2 b + n_3 c \quad \text{...(1.101)}$$

where, n_1, n_2, n_3 are integers, we arrive at a point which is geometrically equivalent to the point we started from → *i.e.,* periodic. Thus, the physical properties of crystalline solids remain unchanged, when we make a translation defined by any vector of the type d. For example, if the potential energy of an electron is given by $V(r)$, then we must have

$$V(r) = V(r + d) \quad \text{...(1.102)}$$

Vectors like d are called direct or real lattice vectors in order to distinguish such vectors from "reciprocal lattice vectors".

Introduction of Fourier Series

In order to discuss the behaviour of an electron in a periodic lattice potential, it is convenient to first consider : how we represent 'periodic function' like equation (1.102) in terms of 3-dimensional Fourier series → this will show why we need to go to Fourier space for band structure calculation. The Fourier series is based on a 'periodic function' following dirichlet conditions. For a 1-dimensional periodic potential, which satisfies the condition (*i.e.,* the same periodicity of the lattice)

$$V(x) = V(x + n_1 a)$$

where, n_1 = integer, and we can always write by making a Fourier expansion as

$$V(x) = \Sigma_n V_n \exp\left(2\pi i \frac{nx}{a}\right) \quad \text{...(1.103)}$$

where, n = integer, and the summation extends over all integers from $-\infty$ to $+\infty$, as in a Fourier series. The coefficients V_n are the Fourier coefficients.

But how do we know that this series indeed satisfies the 'periodicity' requirements ?

This is shown below :

If we replace x in (1.103) by $x + n_1 a$, where n_1 is an integer, we get

$$V(x + n_1 a) = \Sigma_n V_n \exp\left(2\pi i \frac{nx}{a} + 2\pi i\, n\, n_1\right) \quad \text{...(1.104)}$$

However, since nn_1 is also integer, $\exp(2\pi i\, nn_1) = 1$, and the right-hand side of this expression equals $V(x)$. Similarly, the potential in a cubic lattice satisfies the requirement

$$V(x, y, z) = V(x + n_1 a, y + n_2 a, z + n_3 a) \quad \text{...(1.105)}$$

which can also be written in terms of a 3-dimensional Fourier series. Let us now return to the general 3-dimensional lattice for which the 'primitive translations' : *a*, *b*, *c* are not necessarily equal in magnitude, nor perpendicular to each other. It is not immediately obvious how we can actually represent a 'potential' with the periodicity (1.104) in terms of a 3-dimensional Fourier series, *i.e.,* a periodic potential.

But it can be done ! How ?

It can be done rather easily, if we introduce the so-called 'reciprocal lattice'. Therefore, now we need to define the 'reciprocal lattice'. From the vectoral point of view, it is defined by three primitive translations : g_1, g_2, g_3, which satisfy the conditions

$$a_i \cdot b_j = \delta_{ij} = \begin{cases} 1 & \text{if} \quad i = j \\ 0 & \text{if} \quad i \neq j \end{cases} \qquad ...(1.106)$$

Therefore, the vector g_1 is perpendicular to the plane through the direct lattice vector *b* and *c*. The explicit expressions for the *g's* are evidently of the form

$$g_1 = \frac{(b \times c)}{a \cdot (b \times c)}, \quad \text{etc.} \qquad ...(1.107)$$

from which the absolute magnitudes of the *g*'s may be obtained in terms of the primitive translations of the direct lattice. Any vector which is written as :

$$\mathrm{G} = h\, g_1 + k\, g_2 + 1\, g_3 \qquad ...(1.108)$$

where, *h*, *k*, 1 = integers, and the vector is called the 'reciprocal lattice vector'. The end points of these vectors define the reciprocal lattice points. It can be easily shown that the reciprocal lattice of a f.c.c. lattice is a b.c.c. lattice, and vice versa.

Let us now show that the 3-dimensional Fourier series

$$\mathrm{V}(r) = \Sigma_g \mathrm{V}_g \exp(2\pi i\, \mathrm{G} \cdot r) \qquad ...(1.109)$$

exhibits the periodicity requirement (1.102). The symbol V_g stands for $\mathrm{V}_{g1,\, g2,\, g3}$.

The Proof

$$\mathrm{V}(r + d) = \Sigma_g \mathrm{V}_g \exp[2\pi i\, (\mathrm{G} \cdot r + \mathrm{G} \cdot d)]$$

However, G . *d* according to (1.101), (1.106) and (1.108) is equal to $n_1 d_1 + n_2 d_2 + n_3 d_3$ which is an integer. Hence, the right hand side of this last expression is equal to V(*r*), which proves the statement. Hence, we are in a position to represent periodic function in three-dimensions in terms of Fourier series. Now, let us consider the motion of the electrons in a potential of three-dimensional periodicity.

First of all, the Bloch theorem can be extended to three-dimensions. Then the wave functions can be written as follows

$$\psi(r) = \exp(\pm\, ik.r)\, u_k(r) \qquad ...(1.110)$$

where, $u_k(r)$ has the periodicity of the lattice . Hence, in general, we can write it as

$$u_k(r) = \Sigma_g \mathrm{V}_g \exp(2\pi i\, \mathrm{G} \cdot r) \qquad ...(1.111)$$

where, G is a vector in the reciprocal lattice. *It can be shown that two Bloch functions for which the wave vectors differ by 2π times a reciprocal lattice vector are physically equivalent.* For example, let G be a reciprocal lattice vector and let us introduce instead of k, another wave vector $k' = k + 2\pi G$ in (1.110). Then, we can write it as :

$$\psi(r) = \exp(\pm ik'.r)\exp(+2\pi iG.\ r)\ u_k(r) = \exp(\pm ik'.r)\ uk'(r)$$

where, $u_{k'}(r)$ is still periodic, since $\exp(\pm 2\pi iG.\ r)$ is periodic, *i.e.,* we are still left with a Bloch function. Also, we can say that k is not uniquely determined, and that k and $k + 2\pi G$ correspond to physically equivalent states, *i.e.,* the Bloch waves whose wave vector differs by a 'reciprocal lattice vector' are identical. But we must always avoid the occurrence of physically equivalent solutions with different k values.

So, how do we do this ?

The only way to do this → is to restrict the range of k-values. It is most conveniently done by limiting the components of k along the directions of G to the following ranges

$$-\pi g_1 \le k_1 \le +\pi g_1$$
$$-\pi g_2 \le k_1 \le +\pi g_2$$
$$-\pi g_3 \le k_1 \le +\pi g_3 \qquad \text{...(1.112)}$$

In this case, we refer to k as the reduced wave vector. The region in k-space defined by (1.112) is referred as the first "Brillouin Zone" or the "reduced zone".

Some Conceptual Facts

At the Brillouin zone boundaries, the energy E(k) exhibits a 'discontinuity' as also in the case of a one-dimensional lattice. It should be noted that the values of the wave vector: $G.\ k = \frac{\pi n^2}{a}$ are those for which ***the electron suffers a Bragg reflection.*** An electron, which satisfies the Bragg condition cannot obviously 'penetrate' the lattice, since it suffers 'reflections'. Therefore, such an electron does not correspond to a "propagating wave" passing through the crystal, but to a "standing wave". The energy 'discontinuities' or the 'energy gaps' in the E(k) vs. k curve occurring at the Brillouin zone boundaries represent the energy ranges for which it is "IMPOSSIBLE" for an electron to move through the crystal. For example, if such electrons are incident on the crystal from the outside, they are totally reflected and unable to penetrate into the crystal lattice. This should clarify the conceptual nature of the 'Brillouin Zone' and the 'Band Gap', and this concept is obviously very important.

In such a situation, the Schrödinger equation becomes

$$H\psi_k = E(k)\ \psi_k$$

and that for the same problem, which is displaced by G,

$$H\psi_{k+G} = E(k+G)\ \psi_{k+G}$$

Since $\psi_{k+G}(r) = \psi_k(r)$, we have → $H\psi_k = E(k+G)\ \psi_k$, and hence we get

$$E(k) = E(k+G).$$

The behaviour of E(k) is periodic in k-space, it is sufficient to represent it in the first Brillouin zone, *i.e.,* between $-\pi/a$ and $+\pi/a$, or the 'reduced zone'. A typical curve of E(k) vs. **k** is shown in Fig. 1.4, where the allowed and the forbidden bands are clearly seen [7].

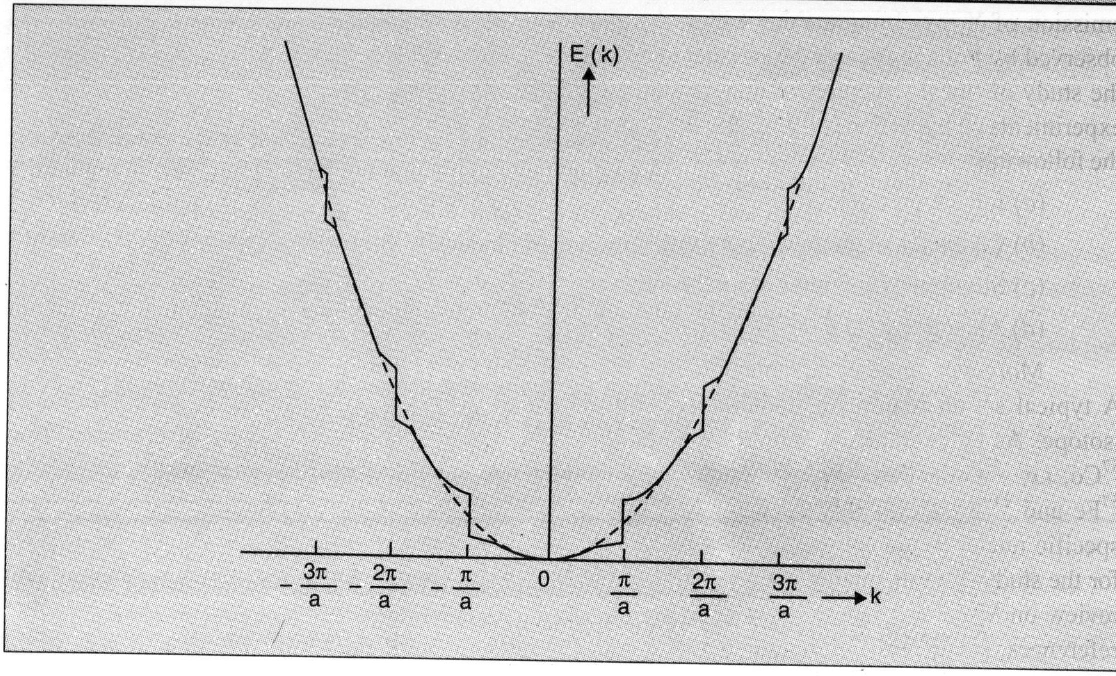

Figure 1.4: The energy [E(k)] vs. the wavenumber (k) curve. The discontinuities in the energy occur for $k = \pm n\pi/a$ (n = 0, 1, 2 . . . . and a = lattice spacing)

1.6. MÖSSBAUER AND ESR SPECTROSCOPY

Both the Mössbauer and ESR spectra measurements have been extremely useful in the study of the magnetic properties of materials, so both of them have a tremendous importance and utility in such type of study, particularly for the study of nano-materials. Both of them give us information on the micro-details in and around an atom. Since the nano-materials may contain only few atoms, *i.e.,* a few nanometers in size, these types of investigations assume a special significance. While Mössbauer spectroscopy deal with the nuclear aspects and their related issues, the study of ESR spectra deals with the electronic transitions between different energy levels, which are sensitive to their surroundings.

Therefore, both these types of studies are inherently 'magnetic' in nature and can give us an wealth of information on the nano-materials, when properly supplemented by magnetization measurements, as will be shown later in the **section — 5.5.** Here, first of all, the theoretical aspects of the Mössbauer spectroscopy will be described and then a discussion on the theoretical issues involved in the ESR spectroscopy will be elaborated. Both these topics are important to understand the magnetic phenomena of nano-sized magnetic materials, as described in the chapter - 5.

1.6.1 Mössbauer Spectroscopy

Preamble

First of all, it should be explained what is Mössbauer spectroscopy and then its importance can be described for the study of nano particles. The Mössbauer effect is basically a nuclear spectroscopy, *i.e.,* a study of nucleus of a Mössbauer atom. This effect is the totality of the resonant absorption or emission of X-rays by a nucleus within the atoms in a solid material. In a silicate glass, it was first observed by Pollack [9] in a Mössbauer spectrum of ^{57}Fe and it started becoming a powerful tool for the study of 'local structure' of non-crystalline solids [10]. As shown later in the **section — 5.5,** the experiments on hyperfine splitting and the dependence on temperature give us useful information about the following :

(*a*) Ionisation state,

(*b*) Character of chemical bonding,

(*c*) Strength of the inter-atomic forces,

(*d*) Anisotropy [11].

Moreover, another advantage is that the experimental set-up is simple and quite inexpensive. A typical set-up requires a 'radio-active source' which decays to an excited state of a Mössbauer isotope. As an example, it should be noted that ^{57}Fe is produced by the capture of electrons from ^{57}Co, *i.e.*, $^{57}Co + \beta = {}^{57}Fe$. Although many isotopes are suitable for Mössbauer experiments, only ^{57}Fe and ^{119}Sn are mainly used for Mössbauer measurements in non-crystalline materials. For a specific nuclei in the solids, the Mössbauer effect is very sensitive, and hence it can be very useful for the study of even smaller concentration of the Mössbauer isotope. On glassy materials, an excellent review on Mössbauer spectroscopy has been made by Kurkjian [12], but there are a number of other references, particularly in the context of local environment inside the glass, which are important to know about the short-range order in such materials [11, 13].

The relaxation of the 'excited nucleus state' into the 'ground state' of the Mössbauer isotope occurs with the emission of a quantum of energy (E_γ), which is typically of the order of 10—100 Kev. When the transmission of this energy (*i.e.*, X-rays) through a 'sample' containing the same isotope as the source is measured by a proportional counter, it is possible to distinguish between the Mössbauer X-ray and unwanted radiations.

When we deal with a 'free' atom, the emission of the quantum of γ-rays gives rise to a 'momentum transfer' to the nucleus by a "recoil" effect and it reduces the energy by some million ev. The natural line-width of such a transition is much smaller, and no 'resonant' absorption can be observed. By contrast, when we deal with a solid material, there is a probability for certain absorption or emission processes that the 'momentum' is taken up by the whole crystal, and a change in the quantum of energy of the γ-rays cannot occur. ***This resonance absorption or emission of the quantum of energy is known as the "Mössbauer Effect".***

Normally, in actual situations, the Mössbauer atom in the absorber can be in a different chemical environment than the Mössbauer atom in the source, and the energy levels of the nuclei of the two Mössbauer atoms (in the absorber and in the source) differ because of the hyperfine interactions (see later). A Mössbauer spectrometer takes "these small differences" in the transition energies into account, ***by vibrating the source*** which gives rise to a continuously varying 'Doppler' shift (originating from the 'thermal motion' of the nuclei) of the emitted radiation and possible resonant absorption in the absorber atom. Hence, the ability of absorption is measured in the Mössbauer spectra as a function of the "velocity" of the source, and hence the Mössbauer parameters are applied to a 'reference material', *i.e.*, the source.

The energy levels of the nucleus and a theoretical Mössbauer spectra of ^{57}Fe is shown in **Fig. 1.5.** There are mainly two important parameters, which are derived from the Mössbauer measurements :

(*a*) The Isomer Shift (δ)

(*b*) The Quadrupole Splitting (Δ)

The former represents the 'energy' arising out of nuclear charge distributions, and the latter expresses the 'electric potentials' at the nuclei resulting from the electrons. But, there is another important parameter, known as the 'hyperfine splitting", which is described in a practical example in case of nano particles of magnetite (see the **section — 5.5** for details)

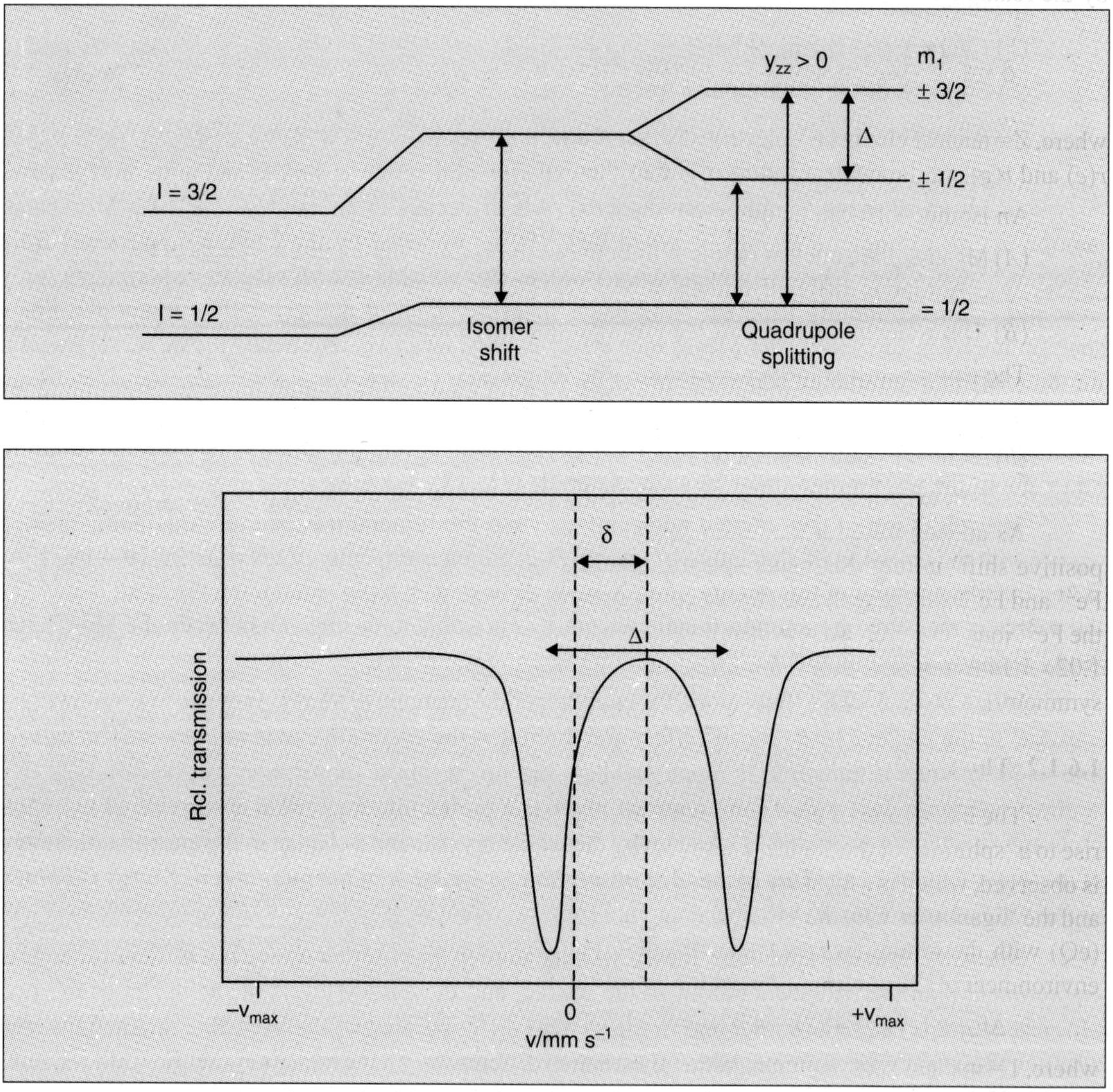

Figure 1.5: The isomer shift and quadrupole splitting resulting from different energy levels of the nuclei for a ^{57}Fe isotope.

1.6.1.1 The Isomer Shift

The value of "isomer shift" originates from the 'electro-static' interaction between the 'charge distribution' of the nucleus and those electrons from which the Schrödinger wave function (ψ) gives a finite 'probability' of being found in the region of the nucleus, *i.e.,* from the enhanced Coulombic interaction of the nuclear charge and the related electron charge. The "isomer shift" (δ) is defined as the 'electric' ⇔ 'monopole' interaction between the nuclear charge and the electric field due to the electrons. Basically, it gives a measure of the 'electron density' in the nucleus [ψ(0)], *i.e.,* the amount of the 'velocity' of the source either towards the absorber or away from it, which is expressed by the relation as

$$\delta = \left(\frac{2\pi}{3}\right) Ze^2 \{|\psi(0)_A|^2 - |\psi(0)_S|^2\}[<r(e)^2> - <r(g)^2>] \quad ...(1.113)$$

where, Z = nuclear charge, e = electron charge, A and S represent the absorber and the source respectively, $r(e)$ and $r(g)$ are the nuclear radius of the excited and the ground states respectively.

An isomer shift can be observed when

(*A*) Mean-square nuclear radius $<r^2>$ between the excited and the ground states of the Mössbauer isotope is different, and also

(*B*) The *s*-electron density of the absorber and the source is different.

The isomer shift is influenced mainly by two effects

(*a*) A direct change of the *s*-electron density of the outer *s*-shells, or

(*b*) A change in the population of the *p*, *d* or *f* valence electrons, which is known as the 'shielding effect'.

As an example, for the ^{57}Fe isotope, the difference of the radii is negative, and thus a 'more positive shift' in the Mössbauer spectra indicates a decrease of the *s*-electron density. Moreover, the Fe^{2+} and Fe^{3+} ions have the electronic configuration d^6 and d^5, which result in a more negative shift for the Fe^{3+} ions. The typical isomer shift values in $2CaO\text{-}3B_2O_3$ glasses containing 2 to 10% iron oxide are: 1.02 - 1.14 (mm/sec) for Fe^{2+} ions, 0.33 - 0.28 (mm/sec) for Fe^{3+} ions in tetrahedral coordination (lower symmetry), and 0.27 - 0.29 (mm/sec) for Fe^{3+} ions in octahedral coordination (higher symmetry) [14].

1.6.1.2 The Quadrupole Splitting

The interaction of non-cubic extra-nuclear 'electric fields' with the nuclear charge density gives rise to a 'splitting' of the nuclear energy levels. Thus, a characteristic two-line spectrum, *i.e.,* a 'doublet', is observed, which is a measure of the 'distortion' from the 'cubic symmetry' of the electron distribution and the 'ligands' around the Mössbauer atom. From the 'interaction' of the 'nuclear quadrupole moment' (eQ) with the inhomogeneous 'electric field gradient' (*eq*) of the electron charge distribution in the environment of the nucleons, we get the "quadrupole splitting" (Δ), which is given by the relation as

$$\Delta E_Q = (e^2qQ)/4I(2I - 1)[3(m_I)^2 - I(I + 1)](1 + \eta^2/3)^{1/2} \quad ...(1.114)$$

where, I = nuclear spin, m_I = magnetic quantum number, and η = asymmetry parameter, which is equal to zero, if there is no distortion from the cubic symmetry, *i.e.,* no asymmetry.

The inhomogeneity of the electric field at the site of the atomic nucleus results from :

(*a*) A 'non-spherical' or 'asymmetry' in the charge distribution, or

(*b*) From different 'ligands', *i.e.,* oxygen ions surrounding the metal cation in an oxide glass, which is popularly known as the 'ligand field environment'.

Now, it is seen that a measurement of the 'quadrupole splitting' can give us "local structural information" of a solid material, *e.g.,* the nano-crystalline particles of magnetite (see later in the **section-5.5**).

The nuclear states with spins $I > \frac{1}{2}$ split into $\frac{1}{2}$ (2I + 1), *i.e.,* into two levels for the ^{57}Fe isotope. In general, the quadrupole splitting in glasses is small for Fe^{3+} ions, *i.e.,* between 0 – 1 mm/sec, which have a nearly spherical charge distribution or symmetry. It is large for Fe^{2+} ions, *i.e.,* 1.5 – 2.5 mm/sec, which have a non-spherical electron wave function. The typical values of Δ in $2CaO\text{-}3B_2O_3$ glasses containing 2 to 10% iron oxide are : 2.26 – 2.21 (mm/sec) for Fe^{2+} ions, 1.91 – 1.47 (mm/sec) for Fe^{3+} ions in tetrahedral coordination (lower symmetry), and 1.12 – 0.91 (mm/sec) for Fe^{3+} ions in octahedral coordination (higher symmetry) [14].

1.6.1.3 The Hyperfine Splitting

The third and the most important parameter for the analysis of Mössbauer spectra of nano-crystalline particles of magnetite (Fe_3O_4) is the 'hyperfine splitting', which is actually a nuclear Zeeman effect. This effect occurs, when there is a magnetic field in and around the nucleus. In a given situation ($H \neq 0$), this magnetic field arises either within the atom itself or within the crystal through 'lattice exchange interactions' [11, 15].

The 'magnetic dipole' hyperfine interaction is described in terms of the Hamiltonian as

$$H_M = -\mu H = -g_N \beta_N I H \qquad ...(1.115)$$

where, μ = nuclear magnetic moment, β_N = nuclear Bohr magneton, and g_N = nuclear gyromagnetic ratio. The energy of the individual nuclear levels is expressed as

$$E_M = -\mu H m_I(1/I) = -g_N \beta_N H m_I \qquad ...(1.116)$$

Now, if the value of the quadrupole splitting is zero, the number of lines in the Mössbauer spectra is rather limited. Then, it is possible to observe six-lines hyperfine splitting spectra.

In many practical applications, like in the case of nano-crystalline particles of magnetite, it may so happen that the magnetic hyperfine interaction (due to nano-magnetic particles) and quadrupole hyperfine interaction (due to the presence of some isolate Fe ions in the system) are superimposed on one another, *i.e.,* the effect of first-order quadrupole perturbation on a magnetic hyperfine spectra. By a very careful Lorentzian line analysis by computer in such a complex situation, these two effects can be separated from a given Mössbauer spectra for the analysis of the nano-crystalline particles of magnetite, as shown in the **section - 5.5** [16-19].

1.6.1.4 Interpretation of the Mössbauer Data

After describing various Mössbauer parameters in the above subsections, it is necessary to give some insights on how to interpret complex Mössbauer spectra in terms of a Hamiltonian. The presence of an internal or external magnetic field completely removes the (2I + 1) degeneracy of the ground and excited states producing $\left(2\times\frac{1}{2}+1=2 \text{ and } 2\times\frac{3}{2}+1=4\right)$, *i.e.,* a total of six sub-levels, where I is the nuclear spin. In the case of iron absorber, *i.e.,* S-state ion like Fe, a pure nuclear Zeeman effect is observed producing a six-line hyperfine splitting (HFS) pattern with the intensities, which are governed by the transition probabilities between different sub-levels for the multipole transitions involved.

The hyperfine interactions depend on both the orbital and spin magnetism of the magnetic ions in crystalline solids. The theory is largely based on the work of Abragam and Pryce [20] in ESR, who introduced the so-called "spin-Hamiltonian" (H_{SH}). This Hamiltonian can also be used in the interpretation of most of the Mössbauer experimental data, which reflect the existence of one or more magnetic hyperfine interactions between the nucleus and its surroundings [15, 21].

In the spin-Hamiltonian containing different energy terms, the magnetic hyperfine interaction is represented as follows

$$H_{SH} = I \,.\, A \,.\, S - g_N \, \beta_N \, H_{ext} \, I \qquad \text{... (1.117)}$$

where, A is the hyperfine tensor, H_{ext} is the externally applied magnetic field.

For large external magnetic field ($g_N \beta_N H_{ext} >> A$), the spins are effectively quantized in the direction of the applied field and the off-diagonal matrix elements are usually neglected. In this case, the nucleus 'senses' an effective field (H_{eff}) that is created, such that the following relation can be written as :

$$I \,.\, A \,.\, <S> - g_N \, \beta_N \, H_{ext} \, I = - g_N \, \beta_N \, H_{eff} \, I \qquad \text{...(1.118)}$$

Therefore, H_{eff} is defined as :

$$H_{eff} = H_{ext} - \frac{\{A \,.\, <S>\}}{g_N \, \beta_N} \qquad \text{... (1.119)}$$

The electron spin enters the equation (1.118) as an 'expectation value', because it is determined by H_{ext}, which tends to hold it fixed in magnitude and direction. Let us define H_{hf} as

$$H_{hf} = - \frac{\{A \,.\, <S>\}}{g_N \, \beta_N} \qquad \text{...(1.120)}$$

Now, we can write the effective field (H_{eff}) as :

$$H_{eff} = H_{ext} + H_{hf} = H_{ext} + \frac{<S>}{S} H_{hf-sat} \qquad \text{... (1.121)}$$

where, $H_{hf\text{-}sat} = \dfrac{A.S}{g_N \beta_N}$ is the saturation value of the hyperfine field. In a given case, $H_{hf\text{-}sat}$ ($H \neq 0$) is not significantly higher than H_{hf} ($H = 0$), which means that the saturation occurs easily, but as shown later in the **section - 5.5,** it may not be case for certain range of sizes of nano-crystalline particles of magnetite embedded within a glassy matrix, which is an interesting system for various applications in nano-magnetics.

An analysis of the Mössbauer spectra of the magnetically ordered (inter-molecular) compounds, *e.g.*, ferrimagnetic materials, such as magnetite, shows that the direction of the Fe ion magnetic moments is the result of the 'competition' between the following terms :

(*a*) Exchange Energy (represented by an effective Molecular Field, H_E),

(*b*) Anisotropy Energy (represented by the Anisotropy Field, H_A), and

(*c*) External Magnetic Field (denoted by H_{ext}), if any.

Thus, it is clear that the magnetically split Mössbauer spectra provide useful information.

In fact, in bulk magnetite, the magnetic HFS pattern is observed at 4^0K without any H_{ext}, as shown later in the **section - 5.5.** With the application of H_{ext}, the anisotropy of small nano particles of magnetite becomes important. The effective hyperfine field (H_{hf}) can increase or decrease depending on the direction of the magnetization. From a detailed analysis of the Mössbauer spectra at various temperatures of measurements with and without any external magnetic field, the "anisotropy" of the ferrimagnetic nano particles of magnetite can also be determined through the estimated value of the hyperfine field (H_{hf}).

1.6.1.5 Collective Magnetic Excitation

This particular topic should be included in the description of the magnetic phenomena in the **section - 1.4,** since it involves the concept of ferromagnetism. However, it is discussed under Mössbauer spectroscopy, since certain aspects of the Mössbauer data can be interpreted by means of the concept of 'collective magnetic excitation', as will be detailed in the **section - 5.5.** As the 'collective electronic excitation' is called 'excitons', in the same way the 'collective magnetic excitation' is called 'magnons' [22].

In a simple ferromagnet, all the spins are parallel in the ground state. Let us consider a case with N number of spins. Each of these spins has a magnitude S on a line or a ring, with the nearest neighbour spins coupled by the Heisenberg interaction. If the 'exchange interaction' takes place between the spin in the pth state (S_p) with that of the spin in the $(p+1)^{th}$ state, *i.e.,* S_{p+1}, then the energy for such an interaction is written as

$$E = -2J \sum_{p=1}^{N} S_p\, S_{p+1} \qquad ...(1.122)$$

Here, J is the 'exchange coupling integral', and $\hbar S$ is the singular momentum of the spin in the pth state. Now, when we deal with the spin S_p as a classical vector, then in the ground state, we can write it as

$$S_p \cdot S_{p+1} = S^2 \qquad ....(1.123)$$

and in this case, the 'exchange energy' in the ground state of the system can be written as :

$$E_0 = -2N\,J\,S^2 \qquad ...(1.124)$$

Now, what is the energy of the 'first excited' state ?

In order to answer this question, first of all, we have to consider an 'excited state' with the direction of just one specific spin reversed. Immediately, we note that this particular reversal of a spin would increase the Heisenberg 'interaction energy' by $8JS^2$, and we can write the 'new interaction energy' as

$$E = E_0 + 8JS^2 \qquad ...(1.125)$$

The above deduction shows that the energy for the next 'excited state' is quite high. However, we could form an 'excited state' with excitation of much lower energy, provided we allow that all the spins share the above 'reversal' process, *i.e.,* all the spins in circular motion with respect to *z*-axis. This elementary 'excitations' of a system of spins possess a 'wavelike form' and are known as "spin waves". When this 'spin wave' is quantized, these excitations are called "magnons", *i.e.,* a 'collective excitation of spins', which are analogous to the 'collective lattice vibrations' or 'phonons'. In other words, while the phonons are 'oscillations' in the relative spatial positions of the atoms in a given lattice, the 'magnons' are the oscillations in the relative orientations of the spins on a given ferromagnetic cubic lattice.

There is an interesting story about the 'magnon dispersion' behaviour, which is not dealt here. It is sufficient to mention that the angular frequency of 'magnons' is proportional to $\boldsymbol{k}^2$, whereas that for phonons, it is proportional to only the wave vector $\boldsymbol{k}$. A classical representation of the 'collective magnetic excitation' or 'spin wave' is shown in **Fig. 1.6.**

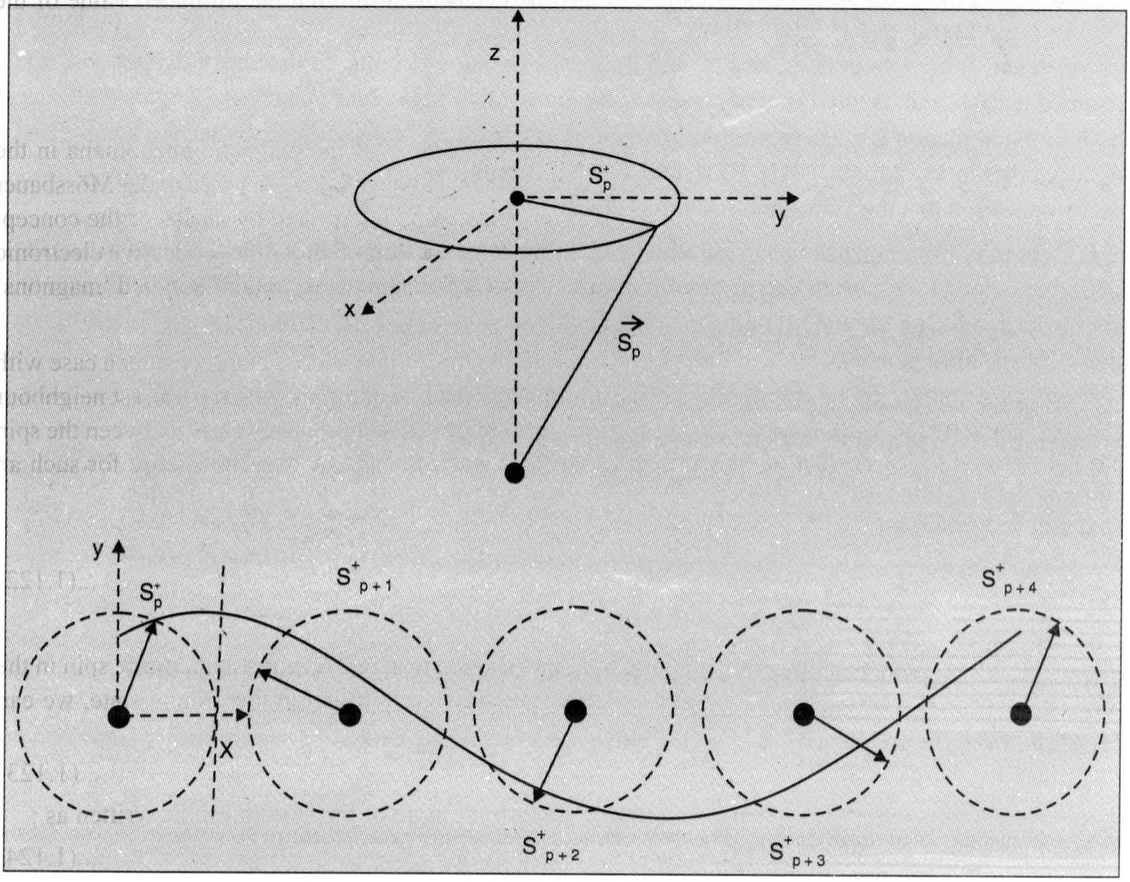

Figure 1.6: Collective magnetic excitation or spin wave in a ferromagnetic substance.

In the context of Mössbauer data, for ferrimagnetic magnetite particles in the nano-range **(6.4 nm to 7.0 nm),** there is a drop in the value of H_{hf} to 410 and 450 KG respectively from a value of 486 KG for the bulk magnetite of larger grain sizes [19]. This was explained in terms of a 'surface effect', but a more plausible model explains this behaviour, *i.e.,* lowering of the hyperfine field value, to a 'collective magnetic excitations' or 'magnons'. This is due to the fluctuations of magnetization directions around the energy minimum corresponding to an 'easy' direction of magnetization , *e.g.,* <111> directions of magnetite [19] (see the details in the **section - 5.5**).

1.6.1.6 Spin Canting

In a magnetic material, there is a magnetic lattice as well as sub-lattices. It is generally known that the '**intra**-sublattice' interactions, which are determined by a negative exchange forces or energy between the spins on the same magnetic sub-lattice, are much weaker than the '**inter**-sublattice' interactions. The molecular field or 'mean field' is denoted by : $B_m = \gamma M$ (γ = mean field constant).

A 'canted spin' arrangement can occur in a 'body centred cubic' **(BCC)** antiferromagnet with four magnetic sub-lattices, where the magnetization in each sublattice is not parallel to the corresponding molecular field. Let us say that these magnetic sub-lattices have magnetizations as M_1, M_2, M_3 and M_4 respectively. In a particular spin configuration, the molecular field in each of these sub-lattices will be normally given by four vector equations, which when solved give rise to the following relations

$$\alpha_1 M_A + M_B = 0 \qquad ...(1.126)$$

and, $$\beta_1 M_B + M_A = 0 \quad ...(1.127)$$

where, A and B are two lattices, and α_1 and β_1 are the two coefficients. In the above deductions, it is assumed that M_1 and M_2, and also M_3 and M_4 are antiparallel in the body-centred arrangement. Since the A ⇔ B interaction is much more stronger than 'intra-sublattice' interactions, we have to assume that M_1 is parallel to M_2, *i.e.* $M_A = M_1 + M_2$, and also M_3 is parallel to M_4, *i.e.*, $M_B = M_3 + M_4$. This has a clear implication that the above two equations (1.126 and 1.127) will have a finite value.

Let us consider that M_3 is not parallel to M_4 in the B lattice so that only $\beta_1 M_B + M_A = 0$, *i.e.*, the second equation (1.127) is not having any finite value. In such a situation, the 'canted spin' arrangement will be obtained, with M_3 and M_4 being no longer collinear, but they are deviating by some "angle" from their collinear arrangement.

For the reason of simplicity, if we consider that $M_3 = M_4$, then the following relations can be written involving the magnetization in the magnetic lattice A as

$$M_A - 2\beta_1 M_3 \cos\theta = 0 \quad ...(1.128)$$

or, it can be written as

$$\cos\theta = \frac{M_A}{2\beta_1 M_3} = \frac{M_A}{\beta_1 M_B} \leq 1 \quad ...(1.129)$$

where, $M_B = 2M_3$, which corresponds to the minimum 'angle' $\theta = 0$, *i.e.*, $\cos 0 = 1$. In this case of 'no deviation' from the collinearity, $\beta_1 = \frac{M_A}{M_B}$. Now, it is clear that a 'canted' spin arrangement may occur in the B lattice, provided $\beta_1 \geq \frac{M_A}{M_B}$ only. The 'canted spin' arrangement involving all the four magnetizations is shown in **Fig. 1.7.**

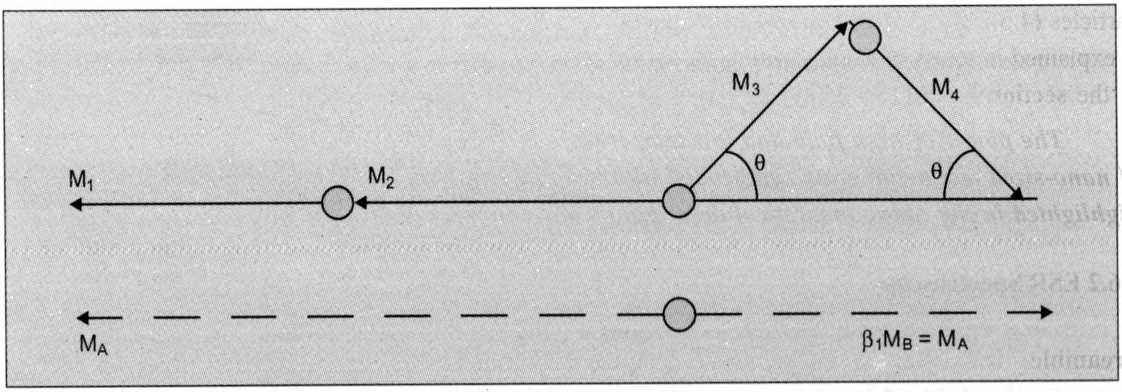

Figure 1.7: A typical diagram for a spin canting model in a BCC crystal

In the context of Mössbauer spectroscopy on nano-crystalline magnetite particles, according to Neel's theory of ferrimagnetism [23], the anti-parallel spin moments have unequal magnitude giving rise to a net resultant moment. In the canted spin arrangements, *i.e.*, the 'deviation' from the collinearity, there will be a disappearance of the 2nd and 5th lines in the six-line spectra of the hyperfine splitting, *i.e.*, $\Delta m_I = 0$ lines. This theory was extended by Yafet and Kittel [24] by taking care of the 'angle' between the two spins in the magnetic sub-lattices, which is based on the above formalism. The details are given in the **section - 5.5.**

The Summary

It is quite tempting to summarize one of the most important subsections of this chapter, since the concepts developed here are not only used in the present book on a detailed analysis of the Mössbauer spectra, but also it gives hints for many more applications in the future. A brief outline is given in the following.

A multi-component basalt glass is an interesting system for studying the 'superparamagnetic' behaviour in a certain nano-range and the ferrimagnetic behaviour of little higher nano-sized magnetite particles, whose growth has been controlled to precipitate even **4.5 nm** to **7.0 nm** size crystallites. Such nano-sized small particles of magnetite are superparamagnetic due to fast relaxation, which needs to be detected. One of the most useful techniques for such a study is Mössbauer spectroscopy both at high fields and at low temperatures. For experiments at 300K, a general two doublet spectra have been observed for samples heat-treated at 650° and 700°C, while six-line hyperfine spectra (HFS) were observed for the samples heated at higher temperatures (*i.e.,* for larger nano-sized crystallites). This latter behaviour has been ascribed to a high degree of anisotropy in the higher nano-sized ferrimagnetic grains.

The observed HFS pattern at 4K with H = 0 in the Mössbauer spectra has been found to be independent of the particle size. The Mössbauer spectra at 4K with $H \neq 0$ show that $\Delta m_I = 0$ lines do not vanish for such nano-sized magnetic particles below a certain 'critical' size within this nano-range, defying Neel's theory. But above this 'critical' size, these lines disappear with the progressive application of the magnetic field. This is a direct proof of Neel's theory above the critical region within such a narrow nano domain. This work shows the 'easiness' of producing nano-sized magnetite grains, which show superparamagnetic behaviour at 300K due to the fast relaxation, whereas the same particles **(4.5 nm – 5.5 nm)** show ferrimagnetic HFS lines at 4K with $\Delta m_I = 0$ lines still visible. This is explained in terms of a spin-canting model for nano-sized magnetite grains within a glass, as shown in the **section - 5.5** [16 - 19].

The power of high field and low temperature Mössbauer spectroscopy in detecting the role of nano-sized magnetite grains embedded within a glassy matrix and their distinction is definitely highlighted in the above theoretical descriptions and practical data later.

1.6.2 ESR Spectroscopy

Preamble

The above description about the Mössbauer spectroscopy involved nuclear phenomena. Here, the Electron Spin Resonance (ESR) involves the electron and its spin. The ESR is a very powerful method and one of the most sensitive techniques for detecting any atomic site and its symmetry. Often, it is coupled with optical spectroscopic measurements. This method aims to study the separation of electronic energy levels of the concerned atom in the presence of an external magnetic field [10].

Here, a brief outline is given on the theory of ESR in order to be able to understand and interpret the ESR spectra of small nano-sized ferrimagnetic magnetite grains in the **section - 5.6.**

1.6.2.1 The Theory of ESR

As an atom or free ion possesses a kinetic energy of K, its permanent dipolar magnetic moment can be written as

$$\mu = \gamma K \qquad ...(1.130)$$

where, γ is the gyromagnetic ratio, which is given by

$$\gamma = -\frac{(g\,e)}{2mc} \qquad ...(1.131)$$

where, e is the electronic charge, m the mass of the electron, c the speed of light, and g is a number whose value depends on the relative contribution of the orbital and spin. For a 'free' electron, $g = g_0 = 2.0023$ (by taking a correction, based on quantum electrodynamics).

If only the orbital moment (L) is present in the atom, then we can write the kinetic energy as

$$K = \hbar L \quad (g = g_L = 1) \qquad ...(1.132)$$

If only the spin moment (S) is present in the atom, then we can write the kinetic energy as

$$K = \hbar S \quad (g = g_S) \qquad ...(1.133)$$

When both L and S are present simultaneously, the value of $g = g_J$ is dependent on the L-S coupling, with $J = L + S$, and this is expressed as

$$g_J = \frac{[J(J+1)(g_L + g_S) + \{L(L+1) - S(S+1)\}\,(g_L + g_S)]}{2J(J+1)} \qquad ...(1.134)$$

The electronic dipolar magnetic moment is written as

$$\mu_J = -\left[\frac{(e\,g_J)}{2mc}\right] \hbar J = g_J\,\beta\,J \qquad ...(1.135)$$

where, β is the Bohr magneton $= \frac{e\hbar}{2mc}$.

In the presence of an applied magnetic field (H), the "Zeeman interaction" is written as

$$E = -\mu_J H = g_J\,\beta\,H\,M_J \qquad ...(1.136)$$

where, M_J is the component of the total angular momentum (J) in the direction of the field. The magnetic dipolar transition is produced following the selection rule : $\Delta M_J = \pm 1$. For an alternating field with frequency ν, which is applied perpendicularly to the field, there is an 'absorption' of energy, which can be written as

$$h\nu = g_J\,\beta\,H \qquad ...(1.137)$$

As the electron spin passes from the parallel position with respect to the magnetic field → towards the anti-parallel direction, an emission occurs in the inverse process. At the 'thermal equilibrium', a lot more number of spins is in the parallel position, which corresponds to a lower energy. Thus, a net absorption takes place. The energy separation ($g_J\,\beta$ H) between the Zeeman levels increases linearly with H. For a value of $g = 2.00$ and H = 3300 Gauss, the frequency corresponds to a value of $\nu \sim 10$ GHz, which conforms to the X-band frequency of the waveguide used in the ESR experiment. In the measurement of the ESR spectra, the frequency is actually kept constant at X-band ($\nu \sim 9.2$ — 9.4 GHz), but the magnetic field is varied to obtain the 'resonance'. The ESR spectra is actually a curve of

$d\chi/dH$ vs. H, where χ is the 'magnetic susceptibility' of the sample, which is put in the powder form in a special quartz tube inside the radio-frequency cavity that is placed between two large water-cooled magnets, whose directions can be varied for alignment.

The observed transition depends in a complex manner on the ions, *i.e.,* paramagnetic centres in the sample, of a given crystal field symmetry showing the effect of spin-orbit coupling and the hyperfine interaction between the electron and the nucleus. In such a case, *e.g.,* for V^{4+} ions in solids, a spin-Hamiltonian is used to interpret the observed ESR spectra. Due to the effect of the spin-orbit coupling (see the **section - 1.4.4**), the value of g of a material differs from that of g_S of a 'free spin'. The value of g depends on the orientation of the 'magnetic field' with respect to the 'crystal field'. Thus, it is a tensorial quantity [23-27].

The simpler form of the spin-Hamiltonia can be written as [20, 28]

$$H_{SH} = g\,H\,S + I\,A\,S \qquad ...(1.138)$$

where, in the first term, the interactions of the 'spin-orbit' and 'Zeeman-levels' are shown and the second term shows the 'hyperfine interaction' with the hyperfine tensor (A). From the ESR spectra, we measure the main components of the g-tensor and A-tensor, whose values characterize the symmetry position of a given ion [10].

1.6.2.2 ESR Spectra of Iron containing Materials

Since we are mainly interested in the nano-crystalline magnetite particles containing iron, its ESR spectra needs some mention here. Normally, a material, *i.e.,* a non-crystalline material like glass containing iron ions, the ESR spectra show two resonances as a function of the magnetic field at $g = 2.0$ and $g = 4.3$ respectively (see the figures in the **section - 5.6**). This type of spectrum can be described by the usual spin Hamiltonian as [20]

$$H_{SH} = g\,\beta\,H\,S + D[(S_x)^2 - (1/3)S(S+1)] + E[(S_x)^2 - (S_y)^2] \qquad ...(1.139)$$

where, D and E are the crystal field parameters for the iron ions, which determine the resonance positions. In the crystal field scenario, there are three Kramer's doublets. The resonance at $g = 4.3$ arises due to the transition within the middle doublet. The other resonance at $g = 2.0$ arises due to the transition in the lowest doublet [29-30]. The $g = 4.3$ resonance is ascribed to the 'tetrahedral' Fe^{3+} ions [31-34], but both 'tetrahedral' and 'octahedral' Fe^{3+} ions with 'rhombic distortions' could produce this resonance [29, 30]. The values of D range from 0.24 to 0.32 cm^{-1} in different borate glasses represents a 'rhombic distortion' of the Fe^{3+} ions [30]. The $g = 2.0$ resonance is normally ascribed to the presence of 'octahedral' Fe^{3+} ions, but it was again disputed [29]. The resonance at $g = 2.0$ is thought to arise from the 'clusters' of paramagnetic' ions containing two or more ions coupled by 'exchange interaction', because the analysis of an interaction Hamiltonian could give rise to a g-value close to 2.0 [35].

As shown in the **section - 5.6**, the line-width analysis complicates the interpretation of the ESR spectra, as compared to that of the Mössbauer data, although both the spectroscopy are highly suitable to study the nano-crystalline particles of magnetite containing both ferrous and ferric ions [16-19]. The increase of line-width as a function of an independent variable indicates 'free' ions with some kind of 'rhombic distortion', *i.e.,* a lower symmetry, but a decrease in the line-width is ascribed to the 'cluster' formation [35]. This sums up a short discussion on the interpretation of ESR spectra for iron containing glasses, which is of some interest to us.

1.7 OPTICAL PHENOMENA

Preamble

In optical phenomena, three topics are of primary importance

(*a*) Absorption for optical properties of materials, *e.g.*, for glasses and some crystals,

(*b*) Stimulated Emission for laser materials, *e.g.*, for Nd-laser glasses and ruby crystal laser,

(*c*) Spontaneous Emission for fluorescent materials.

These properties are normally discussed semi-classically [36, 37]. But, first of all, one of most significant discoveries of the 19th century, by a British scientist named J. C. Maxwell who remains to be legendary in modern science, should be discussed. Obviously, it gives indication towards the "wave like nature of 'light", *i.e.*, the 'electromagnetic wave equation' of Maxwell. The latter has given rise to the study of 'electrodynamics' during the last 100 years or so in order to give us a full understanding of this most important branch of physics. So, it is only proper to deal with this subject very briefly.

1.7.1. Electrodynamics and Light

As it is already known that one very important application of the 'vector relations' consists of the derivation of the electromagnetic wave equations, which are most commonly termed as Maxwell's equations in vacuum. These equations are written as

$$\nabla \cdot \mathbf{B} = 0$$

$$\nabla \cdot \mathbf{E} = 0$$

$$\nabla \times \mathbf{B} = \varepsilon_0 \mu_0 \left(\frac{\partial \mathbf{E}}{\partial t}\right)$$

$$\nabla \times \mathbf{E} = -\left(\frac{\partial \mathbf{E}}{\partial t}\right)$$

where, **B** is the magnetic induction, **E** is the electric field, ε_0 is the dielectric permittivity and μ_0 is the magnetic permeability in mks or SI units in vacuum. Let us decide to eliminate **B** from the third and fourth equations of Maxwell, which can be done by simply taking curl of both sides of the fourth equation and by using the 'time derivative' from the third equation. We can do this, since both time and space derivatives commute.

Now, we can do the above operation as

$$\left(\frac{\partial}{\partial t}\right) \nabla \times \mathbf{B} = \nabla \times \left(\frac{\partial \mathbf{B}}{\partial t}\right)$$

and,
$$\nabla \times (\nabla \times \mathbf{E}) = -\varepsilon_0 \mu_0 \left(\frac{\partial^2 \mathbf{E}}{\partial t^2}\right)$$

Actually, Maxwell's equations are first order "coupled" partial differential equations. In order to decouple them, it is always convenient to take the curl of a curl, and we get

$$\nabla \cdot \nabla \mathbf{E} = \varepsilon_0 \mu_0 \left(\frac{\partial^2 \mathbf{E}}{\partial t^2}\right)$$

or, $$\nabla^2 \mathbf{E} = \varepsilon_0 \mu_0 \left(\frac{\partial^2 \mathbf{E}}{\partial t^2} \right)$$

In the same way, by taking the curl of a curl in the third equation, we get

$$\nabla^2 \mathbf{B} = \varepsilon_0 \mu_0 \left(\frac{\partial^2 \mathbf{B}}{\partial t^2} \right)$$

Now, we see that the first order 'coupled' equations are decoupled, but with a price of getting the second order equations with the Laplacian. In vacuum, where there is no charge or current, the components of **E** and **B** satisfy the following equation as :

$$\nabla^2 f = \frac{1}{V^2} \cdot \frac{\partial^2 f}{\partial t^2}$$

This is the famous electromagnetic vector wave equation, which implies a traveling wave moving with a velocity V. As per Maxwell's equations, we can see that the vacuum supports the propagating electromagnetic waves at a speed, which is written as :

$$V = \frac{1}{\sqrt{\varepsilon_0 \mu_0}} = 3.00 \times 10 \text{ m/s}$$

This is precisely the "speed" of light (c) after we put the values of the dielectric permittivity and magnetic permeability in vacuum. This has an outstanding implication that 'light' is an electromagnetic wave. This shows the power of Vector Calculus and Maxwell's full understanding of it, and consequent judicious use of this methodology towards making an important contribution in the field of physics.

Finally, with the risk of some repetition, we are tempted to say that the "story of light" is nothing but a simple mathematical manipulation of "decoupling" a set of 'coupled first order partial differential equations' into a set of 'second order equations' involving a Laplacian in the Cartesian coordinates, which are little more difficult to solve, but our purpose is well served, since we get the wave equation of 'light'. This mathematical technique, although known to many of us, makes one of the most wonderful discoveries of all times, and we get 'light' on our eyes.

1.7.2. Transition Dipole Bracket

In this semi-classical approach, the electromagnetic radiation field is treated classically, while the molecular eigenstates are quantized. First, by using Maxwell relation of electric and magnetic fields (including the magnetic vector potential A_x) and using Euler-Lagrange equation of motion, a classical Hamiltonian is constructed. This is converted to a Quantum Mechanical Hamiltonian by replacing the classical momentum with the corresponding quantum mechanical operators, which is a standard practice.

After applying a time-dependent perturbation on this QM Hamiltonian, ultimately the 'transition dipole bracket' (X_{m-n}) for an optical transition between two energy states, *i.e.*, optical energy transfer between two eigenstates : E_n (lower energy) to E_m (higher energy), is found out as

$$<\psi_m^{\,0}\,(q, t) \mid H' \mid \psi_n^{\,0}\,(q, t)> = -i\frac{1}{c\hbar}\; A_x(E_m - E_n)\, X_{m-n}\, e^{i(E_m - E_n)t/\hbar} \qquad ...(1.140)$$

with $$X_{m-n} = <\psi_m^{\,0} \mid e\,\Sigma_j\, X_j \mid \psi_n^{\,0}> \qquad ...(1.141)$$

X_{m-n} is the 'transition dipole bracket' with a dimension of a dipole (*i.e.*, charge times length), but it does not represent a permanent dipole, since it couples two zeroth-order eigenstates. Within the bracket, the operator is the 'transition dipole moment operator' for the electro-magnetic radiation field with the electric field of the 'light', which is aligned in the x-direction (similarly, it can be for y and z directions) [7].

1.7.3 Transition Probabilities for Absorption

The radiation field, which is a time-dependent perturbation [$H'(t)$], generates a probability amplitude (C_m) for finding a molecule or an atom (*i.e.*, the system) in an excited state $< \psi_m^{\,0}\,(q, t) \mid$ from an initial eigenstate $\mid \psi_n^{\,0}\,(q, t) >$. These are the eigenstates of the molecular Hamiltonian in the absence of any radiation field.

In a typical 'spectroscopy experiment', the time-dependent perturbation theory can be used in order to calculate C_m as

$$\frac{dC_m}{dt} = \frac{i}{\hbar} < \psi_m^{\,0}\,(q, t) \mid H \mid \psi_n^{\,0}\,(q, t)> \qquad ...(1.142)$$

The bracket is defined in equation (1.140) with the 'transition dipole moment' bracket defined in equation (1.141). For "light" of frequency ν, the 'vector potential' (A_x) is given by

$$A_x = A_x(0)\cos(2\pi\nu t) = \frac{A_x(0)}{2}\,(e^{i2\pi\nu t} + e^{-i\,2\pi\nu t}) \qquad ...(1.143)$$

Then, after including the 'vector potential', the equation (1.142) becomes

$$\frac{dC_m}{dt} = -\frac{1}{2c\hbar^2}\, A_x(0)\, X_{m-n}\,(E_m - E_n)\,(e^{i(E_m - E_n + h\nu)t/\hbar} + e^{i(E_m - E_n - h\nu)t/\hbar}) \qquad ...(1.144)$$

If the system is initially in the state $< \psi_n^{\,0} >$, *i.e.*, $C_m = 0$ at $t = 0$, then by multiplying equation (1.144) by dt and by integrating, let us choose a constant of integration at the initial boundary condition, and we eventually get

$$C_m = \frac{1}{2c\hbar}\, A_x(0)\, X_{m-n}\,(E_m - E_n)\left[\frac{e^{i(E_m - E_n + h\nu)t/\hbar} - 1}{(E_m - E_n + h\nu)} + \frac{e^{i(E_m - E_n - h\nu)t/\hbar} - 1}{(E_m - E_n - h\nu)}\right] \qquad ...(1.145)$$

Now, What do we do?

First, let us consider the case when $E_m > E_n \rightarrow$ this situation indicates the "absorption" of radiation. When $E_m > E_n$, as $h\nu \rightarrow (E_m - E_n)$, the denominator of the first term in the **third bracket** goes to $2h\nu$, while the denominator of the second terms goes to zero, *i.e.*, the second term is relatively large compared to the first term, when the frequency or the energy of the 'light' is at or near the frequency of the transition.

The second term that is kept → is on or near the 'resonance', and the first term that is dropped is effectively 2ν "off" the 'resonance'. Hence, the term that is kept → represents the component of the 'light field' rotating in a sense, which is required to be coupled to the "atomic or molecular" transition moment, which finally induces the "absorption". So, that's how the 'absorption process' takes place in an atomic solid material, and we have to find out the 'probability' of this transition to know more about the absorption process.

The probability of finding the system in the excited state $< \psi_m^0 |$ at time t is $C_m C_m^*$. Let us 'create this product' and use the following trigonometric identities, and then we write it as

$$(e^{ix} - 1)(e^{-ix} - 1) = 2(1 - \cos x) = 4 \sin^2 \frac{x}{2} \qquad \text{...(1.146)}$$

Thus, we can get the above probability (*i.e.,* the above product) as

$$C_m C_m^* = \frac{1}{c^2\hbar^2} \,|\, A_x(0) \,|^2 \,|\, X_{m-n} \,|^2 \,(E_m - E_n)^2 \, \frac{\sin^2 \left[\frac{(E_m - E_n - h\nu)t}{2\hbar} \right]}{[(E_m - E_n - h\nu)]^2} \qquad \text{...(1.147)}$$

Let us take $(E_m - E_n) = E$, where E is the difference in energy between the two eigenstates of the time-independent Hamiltonian, *i.e.,* $\Delta E = E - h\nu$. Thus, ΔE is the quantity of radiation field, which is "off" the 'resonance' from the energy of the transition, *i.e.* we have to eliminate it or make it vanish, if a 'resonance' has to occur. The maximum 'transition probability' has to take place on the 'resonance', *i.e.* $\Delta E = 0$.

As $\Delta E \to 0$, $\sin^2 \left(\frac{\Delta E\, t}{2\hbar} \right) \to \left(\frac{\Delta E\, t}{2\hbar} \right)^2$, and the term $(\Delta E)^2$ in the numerator and denominator of equation (1.147) is canceled out. In this situation, the maximum probability can be given by the relation as

$$| C_m |_{max}^2 = R \frac{t^2}{(2\hbar)^2} \qquad \text{...(1.148)}$$

where, the term R is defined as follows

$$R = \frac{1}{c^2\hbar^2} \,|\, A_x(0) \,|^2 \,|\, X_{m-n} \,|^2 \,(E_m - E_n)^2 \qquad \text{...(1.149)}$$

So, it is clearly seen that the 'maximum probability of transition' is proportional to t^2. This probability has to be only significant in an interval around $\Delta E = 0$ of the width $\approx \frac{4\pi\hbar}{t}$, which is effectively determined by the Heisenberg uncertainty principle. The plot of the 'probability' curve is shown in **Fig. 1.8** that clarifies the situation in the case of an 'absorption' process.

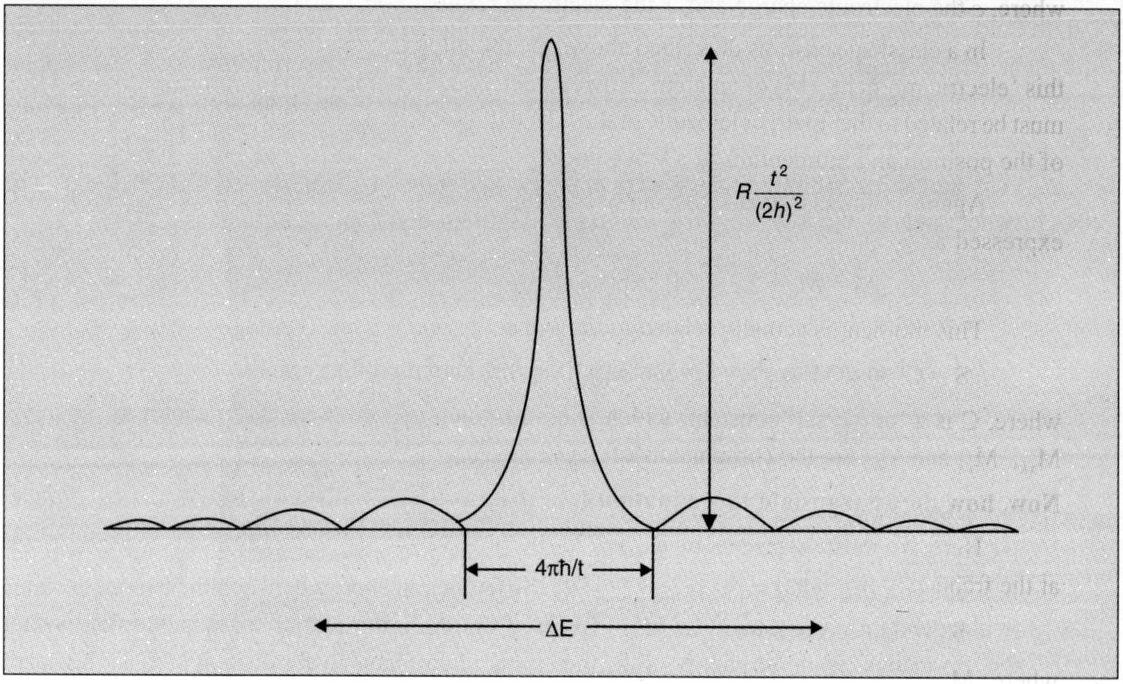

Figure 1.8: The probability curve against ΔE showing a transition to higher energy by an absorption process

For a transition from the initial state to the final state, the total probability of finding the system in the final state is found by an integration of equation (1.147) over ν, and after some mathematical manipulations, when we take $(E_m - E_n) = \hbar\nu_m$, we get

$$C_m^* C_m = \frac{\pi^2 \nu^2_{mn}}{c^2 \hbar^2} \mid A_x(0) \mid^2 \mid X_{m-n} \mid^2 t \qquad ...(1.150)$$

Now, we have to talk about the 'intensity' of this radiation, which is the 'time averaged' magnitude of the **'Poynting Vector' (S)**, and the probability can be written in terms of this intensity as :

$$C_m^* C_m = \frac{2\pi}{c\hbar^2} I_x |X_{m-n}|^2 t \qquad ...(1.151)$$

This clearly shows that the "probability of absorption" per unit time depends linearly on the 'intensity of light'.

Another Look through Poynting Vector

We can find out the 'probability of transition' from another viewpoint, *i.e.,* when we are dealing with a moving electron that radiates energy, this probability is actually associated with the 'matrix elements' of the 'electric moment' as

$$M = -e\,r \qquad \text{...(1.152)}$$

where, e the electronic charge and r the electronic radius.

In a classical sense, as described above, the radiation is related to the second 'time derivative' of this 'electric moment' (M) of an atom. However, in a quantum mechanical sense, the emitted radiation must be related to the 'matrix elements' of the 'electric moment', *i.e.,* in the realm of matrix representation of the position and momentum of a wave packet.

Again, in the classical sense, an oscillating 'electric dipole' with an 'electric moment' can be expressed as

$$M = M_0 + \Sigma_{k=1} M_k \exp(i\omega_k t) \qquad \text{...(1.153)}$$

This moment is actually related to the *Poynting Vector* and radiates power or energy as

$$S = C^2 <|\, d^2M/dt^2|^2> = C^2\, \Sigma_k(|\, M_{xk}|^2 + |\, M_{yk}|^2 + |\, M_{zk}|^2)\,(\omega_k)^4 \qquad \text{...(1.154)}$$

where, C is a 'universal' constant, which is not of much concern here. $< >$ denotes a 'time average', M_{xk}, M_{yk} and M_{zk} are the components of the vector M_k.

Now, how do we translate this equation into the atomic or molecular domain ?

Here, we must replace ω_k by ω_{ik} and M_k by M_{ik}. As a consequence, the 'average power' radiated at the frequency ω_{ik} is given as

$$S_{ik} = C^2\,(\omega_{ik})^4\,(|\, M_{xik}|^2 + |\, M_{yik}|^2 + |\, M_{zik}|^2) \qquad \text{...(1.155)}$$

where, $M_{xik} = M_{xik}(0)\exp(i\omega_{ik}t)$, etc. By taking $-e$ as the electric charge of the electron, φ_i and φ_k as the eigenstates which are quantized, and x, y and z as the positional operators, we can construct the matrix elements by integrating over the volume element (dV) as

$$M_{xik}(0) = -e \int \varphi_i^*\, x\varphi_k\, dV,$$

$$M_{yik}(0) = -e \int \varphi_i^*\, y\varphi_k\, dV, \text{ and}$$

$$M_{zik}(0) = -e \int \varphi_i^*\, z\varphi_k\, dV \qquad \text{...(1.156)}$$

Now, it is quite interesting to note that the radiated frequency corresponds to

$$\omega_{ik} = \frac{(E_i - E_k)}{\hbar} \qquad \text{...(1.157)}$$

which is actually known as the Ritzs combination principle. Thus, a positive frequency means 'emission', and a negative frequency means 'absorption'. This was also explained in the previous section. If the above matrix elements of equation (1.157) are zero, then the transition is 'forbidden'. Hence, this clarifies the situation on different processes, and at the same time, this fixes the 'selection rules' for atomic or molecular spectra to be observed. It should be also noted that $S_{ik} = 0$ for $i = k$, which indicates that only transitions between different states emit radiation.

So, the above description clearly shows the role of the '**Poynting Vector**' in the 'absorption' or 'emission' process in the interaction of radiation with 'matter' in terms of a classical radiation field of Maxwell and the quantized states of an atom or a molecule taking part in the process.

1.7.4. The Stimulated Emission

In the above description, it has been clearly seen that there is an importance of the 'transition dipole bracket' in the absorption of light in terms of the 'probability', and the intensity of light has also been related in equation (1.151) and how it varies with time. The formalism given so far in the above description is quite similar for the 'stimulated emission', except that we have to find the probability of transition from the higher energy state (E_m) to a lower energy state (E_n), *i.e.*, $n \rightarrow m$ transition, which is given per unit time for an isotropic radiation as

$$B_{n \rightarrow m}\, \rho(\nu_{mn}) = \left(\frac{2\pi}{3\hbar^2}\right) | \mu_{mn} |^2 \rho(\nu_{mn}) \qquad ...(1.158)$$

where, μ_{mn} is the 'transition dipole bracket' in *x*, *y*, *z* directions, and ρ is the 'density' of the radiation field, which is defined as the 'time averaged' magnitude of the electric field divided by 4π. By considering the equation (1.145), the problem of absorption of energy from the radiation can be described. First of all, the system starts at the lower energy state (*n*), the probability of a transition to the higher energy state (*m*) is then calculated. If the system is in the higher energy state at the initial stage, the same treatment as above can be given to yield the result on 'stimulated (*i.e.*, induced) emission' of radiation. However, it is known that the process of 'stimulated emission' of radiation involves an action of the radiation field on the system initially in the higher energy state, and it causes a downward transition, *i.e.*, when we add some energy to the radiation field.

Now, the radiation field stimulates or induces a downward transition. The lower state *n* in equation (1.145) is considered to be higher in energy than *m*. This implies that the first exponential term dominates the process of transition. However, we have to preserve the energy levels, *i.e.*, $n \rightarrow m$ transition is taken as the 'absorption' and $m \rightarrow n$ transition is considered to be operative in the 'emission' process. In this situation, as the results are symmetrical for absorption and stimulated emission, we can write the equation for the process as

$$B_{m \rightarrow n}\, \rho(\nu_{mn}) = B_{n \rightarrow m}\, \rho(\nu_{mn}) \qquad ...(1.159)$$

where, $B_{n \rightarrow m}$ is the transition probability for the 'emission' process, as defined in equation (1.158) above. The terms $B_{m \rightarrow n}$ and $B_{n \rightarrow m}$ are known as the "Einstein Coefficient" of absorption and emission respectively. For the probability of a stimulated emission of radiation per unit time, the equation (1.159) shows that it is equal to that of absorption. So that sums up the 'process of stimulated emission' of light vis-a-vis the 'absorption process'.

1.7.5. The Spontaneous Emission

There is a striking difference between the above semi-classical description of '**absorption and stimulated emission of radiation**' (*i.e.* where the field is taken classically but the molecular eigenstates that are involved in the transition process are quantized) and the '**spontaneous emission**'. The latter process cannot be explained from a semi-classical approach.

The 'spontaneous emission' is defined as the emission of radiation from an excited state of an atom or molecule without the presence of a radiation field, which actually causes the stimulated emission. Hence, these excited states of an atom or molecule can be treated quantum mechanically, *i.e.*, quantized. By absorption of a short pulse light, a molecule can be put in the excited state through different routes, *e.g.*, by a chemical reaction, by the bombardment by an energetic electron or by some other energetic means, which ultimately leave the molecule in the state of darkness. But in such a situation, the molecule can still emit a 'photon', *i.e.*, a quantum of light energy. This particular emission is called the "fluorescence" which is the spin-allowed spontaneous emission and it takes place immediately, whereas if it takes place after a certain time lag, it is known as the "phosphorescence" that is a spin-forbidden spontaneous emission. Both these processes take place without the presence of an applied radiation field, which normally gives rise to a stimulated emission of radiation.

For any substance, if N_m is the number of atoms or molecules in the higher energy state (E_m), and that for a lower energy state (E_n)) is N_n, then the Boltzman population of atoms or molecules at a temperature T is given by

$$\frac{N_m}{N_n} = \frac{\exp\left(\frac{-E_m}{kT}\right)}{\exp\left(\frac{-E_n}{kT}\right)} = \exp\left(\frac{-h\nu_{mn}}{kT}\right) \qquad ...(1.160)$$

where, k is the Boltzman constant. The number of upward transitions have to be equal to that of the downward transitions, provided the system is in equilibrium. Hence, it can be written as

$$N_m\,[A_{m\to n} + B_{m\to n}\,\rho(\nu_{mn})] = N_n B_{n\to m}\,\rho(\nu_{mn}) \qquad ...(1.161)$$

Here, it has to be clearly noted that the coefficient $A_{m\to n}$ is not multiplied by the radiation density [$\rho(\nu_{mn})$], since the 'spontaneous emission' takes place without the presence of any radiation field, but both the B coefficients are multiplied by the radiation density (ρ), which is the 'time averaged' magnitude of the electric field. Actually, at the equilibrium, there is a necessity of the presence of the coefficient $A_{m\to n}$ in order to have a lesser 'populations' at the higher energy state than that in the lower energy state. By considering the above two equations (1.160 and 1.161), and solving the resulting equation for $\rho(\nu_{mn})$, we get

$$\rho(\nu_{mn}) = \left\{\frac{A_{m\to n}\exp\left(\frac{-h\nu_{mn}}{kT}\right)}{\left[\left[-B_{m\to n}\exp\left(-\frac{h\nu_{mn}}{kT}\right) + B_{n\to m}\right]\right]}\right\} \qquad ...(1.162)$$

It is already known that $B_{n\to m} = B_{m\to n}$, and in this case the equation (1.162) reduces to :

$$\rho(\nu_{mn}) = \frac{\left(\frac{A_{m\to n}}{B_{m\to n}}\right)}{\left[\exp\left(\frac{h\nu_{mn}}{kT}\right) - 1\right]} \qquad ...(1.163)$$

Now, we can consider a blackbody radiation density by taking the substance as a 'blackbody', and solving for the Einstein coefficient $A_{m\to n}$ and also by using the equation (1.158), it is possible to find out the probability per unit time for an excited state of an atom or molecule to go through a process of 'spontaneous emission' as :

$$A_{m\to n} = \left\{\frac{[32\pi^3(\nu_{mn})^3]}{3c^3\hbar}\right\} |\,\mu_{mn}|^2 \qquad ...(1.164)$$

In the context of the 'dipolar approximation', the above equation is derived, since this approximation is also used in the derivation of another important Einstein coefficient, which explains the 'stimulated emission', as done in the previous section. Here also, like in the two previous cases, the 'spontaneous emission' is dependent on the square of the absolute value of the 'transition dipole bracket'. It should be clearly noted that there is no probability of absorption or stimulated emission if the intensity is considered zero, but there will be always 'spontaneous emission', since it does not depend on the 'intensity' factor, as in equation (1.151) [36, 37].

1.7.6. The Optical Transition

The optical transitions are important in many semiconductors, and it has great importance and applications in laser materials. Since all these materials are of some significance, these transitions need some mention here. The optical transitions can take place in two different ways

(*a*) In an unassisted optical transition, the reduced wave vector (***k***) does not change, and

(*b*) In the second type, there are phonon-assisted transitions (phonons are quantized vibrations).

In the second type, in the absorption process, the electron absorbs a quantum of energy, and one or more phonons have to go to a higher energy state with different reduced wave vector (***k***). In the emission process, the electrons emit a quantum of energy , and one or more phonons have to go to a lower unoccupied energy state again with a reduced wave vector (***k***). However, the phonon-assisted transitions require the availability of phonons, these have the lesser probability than the unassisted transitions.

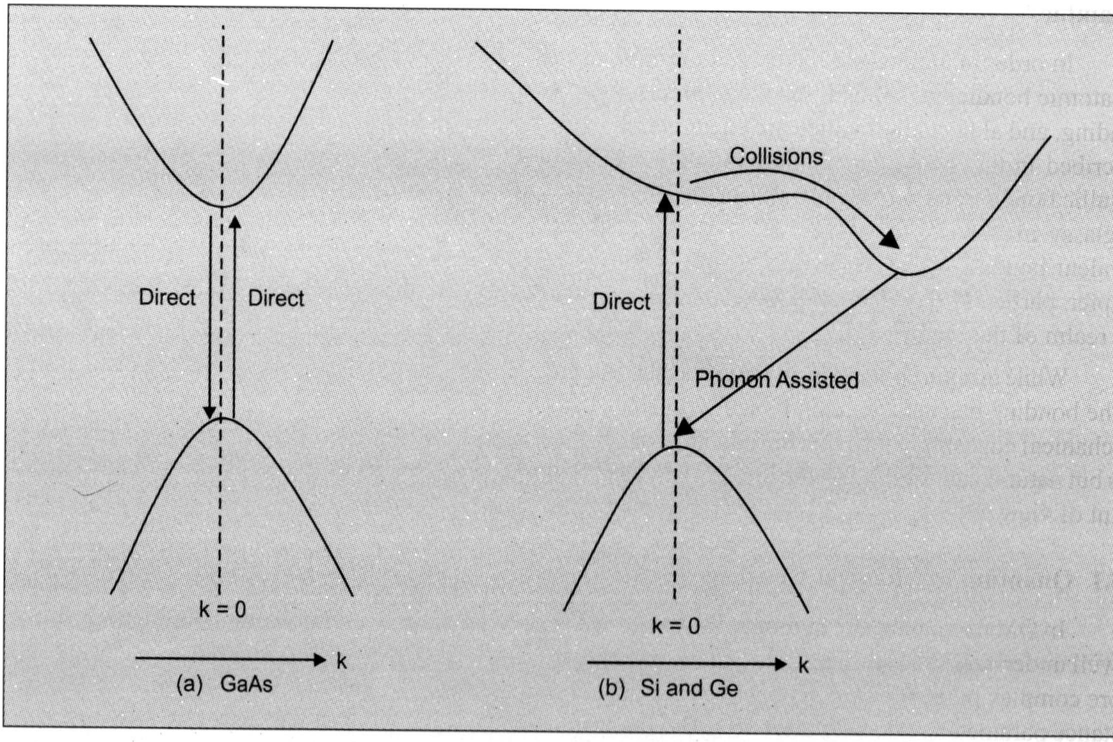

Figure 1.9: The absorption and emission processes with (*a*) Direct transitions, (*b*) The absorption in direct transition and indirect transition in the emission process

Now, let us assume that the valence and conduction bands of a semiconductor are positioned in such a manner that their respective band maximum and minimum lies at $\boldsymbol{k} = 0$. In this situation, as shown in **Fig. 1.9(*a*)**, the unassisted transitions take place back and forth such that the absorption process involves a quantum of energy, which can also be re-emitted. In such a case of re-emission, the 'optical gap' and the 'band gap' (both of which involve a direct transition) then coincide, which is very clear from **Fig. 1.9(*a*)**. Such semiconductor materials will have applications for 'solid state lasers', such as GaAs and similar materials fall into this category.

However, when the valence band has its maximum at $\boldsymbol{k} = 0$, but the conduction band has its minimum at $\boldsymbol{k} \neq 0$, then the most probable absorption process takes places as : $\boldsymbol{k} = 0 \rightarrow \boldsymbol{k} = 0$.

In this scenario, the valence band electrons have to lose energy rapidly till they arrive at the bottom position of the conduction band, when these transitions will emit 'light' simply by a phonon-assisted process. Hence, the absorbed quantum of energy cannot obviously be re-emitted, as shown in **Fig. 1.9(*b*)**. These semiconductor materials cannot of course be used as 'solid state lasers', since there is no re-emission process involved, as in the previous case. It has to be noted that in this case, the 'band gap', which involves a direct transition, and the 'optical gap', which induces an indirect transition, do not match at the same $\boldsymbol{k}$ value. The silicon (Si) and germanium (Ge) fall in this category of semiconductors.

1.8. BONDING IN SOLIDS

Preamble

In order to understand the behaviour of nano-materials, or nano-sized particles in any material, the atomic bonding in solids is the most interesting topic without any doubt. There are various types of bonding, and almost all of which are encountered in the type of materials, whose various properties are described in this book, *e.g.*, ionic bonding (in **chapters-3** and **5**), covalent bonding (in **chapter-2**), metallic bonding in nano-metallic particles of bismuth and selenium grains embedded in an amorphous or glassy matrix (in **chapter-6**) etc. But out of these different types of bonding in solid materials, covalent bonding is a very important type, since it has some relations with the other topics of this chapter, particularly with quantum mechanics, *i.e.*, the elementary concepts that are important to grasp the realm of the 'nano-world'.

While in ionic bonding in any material, the electrons are localized on the respective sites participating in the bonding mechanism, *e.g.*, typically in NaCl salts, the covalently bonded solids involve quantum mechanical concepts. Since this introductory chapter deals with various aspects of quantum mechanics, it is but natural that we discuss the idea of "covalent bonding" very briefly from the quantum mechanical point of view.

1.8.1 Quantum Mechanical Covalency

In QM covalency, the hydrogen molecule is a very good example. If the bonding in this molecule is well understood, the similar ideas can be extrapolated for finding out the covalent bonding scheme in more complex molecules, *i.e.*, many body problems. But, first of all, let us start with the labeling of the distance parameters, *i.e.*, between two hydrogen nuclei and between two extra electrons involved in the bonding scheme. This is shown in **Fig. 1.10**, *i.e.*, the hydrogen nuclei or the protons *a* and *b*, which are separated by the inter-nuclear distance R_{ab}. Here, the electrons are simply labeled as 1 and 2, and the distance between them being denoted by R_{12}. The distance between the electron 1 and the hydrogen nucleus *a* is R_{a1}, that between 2 and the same nucleus as R_{a2}, etc.

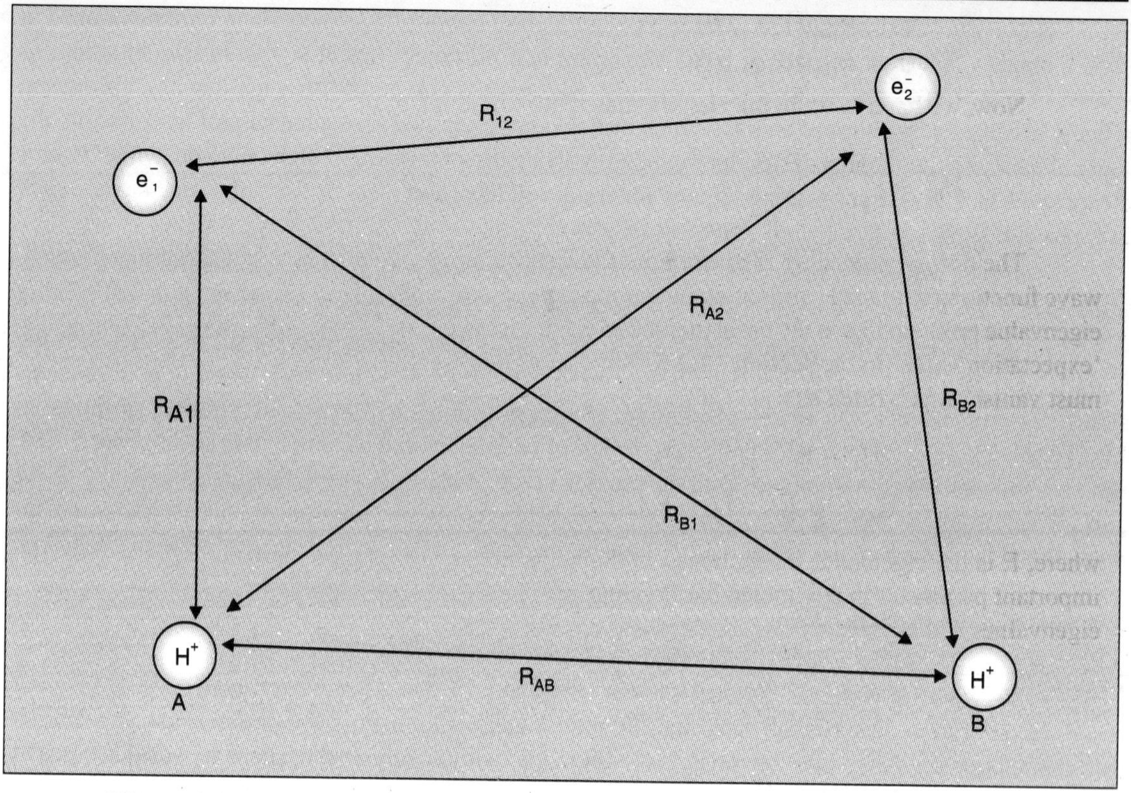

Figure 1.10: Distance parameters in the bonding scheme for a hydrogen molecule

Normally, we invoke Born-Openheimer approximation, when we involve nuclear wave-functions in the discussion, to take care of the nuclear interactions. As per this approximation, the electron wave function can be written as

$$(\nabla_1^2 + \nabla_2^2)\psi + \left(\frac{2m}{\hbar^2}\right)\left[E + \frac{e^2}{R_{a1}} + \frac{e^2}{R_{b1}} + \frac{e^2}{R_{a2}} + \frac{e^2}{R_{b2}} - \frac{e^2}{R_{12}} - \frac{e^2}{R_{ab}}\right]\psi = 0 \qquad ...(1.165)$$

where, m is the mass of the electron and the other terms have their usual meaning, but different terms need to be described in order to construct the Hamiltonian of the whole system. Here, the first term. *i.e.* kinetic energy, involves the Laplacian operators for both the electrons acting on the wave function. The eigenvalue of the system is E. Within the square bracket, the 2nd – 5th terms are the Coulombic interaction terms for the electrons 1 and 2 with the nuclei a and b. A term like $4\pi\varepsilon_0$ (ε_0 = dielectric permittivity of the vacuum) is dropped from the electro-static attraction/repulsion energy in order to make the equation simpler. Finally, the sixth term is the electron-electron repulsion with the last term being the repulsion between two hydrogen nuclei.

When the inter-nuclear separation (R_{ab}) is very large, the energy of the system is $2E_H$ (*i.e.,* the normal ground-state energy of two hydrogen atoms), since it will simply consist of two non-interacting hydrogen atoms. In this case, the system wave functions in terms of two normalized basis functions [$\varphi_{1sa}(I)$ and $\varphi_{1sb}(I)$] with degenerate states can be expressed as

$$\psi_1 = \varphi_{1sa}(\mathrm{I})\ \varphi_{1sb}(\mathrm{II})$$
$$\psi_2 = \varphi_{1sa}(\mathrm{II})\ \varphi_{1sb}(\mathrm{I}) \qquad ...(1.166)$$

Now, we have to write the Hamiltonian matrix in terms of the above two basis functions as :

$$H_{SH} = \begin{vmatrix} H_{Ii} & H_{I2} \\ H_{II1} & H_{II2} \end{vmatrix} \qquad ...(1.167)$$

The normal procedure is to work out the respective matrix element with double integral on two wave functions with the Hamiltonian operator acting on one of these wave functions, and taking it as an eigenvalue problem like in the usual method in quantum mechanics, as described in the calculation of the 'expectation value' in the **section - 1.2.5**. Now, let us construct the following secular determinant that must vanish. It is written as :

$$\begin{vmatrix} H_{I1} - E & H_{I2} - \Delta^2 E \\ H_{II1} - \Delta^2 E & H_{II2} - E \end{vmatrix} = 0 \qquad ...(1.168)$$

where, E is the eigenvalue as explained in the beginning, and Δ is the 'overlap integral', which is an important parameter in any molecular bonding scheme. Now, let us solve this determinant for the two eigenvalues and eigenfunctions, *i.e.,* symmetric (S) and anti-symmetric (A), as

$$E_S = \frac{(H_{I1} + H_{I2})}{(1 + \Delta^2)}$$

$$E_A = \frac{(H_{I1} - H_{I2})}{(1 - \Delta^2)}$$

$$\psi_S = \frac{1}{(\sqrt{2} + 2\Delta^2)} \cdot [\varphi_{1sa}(\mathrm{I})\ \varphi_{1sb}(\mathrm{II}) + \varphi_{1sb}(\mathrm{I})\ \varphi_{1sa}(\mathrm{II})]$$

$$\psi_A = \frac{1}{(\sqrt{2} - 2\Delta^2)} \cdot [\varphi_{1sa}(\mathrm{I})\ \varphi_{1sb}(\mathrm{II}) - \varphi_{1sb}(\mathrm{I})\ \varphi_{1sa}(\mathrm{II})] \qquad ...(1.169)$$

Now, we have to find two 'hydrogen atom Schrödinger equations' for one hydrogen atom around nucleus *a* and another around the nucleus *b*. From these equations, we have to find out the 'terms' in the Hamiltonian, which consist of the above 'basis functions' being the eigenfunctions of these 'terms' leading to the energies of two hydrogen atoms. Therefore, these 'terms' should be replaced by $2E_H$. These terms in the first Hamiltonian are written as

$$H_{I1} = 2E_H + 2J + J_1 + \frac{e^2}{a_0 C} \qquad ...(1.170)$$

where, J is a 'coulomb integral' consisting of two basis functions and the coulomb energy as the operator acting on one of them over the volume element, C is the inter-nuclear separation in units of Bohr radius $(a_0) = \dfrac{R_{ab}}{a_0}$, and J_1 is an extra term that is estimated analytically.

After the above step, we can write the 'terms' in the second Hamiltonian as

$$H_{I2} = 2\Delta^2 E_H + 2\Delta E_{ex} + E_{ex1} + \Delta^2 \left(\frac{e^2}{a_0 C}\right) \quad ...(1.171)$$

where, E_{ex} is the 'exchange integral' and is a very important term in the molecular orbital scheme of bonding in covalent solids, and it's presence is the reason why 'covalency' is called quantum mechanical. E_{ex1} is an extra term that can be estimated analytically.

Now, we have to write the ultimate formula for the energies involving symmetric and anti-symmetric bonding for the H_2 molecule as

$$E_S = 2E_H + \frac{e^2}{a_0 C} + \frac{(2J + J_1 + 2\Delta E_{ex} + E_{ex1})}{(1 + \Delta^2)}$$

$$E_A = 2E_H + \frac{e^2}{a_0 C} + \frac{(2J + J_1 - 2\Delta E_{ex} - E_{ex1})}{(1 - \Delta^2)}$$

The different forms of energy, *i.e.*, E_S, E_A, $E_N = H_{I1}$ (the classical mechanics energy), are plotted against the inter-nuclear separation in **Fig. 1.11.**

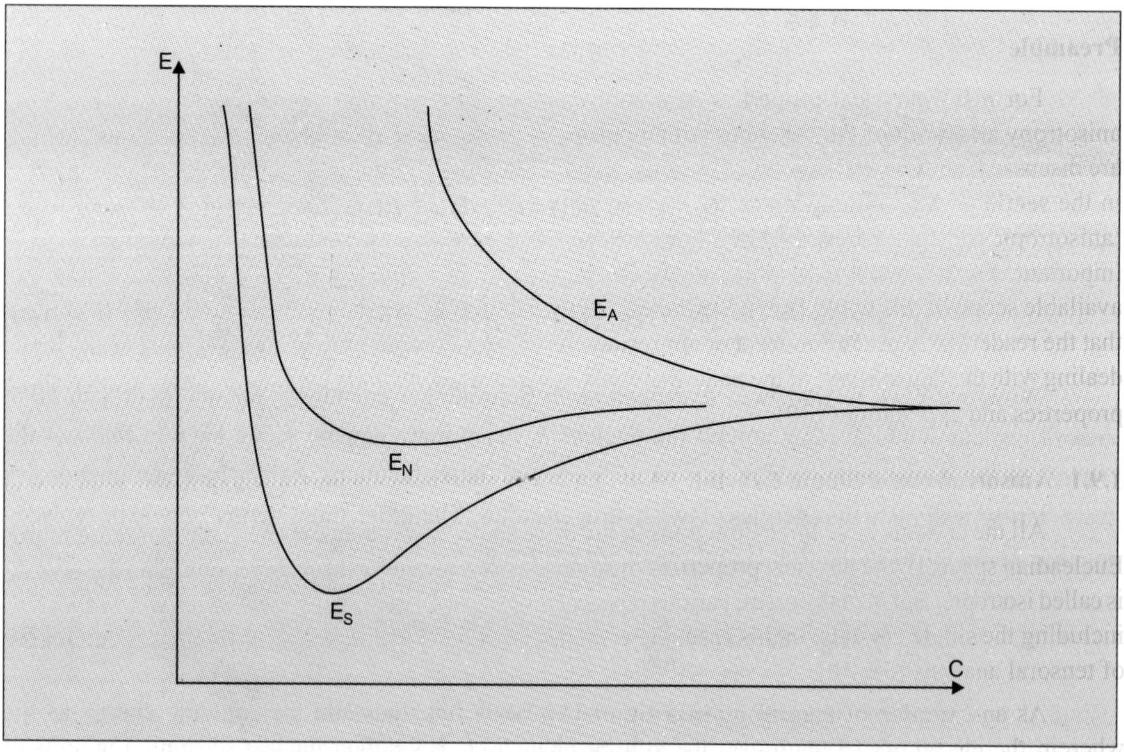

Figure 1.11: Energy diagram for various forms of energy of a hydrogen molecule

There is no interaction between the two hydrogen nuclei, when the separation between them is large, all the three curves should have the same value of the energy, as shown in **Fig. 1.11**. But, this situation changes, when the separation starts getting reduced. In this scenario, the energy of the anti-symmetric bonding increases continuously with the inter-nuclear distance, and hence it is not a stable form of molecular bonding. But, E_N shows a shallow minimum, when the inter-nuclear distance between the two hydrogen nuclei starts getting reduced further, thereby showing some kind of stability, although it does not conform to the experimental value of the bonding energy of the hydrogen molecule.

If this distance is still reduced, *i.e.,* two hydrogen atoms are brought still closer to each other, then the value of the symmetric bonding shows a deep minimum indicating a high degree of stability of the covalent bonding. The proof of this stability lies in the fact that this energy of symmetric bonding increases thereafter, if the distance is further reduced. Hence, we define the point of stability, at which E_S shows a minimum, as the equilibrium "bond length" with the depth of the minimum signifying the equilibrium 'bond strength'.

For a many-body problem, with more than one electron, the scheme of bonding is quite similar, with the exception that many-electron interactions have to be taken into account to construct a proper Hamiltonian under the Born-Oppenheimer approximation in order to calculate the total energy of the system.

1.9 ANISOTROPY

Preamble

For many physical properties of a solid, the anisotropic properties are very important. This anisotropy arises out of the "changes" of any property along 'three different spatial directions', which are discussed in terms of 'tensors'. This anisotropy is described with respect to the magnetic properties in the **section - 5.5**, without invoking the concept of tensors, but it is discussed only in terms of an 'anisotropic constant' for nano-crystalline particles of magnetite, which shows that this concept is quite important. For other physical properties to be discussed, this concept is again not used within the available scope of this book. But, nevertheless, this particular aspect should be explored very briefly so that the readers may use the concept or appreciate the problem of anisotropy, if encountered in any topic dealing with the 'anisotropy' of the nano-materials or materials containing nano-particles with interesting properties and applications [38].

1.9.1 Anisotropy in a Single Crystal

All the crystals have three directions in the three spatial coordinates in the Cartesian sense in the Eucleadian space. If the physical properties of solids do not vary along these directions, then the crystal is called isotropic. But in many cases, various physical properties do vary in different Cartesian coordinates, including the single crystals. In this case, the crystals are called anisotropic, which is the subject matter of tensoral analysis [39, 40].

As an example of anisotropy in a single crystal, let us consider the 'electrical conductivity' wherein the electric field (E) gives rise to a current (I), which is a vector that will have the same direction as the electric field vector. From the linear relation between the cause and effect (*i.e.,* the causality), the current components relative to the arbitrarily chosen Cartesian coordinate system are given as

$$I_x = \sigma_{xx}E_x + \sigma_{xy}E_y + \sigma_{xz}E_z$$
$$I_y = \sigma_{yx}E_x + \sigma_{yy}E_y + \sigma_{yz}E_z$$
$$I_z = \sigma_{zx}E_x + \sigma_{zy}E_y + \sigma_{zz}E_z$$

where, the quantities σ_{ik} are the components of the 'conductivity tensor', which is a symmetric tensor, *i.e.,* $\sigma_{ik} = \sigma_{ki}$. We can make use of this symmetry property and we can multiply the above three equations by E_x, E_y, E_z respectively, and upon adding, we simply get the relation as :

$$I_x E_x + I_y E_y + I_z E_z = \sigma_{xx}E_x^2 + \sigma_{yy}E_y^2 + \sigma_{zz}E_z^2 + 2\sigma_{xy} E_x E_y + 2\sigma_{yz} E_y E_z + 2\sigma_{zx} E_z E_x$$

It is seen that the right hand side represents a quadratic surface. If we choose the coordinates along the 'principal' axes of the surface, then the mixed terms vanish, and we get the new coordinate system as

$$I_x = \sigma_1 E_x$$
$$I_y = \sigma_2 E_y$$
$$I_z = \sigma_3 E_z$$

where, σ_1, σ_2 and σ_3 are the 'principal' conductivities. Therefore, the electrical conductivities of any crystal (even possessing lower symmetry) may be characterized by the above three conductivities, or rather by three specific resistivities. It has to be clearly noted that both the current and the field have same direction only when the applied field falls along any one of the three 'principal' axes.

In cubic crystal, the three quantities are equal, the conductivity does not vary with the direction. In hexagonal, trigonal and tetragonal crystals, the conductivity depends only on the angle (ϕ) between the direction in which σ is measured and the hexagonal, trigonal or tetragonal axis, since in these crystals, two out of the three quantities are equal. In such a situation, we can write the conductivity equation as

$$\sigma(\phi) = \sigma_{\perp}/\sin^2\phi + \sigma_{\parallel}/\cos^2\phi$$

where, the signs with σ refer to the directions perpendicular and parallel to the axis. Now, starting with the above relations, let us describe different properties of solids in terms of the effect of two terms at a time, *i.e.,* scalars, vectors and tensors.

1. **Vector - Vector Effect.** The above effect can be called "vector-vector" effect, since an electric current (*i.e.,* a vector quantity) is produced by an applied electric field, which is also a vector quantity. The relations deduced above can also be applied to the other physical properties, like thermal conductivity. Here, a thermal current vector is created by a 'thermal gradient', which is also a vector quantity. Another example is the 'diffusion' under the influence of a 'concentration gradient', and here both the variables are vector quantities.
2. **Scalar-Tensor Effect.** Similar relation as above can be deduced for this effect as well. As an example, the case of a "deformation" (*i.e.,* a tensor quantity) of a solid can be cited, which results from a change in temperature, which is a scalar quantity. This may be characterized by the three 'principal' expansion coefficients : α_1, α_2 and α_3. Here again, in cubic crystals, $\alpha_1 = \alpha_2 = \alpha_3$, and such crystals are obviously isotropic in this context. The angular dependence of α for hexagonal, trigonal and tetragonal crystals is also given by the last equation, as shown above, for the electrical conductivity.

3. **Vector-Tensor Effect.** A typical example is piezoelectricity, *i.e.,* the creation of an electrical polarization, which is equal to the atomic displacement or deformation (*i.e.,* a tensor quantity) with the application of an electric field vector or vice-versa. This is an important property with so many important applications in human life. It is better to visualize a piezoelectric material (*e.g.,* a gas lighter) as a vector (cause) and a tensor (effect).
4. **Tensor - Tensor Effect.** In the mechanics of solids, the most important property studied so far is the 'elastic deformation or strain', which is a tensor quantity, but this effect can usually take place under the influence of some form of 'mechanical stress' that has to be applied, which is again a tensor quantity. This tensor-tensor effect may require many more components (stress or strain components) than appeared in a rather simpler case of a vector-vector effects, as described above.

Well, this particular short discussion sums up the applications of "tensor" for some of the important physical properties of solids along with vector and scalar quantities to give a more or less comprehensive outlook on this type of mathematical technique, although there are numerous other examples on the applications of "tensor" in materials science and solid state physics with a very interesting mathematical treatment. All these aspects could be important for nano-materials.

REFERENCES

1. L. Pauling, 'General Chemistry ', Dover, New York, (1988) [A book by a Legend].
2. R. E. Serway, 'Physics for Scientists and Engineers', 3rd Ed., Saunders, Philadelphia, (1990).
3. R. A. Alberty and R. J. Silbey, 'Physical Chemistry', Wiley, New York, (1992).
4. C. W. Sherwin, 'Introduction to Quantum Mechanics', H R & A, New York, (1959).
5. A. W. Adamson, 'A Textbook of Physical Chemistry', Academic Press, Orlando, (1986).
6. E. B. Wilson, J. C. Decius and P. C. Cross, 'Molecular Vibrations', McGraw Hill, New York, (1955) [A standard book for everybody].
7. C. Kittel, 'Introduction to Solid State Physics', Wiley (E. U.) Edition, New Delhi, (1971).
8. A. J. Dehkker, 'Solid State Physics', McMillan, London, (1969).
9. H. Pollak, M. DeCoster and S. Amelinckx, 'Mössbauer Effect in Amorphous Substances', Ed. D. M. J. Compton and D. Sehuen, Wiley, New York, (1962) pp. 298.
10. J. Wong and C. A. Angell, 'Glass Structure by Spectroscopy', Dekker, New York, (1976).
11. J. M. D. Coey, J. de Physique, 35 (1974) 89.
12. C. R. Kurkjian, 'Mössbauer Spectroscopy in Inorganic Glasses', J. Non-Crystallne Solids, 3 (1970) 157.
13. M. D. Dyare, 'A Review of Mössbauer Data', Am. Mineralogist, 70 (1985) 304.
14. A. K. Bandyopadhyay, M. Ribes, F. Pernot and J. Zarzycki, 'Mössbauer and ESR Studies on Mixed Transition Metal Oxide Glasses', Phys. Chem Glasses, 23 (1982) 31.
15. J. Chappert, 'High Field Mössbauer Spectroscopy', J. de Physique, 35 (1974) C6-71.
16. A. K. Bandyopadhyay, J. Zarzycki, P. Auric and J. Chappert, 'Magnetic Property of a Basalt Glass and Glass Ceramics', J. Non-Crystalline Solids, 40 (1980) 353.

    [Proceedings of the 5th Univ. Conf. on Glass Sc., Troy (USA), Eds. M. Tomozawa, R. A. Levy, R. K. MacCrone and R. H. Doremus, North Holland, Amsterdam, (1980)].

17. P. Auric, J. Chappert, A. K. Bandyopadhyay and J. Zarzycki, 'Mössbauer Study of Crystallite Formation in Basalt Glass Ceramics', J. de Physique, C1-41 (1981) 277.
[Proceedings of the Intl. Conf. on Mössbauer Spectroscopy, Portoroz (Yugloslavia), (1980)].
18. P. Auric, J. Chappert, A. K. Bandyopadhyay and J. Zarzycki, "Superparamagnetism and ferrimagnetism of the small particles of magnetite in a silicate matrix", J. Non-Crystalline Solids, 50 (1982) 97.
19. A. K. Bandyopadhyay and P. Auric, 'Hyperfine Splitting and Anisotropy of Magnetite Particles within a Glass Matrix by Mössbauer Spectroscopy', Trans. Ind. Ceram. Soc., 46 (1987) 136.
20. A. Abragam and M. H. L. Pryce, 'Theory of Nuclear Hyperfine Structure of EPR Spectra in Crystals', Proc. Royal Soc., A205 (1974) 135.
21. J. Chappert and R. B. Frankel, 'Mössbauer Study of Ferrimagnetic Ordering in Ni-ferrite and Cr-substituted Ni-ferrites', Phys. Rev. Letters, 19 (1952) 570.
22. M. S. Rogalski and S. B. Palmer, 'Solid State Physics', Gordon and Breach, Amsterdam, (2000).
23. L. Neel, 'Influence of Thermal Fluctuations on the Magnetization of Ultrafine Ferrimagnetic Grains', C. R. Acad. Sci., 228 (1949) 664.
24. Y. Yafet and C. Kittel, 'Antiferromagnetic Arrangements in Ferrites', Phys. Rev., 87 (1952) 290.
25. A. K. Bandyopadhyay, 'A study of Interaction between Copper and Manganese in a Soda Borate Glass by ESR', J. Materials Sci., 15 (1980) 1605.
26. A. K. Bandyopadhyay, P. Auric, J. Phalippou and J. Zarzycki, 'Spectroscopic behaviour and internal friction of glasses containing iron having different redox ratios', J. Materials Sci., 15 (1980) 2081.
27. A. K. Bandyopadhyay, 'Optical and ESR Investigation of Borate Glasses containing Single and Mixed Transition Metal Oxides', J. Materials Sci., 16 (1981) 189.
28. W. Low, Solid State Phys. Suppl., (1960) 2.
29. C. R. Kurkjian and E. A. Sigety, Phys. Chem. Glasses, 9 (1968) 73.
30. D. Loveridge and S. Parke, Phys. Chem. Glasses, 12 (1971) 19.
31. R. F. Tucker, 'Advances in Glass Technology', Plenum Press, New York, (1962) pp. 103.
32. C. Hirayama, J. G. Castle Jr. and M. Kuriyama, Phys. Chem Glasses, 9 (1968) 109.
33. A. Bishay and L. Makar, J. Am. Ceram. Soc., 52 (1968) 605.
34. M. Firsov, V. M. Firsov, N. V. Petrovykh and P. I. Litvinov, Proc. of the 10th International Congress on Glass, Kyoto (Japan), (1974) pp. 73.
35. D. W. Moon, J. M. Aitken, R. K. MacCrone and G. S. Cieloszyk, Phys. Chem. Glasses, 16 (1975) 91.
36. F. Abeles (Ed.), 'Intl. Colloq. on Optical Properties and Electronic Structure of Metals and Alloys', North Holland, Amsterdam, (1966).
37. S. Nudelman and S. S. Mitra (Eds.), 'Optical Properties of Solids', Plenum, New York, (1969).
38. R. S. Bera, A. K. Bandyopadhyay and P. C. Ray, 'Mathematical Physics for Engineers', New Age International, New Delhi, (2006) In Press.
39. R. E. Newnham, 'Properties of Materials', Oxford Univ. Press, Oxford, (2002).
40. G. B. Arfken and H. J. Weber, 'Mathematical Methods for Physicists', Acedimic Press, San Diego, USA, (2001).

Chapter 2

Silicon Carbide

PREAMBLE

Under normal conditions, the full densification of silicon carbide is difficult due to its covalency. In order to achieve full densification of silicon carbide, normal Acheson-type α-silicon carbide cannot be used due to various problems, particularly due to impurities. One way out is to prepare "nano particles", which are discussed later. These 'nano particles' of silicon carbide can be sintered under suitable conditions of temperature and atmosphere of sintering. These high density sintered SiC products are ideal candidates for special high performance applications, such as ceramic engines, and many more engineering applications, including high temperature structural applications. It is quite tempting to talk about various applications of a given material, before talking about its preparation and characterization. Hence, the first part of the chapter is devoted to the applications of silicon carbide in hi-tech areas.

2.1. APPLICATIONS OF SILICON CARBIDES

There has been an increasing demand for high performance materials, which can withstand severe conditions such as abrasion, high temperature, pressure and atmosphere in various applications as follows :

1. High Temperature Heat Engines,
2. Nuclear Fusion Reactors,
3. Chemical Processing Industry, and
4. Aeronautical and Space Industries.

In order to meet these demands, it requires the introduction of expensive and 'scarce metallic alloys' which also frequently need additional 'completed cooling arrangements'. Hence, the possibility of using silicon carbide can be explored for important applications in various industries as mentioned above [1-3].

2.1.1. Important Properties

Silicon carbide seems to have the potential to satisfy the requirements in most of the applications as 'high temperature structural material' for the following important properties :

(*a*) Superior Oxidation Resistance,

(*b*) Superior Creep Resistance,

(*c*) High Hardness,

(*d*) Good Mechanical Strength,

(*e*) Very High Young's Modulus,

(*f*) Good Corrosion & Erosion Resistance, and

(*g*) Relatively Low Weight.

It is important to mention here that the raw materials of silicon carbide are relatively inexpensive, and can be made in complex shapes, which may be engineered by conventional fabrication processes, such as dry pressing, extrusion and injection moulding. Therefore, the final products are cost-competitive besides offering the advantages of superior technical performance over the other materials.

2.1.2. Ceramic Engines

The application of silicon carbide in 'ceramic engine' offers a higher efficiency of the engine, which could be operated at a higher operating temperature, thereby ensuring the complete combustion of the fuel. This ensures that the full energy obtainable from a given fuel is harnessed in the automobile combustion process and at the same time reducing the environmental pollution. Moreover, the material has only one-third of the density of the 'super-alloys' giving it an additional advantage, when used in designing and application in an automobile system. In ceramic heat engines, there are many facts in the use of the silicon carbide material, which has the potential for applications as follows

(*a*) Turbo-Charger Rotor,

(*b*) Piston Liner,

(*c*) Valve Trim Components in Gasoline Engines, such as

1. Rocker Arm Pads,
2. Push Rod Tips,
3. Pre-Combustion Caps,
4. Wrist Pins,
5. Lifters Roller Followers.

2.1.3. Other Engineering Applications

The silicon carbides also find many other applications in diverse areas as follows

A. Stationary and Rotating Components in 'Automotive Gas Engines', such as

1. Turbine Wheel Combustion Chamber,
2. Rotor of Various Dimensions, and
3. Nose Cones.

B. Inherent Hardness, Chemical Inertness and a High Abrasion Resistance make silicon carbides commercially applicable in 'Chemical Processing Industry', such as

1. Seals and Valves,
2. Nozzle for Sand Blasters
3. Lens Modes, and
4. Wear Plates for Spray Drying and Wear Dies

C. The other Applications for these 'Wear and Erosion Resistant' silicon carbides include

1. Aircraft Journal Bearing,
2. Thrust Bearing,

3. Ball Bearings,
4. Oil Patch Plungers,
5. Pump Impellers,
6. Pump Components,
7. Various Fixtures,
8. Mould Billets, and
9. Extrusion Dies.

D. Due to high Thermal-Shock Resistance and Creep Resistance Characteristics, Silicon Carbides are also used as Nozzles in many other applications, such as

1. Heat Exchanger Tubes,
2. Diffusion Furnace Components, and
3. Furnace Rollers.

E. The silicon carbides already find applications as 'Heating Elements' because of its Self-Heating and Self-Emitting Glow Characteristics.

F. The silicon carbides also hold promise for its use as "Electronic Substrates' for Semiconductor Industry. By having a 'Thermal Conductivity Value' several times higher than that of alumina, and silicon carbide as a 'substrate' can handle the 'Heat Dissipation' task of the 'Electronic Chips' and also protect the 'Chips Wall'.

The development of highly dense silicon carbide requires a basic understanding of the mechanism of sintering by the addition of various additives or dopants, and their correlation with the microstructure, to eventually appreciate the inherent mechanism for the development of superior thermo-mechanical properties. With so many important properties with diverse applications, as mentioned above, the study of sintering and fracture mechanical behaviour **(see chapter - 4)** of silicon carbide assumes a great importance and a considerable significance.

2.2. INTRODUCTION

As mentioned in the beginning of this chapter that one way out to make fully densified silicon carbide is to prepare 'nano particles', which have been attempted through various routes by many workers. However, no work has been reported on 'attrition milling' and subsequent purification for the preparation of nano-crystalline particles of silicon carbide as a precursor material for full densification. There are different important variables associated with the grinding process, and it is not at all easy to prepare nano particles through the attrition grinding route. In the next section, a complete investigation is described on the preparation of nano-crystalline particles of both α- and β-silicon carbides by 'attrition milling' route through the study of various parameters of grinding in details, and subsequently optimizing the grinding process to achieve the goal [4, 5].

Since the 'sintering' of silicon carbide and subsequent development of highly dense material is a challenging problem to researchers, the sintering behaviour of nano particles of silicon carbide is described in the subsequent sections in order to give a complete picture of this important material with so many interesting properties for diverse applications. The effect of different dopants and carbon, and the atmosphere on the sintering behaviour are also described.

It has been observed that the powder compact of very fine pure crystalline solids does not densify during the heat-treatment without any applied pressure. The reason for this lack of sintering characteristic involves kinetic as well as thermodynamic considerations. The bonding in silicon carbide is predominantly covalent with 10% ionic character, which results in the directional inter-atomic forces and low coordination number in silicon carbide. Therefore, the activation energy of self-diffusion is extremely high, and consequently the self-diffusion coefficient is low and thus, the formation and mobility of structural defects are very low [2, 3].

2.2.1. Ultrafine Particles of Silicon Carbide

The synthesis of ultrafine powders has drawn a considerable attention, because these powders after sintering have physical and chemical properties, which are superior to those made from the materials with a normal size distribution. In particular, the emphasis is on the preparation of ultrafine ceramic powders for their applications in the development of sintered products having higher density and lower sintering temperature [6]. One such industrially important ceramic material is silicon carbide. Recently, sub-micrometer-sized silicon carbide has received more attention because of a significant improvement in mechanical properties of ceramic nano-composites, compared to the micrometer-sized particulates [7 - 9].

Singh *et. al.* [10] have prepared nano crystalline silicon carbide particles from thermally pre-treated rice husk by a 'thermal plasma' process. Chen *et. al.* [11] also prepared nano crystalline silicon carbide by 'chemical vapour deposition' (CVD) route. Martin et. al. [12] reported to have successfully prepared nano crystalline silicon carbide by 'carbo-thermal reduction' of silica sol and sugar. The preparation of silicon carbide particles with nanometer-sized grains prepared from chlorine containing polysilane/ polycarbosilanes (PS/PCS) has also been reported [13 -15]. In these investigations, the polysilane-to-polycarbosilane conversion leads to a conversion of organo-silicon compounds to a 'random silicon carbide network', and the cross-linking of the molecules has been investigated. The silicon carbide derived from polymeric precursors shows inhomogeneities in the nanometer range. No literature is generally found for the preparation of nano crystalline silicon carbide particles by 'attrition milling' of Acheson Type α- and/or β-silicon carbide powders.

2.2.2. Problem of Preparing Nano Particles

The preparation of nano-crystalline α-silicon carbide for studying the sintering mechanism is also a very difficult problem. The properly distributed nano crystalline particles within the very narrow range by grinding necessitates a proper understanding of the process of grinding of silicon carbide. At the initial stage, large cracks in original particles propagate, generating smaller particles with very fine cracks. Thus, the probability of particles having cracks progressively decreases with grinding leading to the generation of newer stronger particles. Hence, continuously higher and higher fracture stress is required to grind the particle. During the grinding, it becomes progressively difficult to obtain further reduction in particle size in nano crystalline region [4].

In the nano crystalline region, the nano particles develop the tendency to agglomerate and the physical equilibrium between the aggregates and fragmentation is established, resulting in a decrease in 'storing of stress energy'. Consequently, it results in an increased stress for initiating fracture and the generation of nano particles. There are two principal factors which need to be controlled to obtain nano particles :

1. Ratio of the weight of the metal balls of grinding media to the total weight of the particles to be processed by the grinding route.
2. Semi-fluid/viscous characteristic of the grinding media containing fine particles.

Another factor of importance is the total time of grinding, which also needs to be controlled in order to optimize the process of grinding to obtain nano particles [4, 5].

The purity of the nano particles is also very important. The commercially available α-silicon carbide is made by Acheson process, in which a mixture of sand and crushed coke with common salt and saw dust is electrically heated in a 'resistance furnace' above 2600°C. The reduction reaction of high-grade silica results in a-silicon carbide grains. The overall reaction is

$$SiO_2 + 3C = SiC + 2CO$$

The sawdust is added to increase the porosity of the furnace charge, thus increasing the circulation of the reacting gases, and facilitating the removal of carbon dioxide.

As these raw materials always have some impurities, the Acheson α-silicon carbide also contains impurities. Some of these impurities have deleterious effect on the sintering of silicon carbide. Moreover, the product of silicon carbide obtained from Acheson (resistance) furnace are lumps and grits. These are ground to different sizes by various mechanical milling processes, thereby further adding the impurities to the ground particles. This available powder does not sinter. This powder needs to be cleaned. Moreover, after purification, this powder does not give a high density. This needs to be further ground. So, further impurities are incorporated into the nano crystalline powder during the preparation from fine particles by the grinding route.

Finally, the very method of manufacture of silicon carbide keeps some excess carbon in it. Thus, the purification is very much essential to remove all these impurities to make the nano-crystalline silicon carbide particles sinter.

The purpose of any investigation should be to develop a scientific basis and a better understanding of the details of the grinding process by 'attrition milling' in order to be able to prepare nano-crystalline particles of both α- and β-silicon carbides by optimizing different parameters of the process of grinding. In this way, the optimized 'grinding parameters' can be used to prepare the nano particles by attrition grinding route for the purpose of sintering [2-5].

2.2.3. Sintering of Ceramics

The ceramic materials are refractory, *i.e.* it can withstand high temperature, and they are also brittle. Hence, they pose problems in their preparation, whereas the metals and polymers are melted, cast and sometimes machined to get the right type of dimensions. The preparation of ceramic materials mainly consists of grinding the raw materials, mixing with suitable binders and then pressing into different shapes and sizes to form the so-called 'green bodies', which are low in strength so that they cannot be used for any application. The initial density of the green body is very important to achieve a higher sintered density.

After this stage, they have to be fired or sintered at sufficiently high temperature to complete different desirable 'phase transformations' and also develop necessary 'atomic bonding or cohesion' so that they have high strength in order to be suitable for a given application. This process of sintering is a subject by itself, so it cannot be discussed in details here. Moreover, there are various types of sintering, *i.e.* gas-phase sintering (ZnO), liquid-phase sintering (ceramic whitewares and some refractories) and solid state sintering that consists of a host of ceramic materials with good mechanical and other properties of importance. Here, only the latter will be discussed very briefly for the acquaintance on the subject, which is only necessary for the understanding of sintering of the nano materials.

Generally speaking, the grain shape of ceramic raw materials is considered to be composed of spherical ensembie. When these 'spheres' are packed in a given volume, then there will be naturally a lot of 'pores' between the grains of different sizes. The elimination of these 'pores' in order to make the spherical particles come closer together with high cohesion is called sintering, when almost the theoretical density of a given material can be achieved. The heating process involved in sintering gives the necessary energy, but there must be some driving force for 'pore elimination', with a consequent change in the surface dynamics.

What is the macroscopic driving force for sintering ?

The reduction of 'excess energy' associated with the surfaces is the main driving force. This reduction can take place mainly by two processes :

1. Total surface area is decreased by increasing the particle sizes leading to a 'coarsening stage',
2. Solid/Vapour interfaces are eliminated with the creation of 'grain boundaries', which is followed by the 'grain growth' leading to → 'densification or sintering'.

Actually, in a given ceramic material undergoing sintering, the above two processes are in 'competition'. However, our main goal is to reduce the former and optimize the latter, *i.e.* the densification. At a given sintering temperature, if the atomic processes leading to densification dominate, the pores start getting smaller and eventually disappear with sintering time, and the sintered compact shrink in size. However, if the atomic processes are such that the 'coarsening' is faster, both the pores and grains coarsen and get larger with time, *i.e.* excessive grain growth, that must be avoided at all cost.

The necessary condition for solid-state sintering to occur is that the grain boundary energy is less than twice the solid/vapour surface energy so that the 'solid dihedral angle' between the grains has to be less than 180° for the densification to occur. This is explained clearly in terms of an equation in the **section - 2.4.1,** with the details for sintering of nano particles of SiC.

For an effective sintering process in solid nano materials, the mass has to be transferred mainly by three different atomic diffusion mechanisms as :

1. Surface diffusion,
2. Volume diffusion, and
3. Grain-boundary diffusion.

This obviously excludes the other processes like evaporation-condensation, and viscous and creep flow in sintering. Now, it is important to know which of the above diffusion processes is operative in the sintering mechanism. This is described in terms of a 'model' that is relevant for the sintering of nano particles of SiC in the **section - 2.7.3,** alongwith the role of dopants in creating the necessary driving force for full densification. It has to be remembered that when the particles are small, say in the nano range, the total surface area is very large, which consequently gives a high surface energy for these nano particles. Hence, the driving force for reducing this surface is large enough to drive the sintering process to be completed faster.

Therefore, it is prudent to go ahead to full densification by the preparation of the nano particles of SiC, which is otherwise difficult for a covalent solid like SiC. This is what has been actually done in the present case. All the experimental data which have been presented here are interpreted with the help of such atomic diffusion processes in the entire discussion on sintering of nano particles of SiC in the following subsections. Moreover, in order to make our knowledge clear, the role of different dopants or additives, the role of adding carbon and the dependence of different atmospheres of sintering have all been delineated in order to elucidate the problem of sintering further. This should help the readers to understand the mechanism of solid state sintering through 'practical' data on an actual material like SiC.

2.3. NANO MATERIAL PREPARATION

A commercially available Acheson type α-silicon carbide of grade 1000 from M/s Grindwell Norton Ltd, India, with a silicon carbide content of 98.7%, was used. The specific surface area of this starting powder (1.5 m^2/gm) was measured by a Surface Analyser of M/s Micromeritics, U.S.A. The initial particle size was measured in a Zetasizer (Model 1000-HS) of M/s Malvern Instruments Limited, U.K. The sub-micron grade of β-silicon carbide from M/s Superior Graphite Company, U.S.A., was used for attrition milling operation. After the optimum milling operation, the nano particles were purified by a special process, as described later. The initial particle size of the as-received α-silicon carbide was 0.39 μm, and that for β-silicon carbide was 0.52 μm, which are normally specified by the 'suppliers' of any powder materials.

2.3.1. Attrition Milling

In order to describe various steps needed to prepare nano particles of silicon carbide by 'attrition milling' route, it is important to have an idea on the working of an attrition mill. Apart from knowing the 'operation', it is also necessary to explore different types of attrition mill for various types of materials with a variety of sizes of the mill in order to be able to up-scale the actual production from the laboratory level to commercial production of nano particles.

2.3.1.1. How Batch Attritors Work?

The generic name of an 'Attritor' is very often referred to as "stirred ball mill." The operation of an Attritor is simple and effective. The material to be ground and the grinding media are placed in a stationary (jacketed) tank. This media is spherical, normally measuring about 0.24 to 0.95 cm in diameter. The type of grinding media commonly used are : carbon steel, stainless steel, chrome steel, tungsten carbide, ceramics like alumina or zirconia.

The material and the media are then agitated by a shaft with arms, which are rotating at high speed. This causes the media to exert both shearing and impact forces on the material, resulting in optimum size reduction and dispersion.

No pre-mixing is necessary. The material can be directly fed into the jacketed grinding tank of the Attritor. Sometimes, there are certain 'additions' to the formulation of the main material requiring only slight grinding, which should be easily added later.

The Attritors are equipped with specially designed high starting torque, two-speed, electrically driven motors. The high starting torque motor means no clutches, etc., are needed in the grinding operation. There is a usefulness of a two-speed motor. The lower speed is used for charging and discharging, and the high speed is used for grinding. Depending upon the application, the electricals are of the type 'totally enclosed fan cooled' (TEFC) or explosion-proof.

The batch Attritors are equipped with a built-in pumping system that maintains the circulation during grinding operation for accelerated attrition and uniformity. This pump can also be used for material discharging. The final result of the grinding action is a fine and even particle dispersion.

There are several advantages of an Attritor Mill, such as : (*a*) Compact design, (*b*) Rugged construction, (*c*) No special foundation required, (*d*) Simple to operate, (*e*) 10 times faster than ball mills, (*f*) Energy efficient, (*g*) No pre-mixing necessary, (*h*) Minimum maintenance required.

The batch Attritors offer the advantage that the material can be inspected any time during the grinding cycle. Therefore, the adjustments or formulation additions can be accomplished without stopping the machine. The Attritors are ideal for long-term, heavy-duty, semi- and fully-automated production operations. They may also be used in conjunction with programmable controllers in highly sophisticated production environments.

The batch Attritors range in size from about 72 litres to 2700 litres 'gross tank capacity' and from a 5 HP motor to a 1500 HP motor. These versatile and easy-to-maintain grinding mills have been successfully used in a variety of industries from chocolates to tungsten carbide. Moreover, the Attritors can be produced to meet the requirements of even the most stringent applications such as the 'electronic ceramic products'. For this application, the Attritors are available with metal-free systems to assure a contamination-free environment.

Certain special nano materials, such as high density 'tungsten carbide', require a specially designed heavy-duty batch Attritor, which are generally available with tank capacities ranging from 54 litres to 1100 litres, and 10 HP to 125 HP motor. These powerful grinding mills feature a 'tapered tank' to reduce load on the bottom arms for easy starting. A special lifter arm raises the media and product off the bottom of the tank. The arms are sleeved with tungsten carbide to reduce wear and contamination. The tank has a ball discharge valve and chute at the bottom for easy discharge of media. A water-cooled cover with a floating teflon seal minimizes solvent loss. This type of Attritor can be fitted with an optional 'torque meter' to precisely measure energy input for determining the 'processing time'. This is to maintain 'efficiency' in the process.

A worm gear assembly attached to the tank allows the tank to be tilted. In this fashion, the shaft and arms can be easily removed and/or inspected, and the grinding media changed, if necessary. There is a pumping system, which is used to circulate material during grinding and to discharge the ground materials.

2.3.1.2. Preparation of Nano Particles

For the preparation of nano particles, a commercial attrition mill (Model PRIS) of M/s Metisch Feinmath Technik GmbH, Germany, consisting of a water cooled 500 mL steel vessel and a motor driven steel agitator was used. In contrast to the conventional ball mills, in the attrition mill, the special steel balls are agitated by a series of stirring arms mounted to a motor shaft, as explained in the previous section. This attritor is very efficient in reducing the particle size, when various grinding parameters are optimized. A picture of an attrition mill used in the preparation of 'nano particles' of SiC is shown in **Figure - 2.1.**

The particles were ground in the above attrition mill in five different 'fluid media'. These five fluid media were ethylene glycol, *n*-hexane, xylene, butanol and water with surface active agent. The ratio of the steel balls to silicon carbide to get nano particles was also studied in these fluid media for different grinding time in hours. Experimentally, for α-silicon carbide, the findings of the ratio of the steel balls to silicon carbide and the required grinding time in hours, which were found to be the best, were ultimately used in order to obtain nano crystalline particles from sub-micron sized silicon carbide in a particular fluid medium.

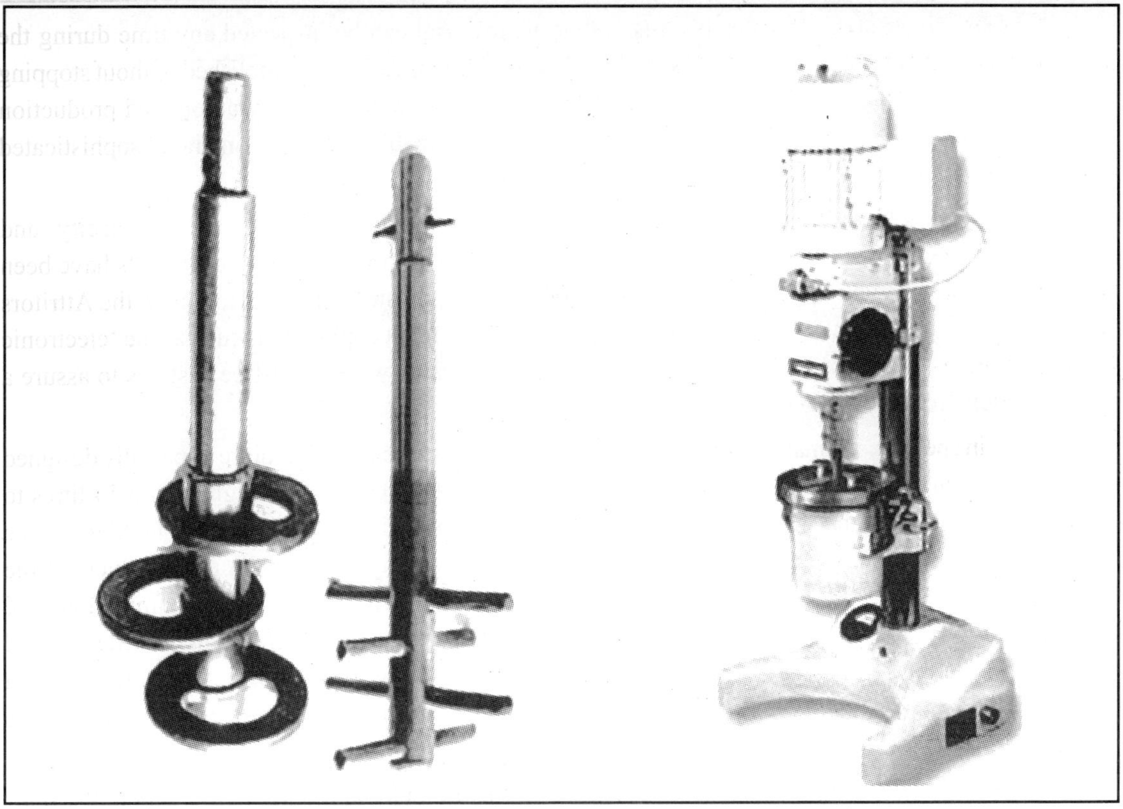

Figure 2.1 : A typical diagram of an attrition milling machine.

These findings were also used for getting nano particles of β-silicon carbide. The particle size distributions, after grinding in different fluid media, with different constant ratio of the weight of the steel balls to silicon carbide, were measured by using a particle size analyser (Model : Autosizer-2C of M/s Malvern Instruments Limited, U.K.) [4, 5].

The particles were examined in a 'Transmission Electron Microscope' (Model : TEM 400CX of M/s JEOL Ltd., Japan), which was operated at an accelerating voltage of 100 KV, before and after grinding in order to check the nature of the particles.

2.3.1.3. Nano Particles of SiC

First of all, the initial conditions of the as-received silicon carbide materials are shown in terms of their particle size, and other data, related to the effect of different variables on the grinding efficiency of the material. **Figure. 2.2** shows the distribution of the initial particle size of the as-received α-silicon carbide. It is seen that the particle size distribution is quite narrow. Therefore, this work is concerned with quite a narrow range of initial particle size. **Figure 2.3** shows the transmission electron micrograph of the initial sub-micron size particles of α-silicon carbide, which clearly shows a sort of 'spherical nature' of the nano-particles.

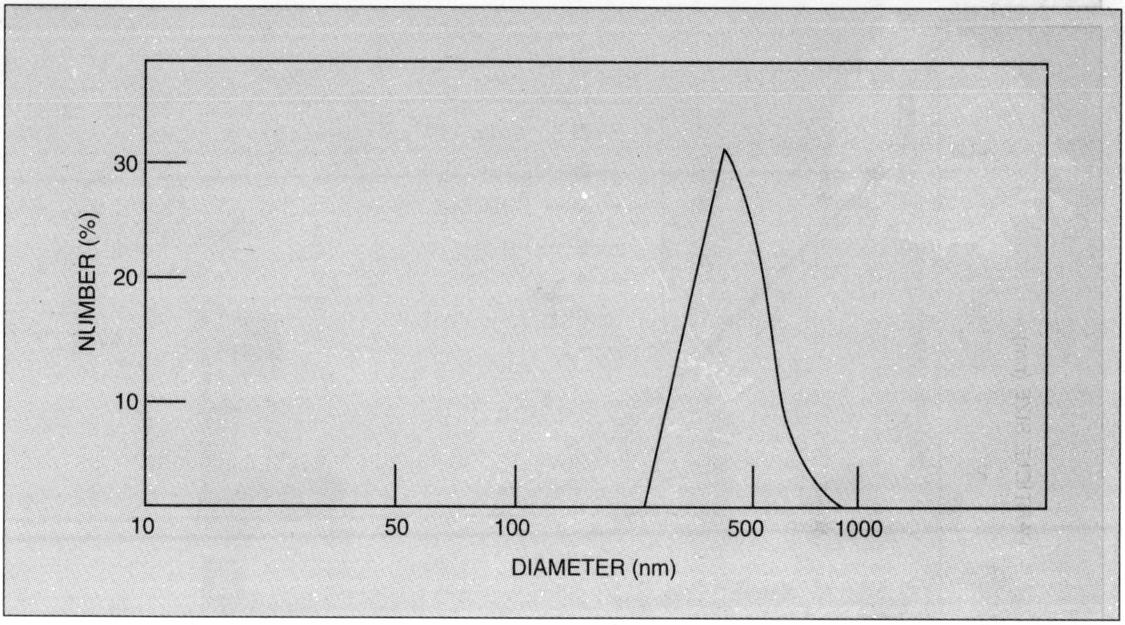

Figure 2.2: Particle size distribution of as-received α-SiC.

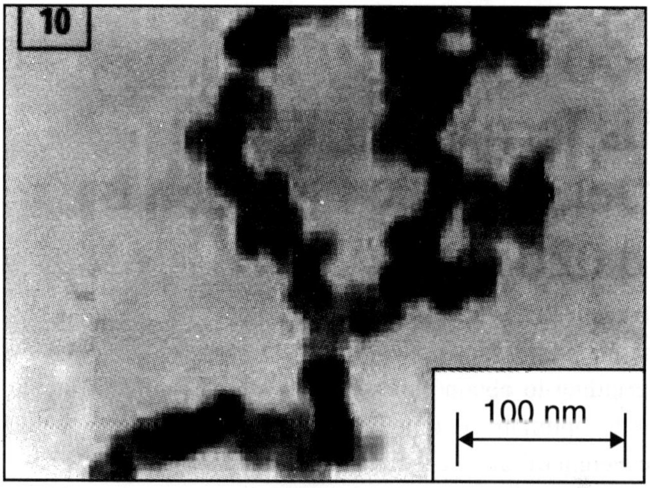

Figure 2.3: TEM photo of the initial particles of α-SiC.

2.3.2. Optimization of the Attrition Milling

Figure 2.4 shows the effect of the ratio of the weight of the steel balls to that of α-silicon carbide, which is required to obtain the nano crystalline particles. It is seen that as the ratio increases, the particle size of α-silicon carbide sharply decreases. Several curves are shown in **Figure 2.4,** wherein each curve represents a different fluid medium of grinding, like water with surfactants, ethylene glycol, N-hexane, butanol and xylene, for a constant time of grinding of 12 hours.

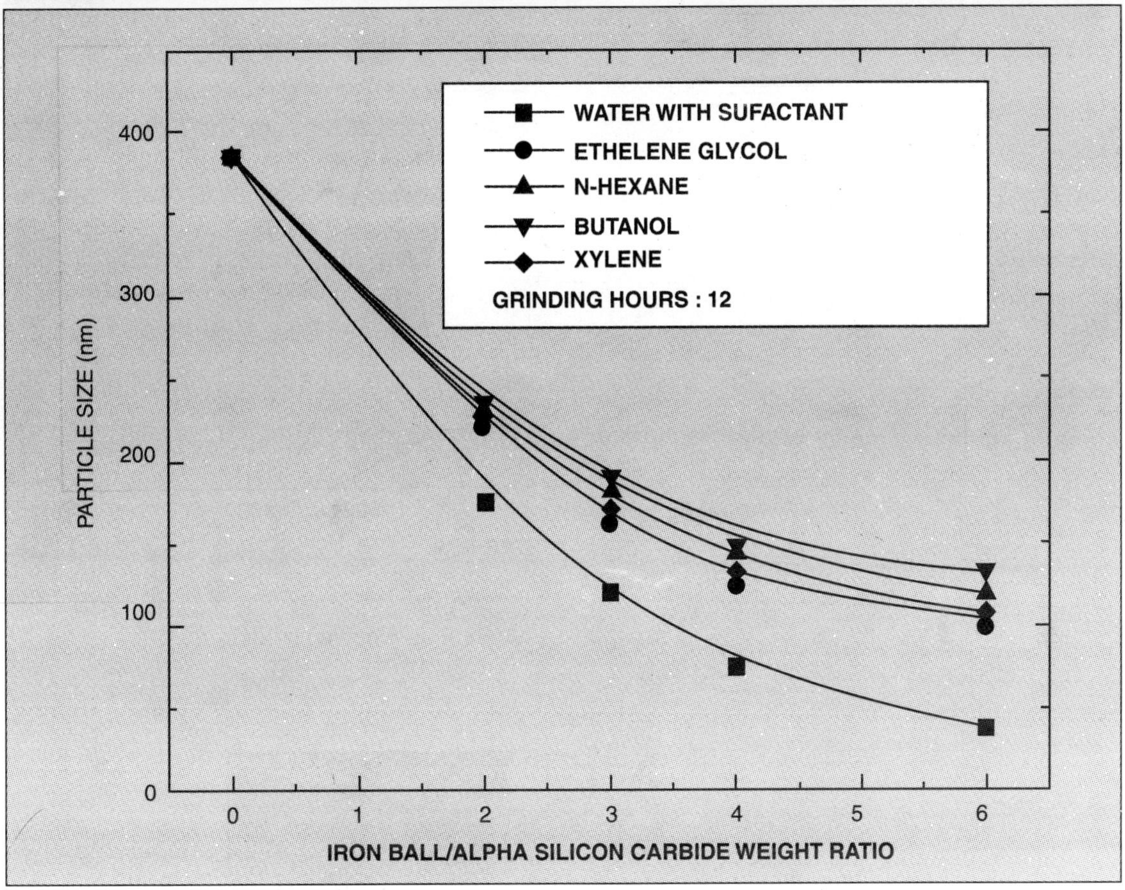

Figure 2.4 : Particles sizes in five different fluid media of grinding against the ratio of the weight of the steel balls to that of SiC.

The decrease of particle size with the 'ratio' is more pronounced for water with surfactants, compared to that of the latter four fluid media, wherein the difference is negligible in terms of grinding efficiency in reducing the particle size. This is an important result in that it helps to optimize even the 'fluid medium' for grinding to obtain nano-sized SiC materials. At higher ratio, the curves tend to saturate showing that the optimum condition for this 'ratio' has reached. It has also been found that at the ratio of 6:1 of the weight of the steel balls to that of α-silicon carbide, it is possible to obtain nano-crystalline particles. The average particle size in the fluid medium of water with surfactants at the ratio of 6:1 is found to be **37 nm.**

Figure 2.5 shows the particle size against the grinding time in hours, which are required to obtain nano-sized α-silicon carbide particles. It is seen that for a fluid medium of water with surfactants, the particle size continuously decreases with increasing time of grinding, as expected. The effect of grinding time is more pronounced for a fluid medium of water with surfactants, compared to that of the other four fluid media, as also observed in **Figure 2.4,** for the effect of the grinding ratio. It is also seen that the nano-sized particles of α-silicon carbide of an average size of **37 nm** are obtained for a minimum of grinding time of 12 hours.

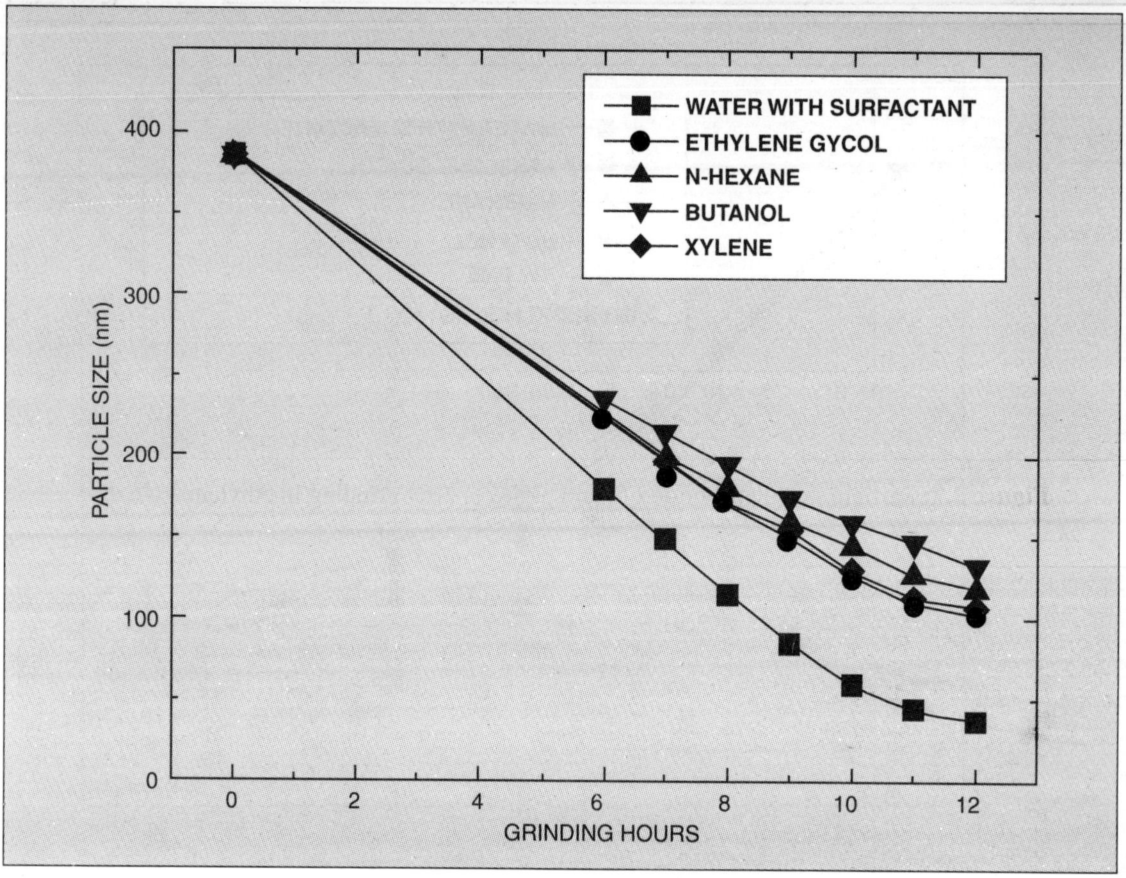

Figure 2.5: Particles sizes in five different fluid media of grinding against the time of grinding in hours.

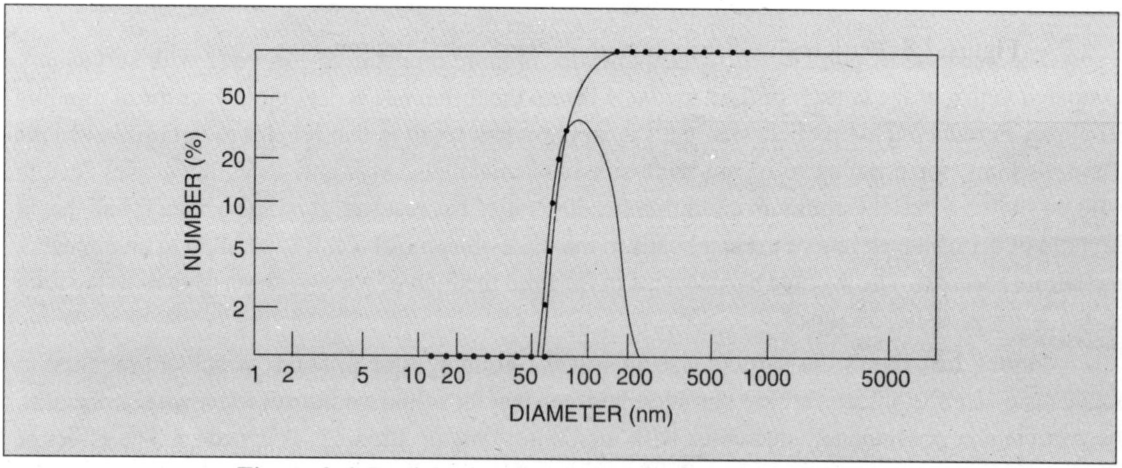

Figure 2.6: Particle size distribution of nano particles after grinding in water with surfactants.

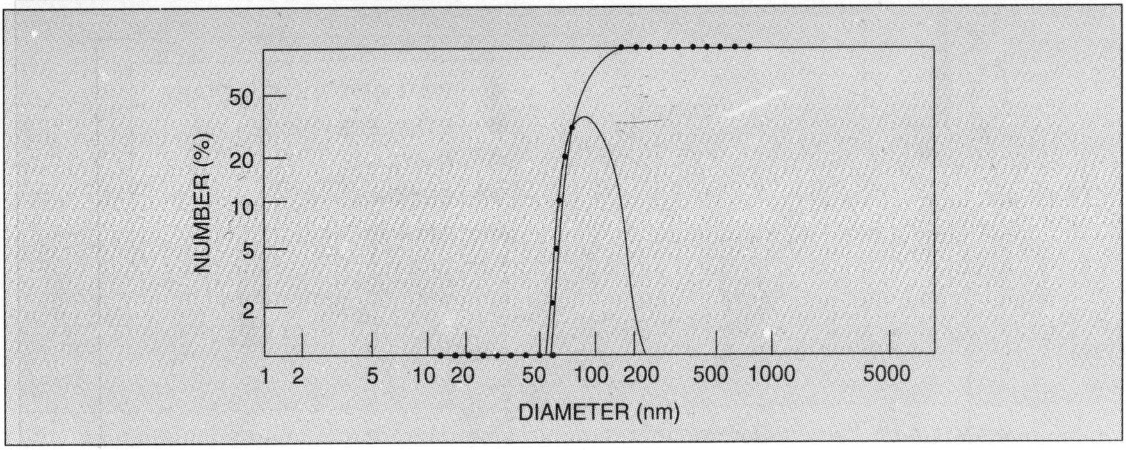

Figure 2.7: Particle size distribution of nano particles after grinding in ethylene glycol.

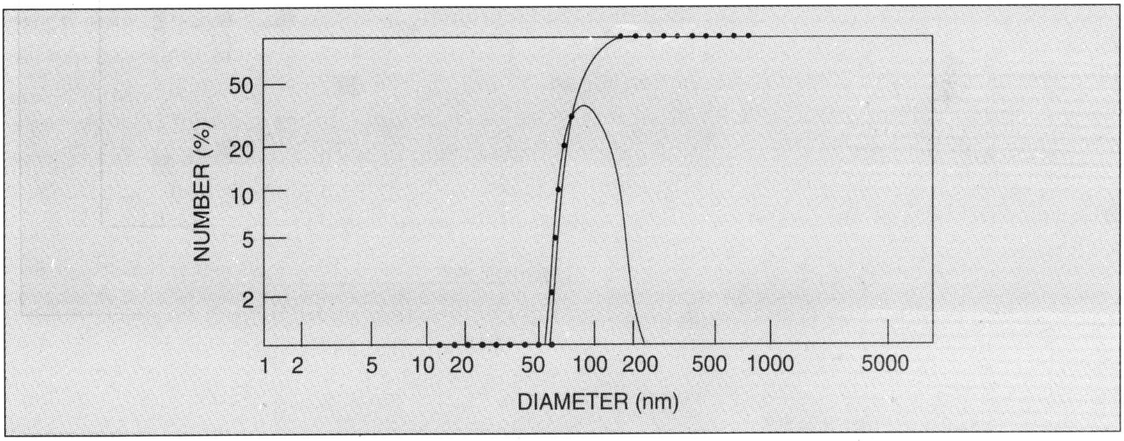

Figure 2.8: Particle size distribution of nano particles after grinding in N-xylene.

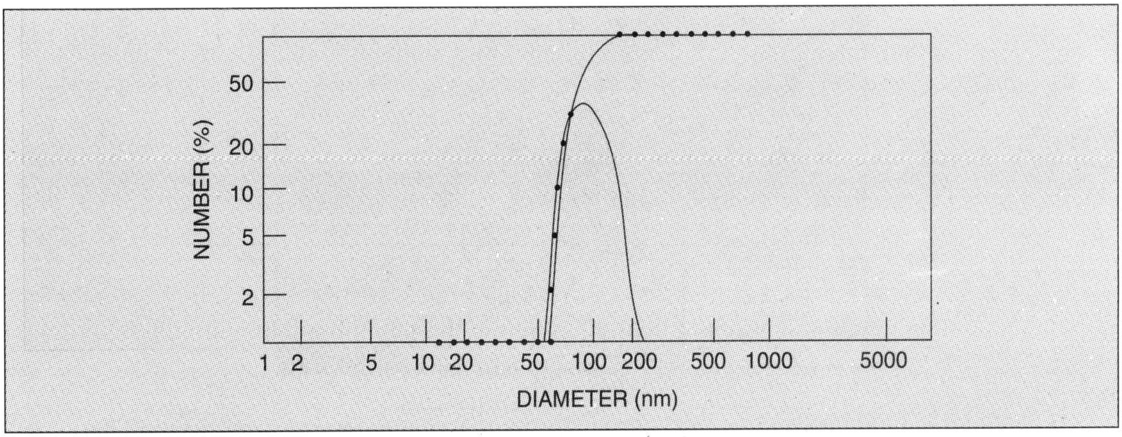

Figure 2.9: Particle size distribution of nano particles after grinding in N-hexane.

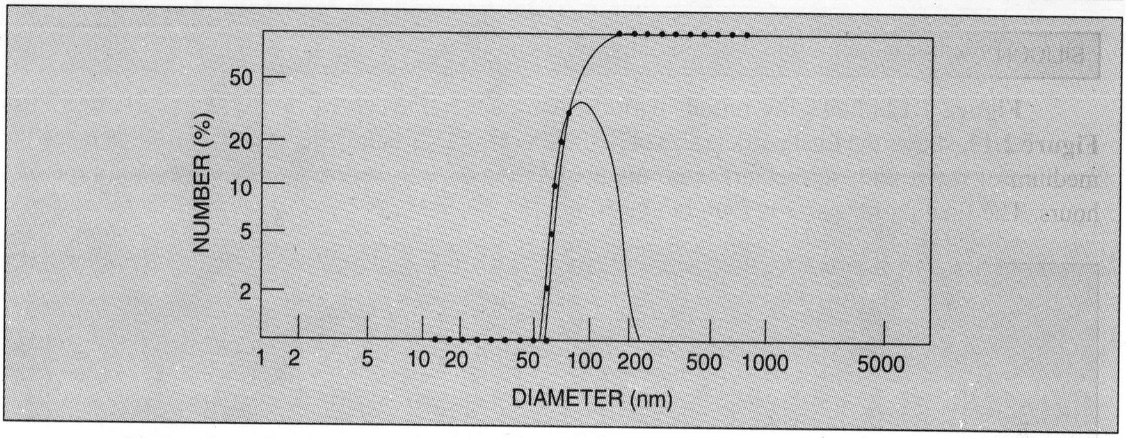

Figure 2.10: Particle size distribution of nano particles after grinding in butanol.

Figures 2.6 – 2.10 show the distribution of nano-sized α-silicon carbide particles for five different fluid media of grinding, after the grinding operation in the attrition mill. These five figures reflect the results on the effect of variation of the fluid medium on grinding to obtain nano-size particles. It also shows that the particle size distribution is quite narrow. From these figures, it can be concluded that the minimum nano-size particles of α-silicon carbide of **37 nm** can be obtained by attrition grinding of initial α-silicon carbide particles with an average size of 0.39 μm, in the medium of water with surfactants with the ratio of the weight of the steel balls to that of α-silicon carbide of 6:1 for an optimum grinding time of 12 hours. Finally, **Figure 2.11** shows the transmission electron micrograph of the final particle size of these nano crystalline particles of α-silicon carbide.

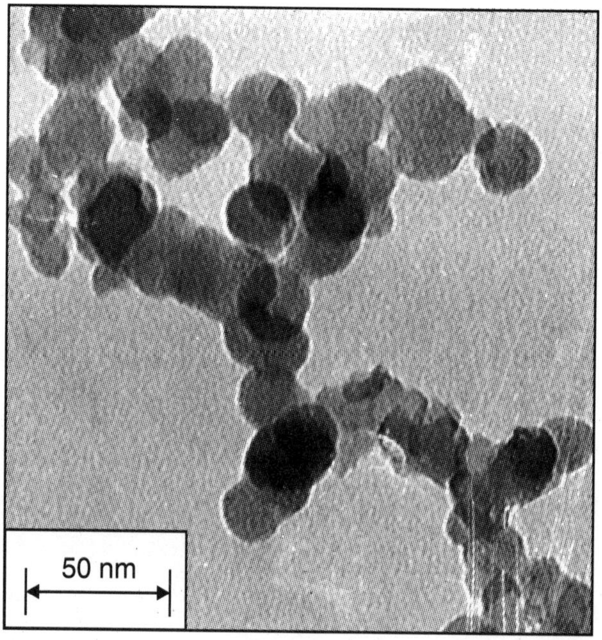

Figure 2.11: TEM photo of the final nano particles of SiC after optimum attrition milling

Figure 2.12 shows the initial particle size of the as-received β-silicon carbide (0.52 μm). **Figure 2.13.** shows the final particle size of β-silicon carbide particles after attrition grinding in the fluid medium of water with surfactants, with the above ratio of 6:1, for an optimum grinding time of 12 hours. The final particle size is found to be 50 nm [4, 5].

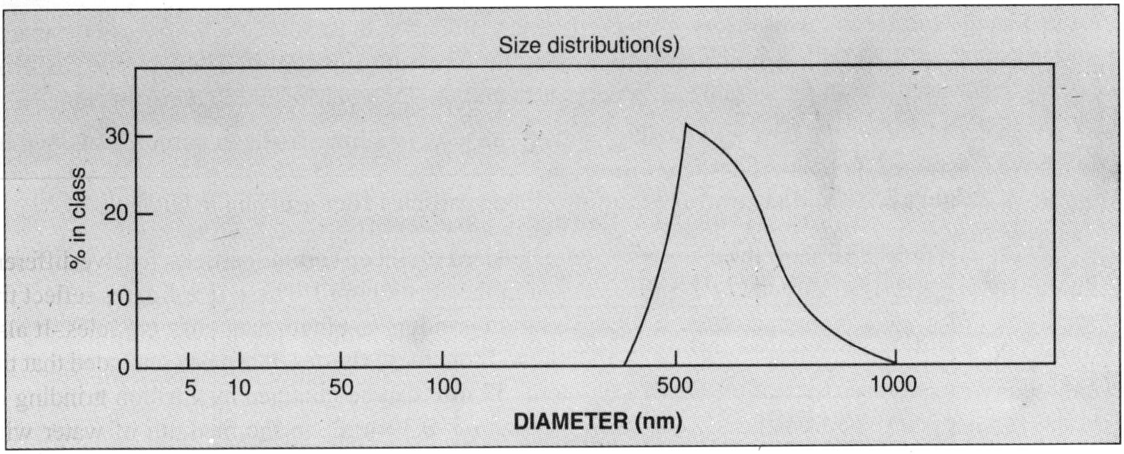

Figure 2.12: Initial particle sizes of as-received β-SiC.

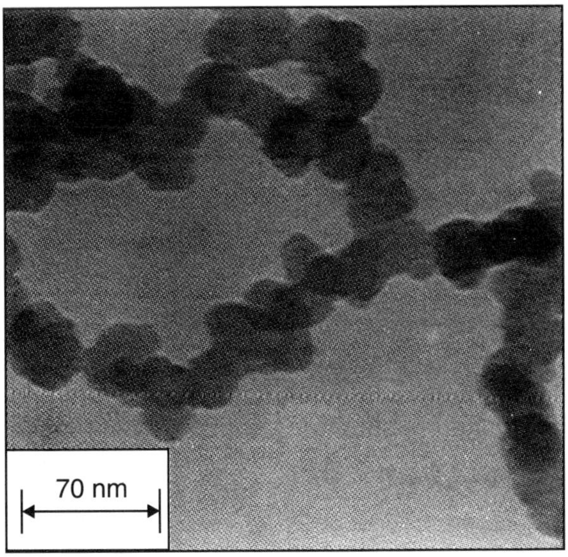

Figure 2.13: Final particle sizes of β-SiC for an optimum attrition grinding.

2.3.3. Purification of Nano Particles

These nano particles of silicon carbide were prepared in order to be able to sinter to full density. Hence, a special care is needed to purify the nano materials. During grinding, the silicon carbide powder is contaminated by iron. These iron wears along with the other impurities were remøved by treating them with hydrochloric acid. An excess carbon, which was present in the silicon carbide powder during the preparation, was burnt off in air at 700°C. Since a slight oxidation of the silicon carbide powder occurs under these conditions, the powder was treated with hydrofluoric acid to remove the silica formed on the surface of the powders. This purification is done with utmost care in order to be able to start the sintering work with as pure materials as possible. These nano particles of both α- and β-silicon carbides after purification are suitable as 'precursor materials' for sintering to full densification [3].

Table 2.1 shows the powder characteristics of nano-crystalline α-silicon carbide obtained by the above mentioned process after the purification.

Table 2.1 : Powder Characteristics

PHASE COMPOSITION	Mainly α-SiC
Total Carbon (wt%)	30.27
Free Carbon (wt%)	i.00
Oxygen (wt%)	0.33
Impurities :	
Al (ppm)	125
Fe (ppm)	130
Ca (ppm)	150
Specific Surface Area (m^2/gm)	50.50
Equivalent Partile Size (nm)	37.00

2.4. SINTERING OF SIC

In order to effectively sinter it, not only high temperature is needed, but also there is a need of various dopants like boron carbide and aluminium nitride mixed with carbon. This creates vacancies of carbon and silicon, which increases their bulk diffusion, thereby making the solid-state sintering possible without any liquid/glassy phase at the grain boundary, as revealed by the Transmission Electron Microscopy (TEM), the Scanning Electron Microscopy (SEM) [2, 3, 16-20] and the Electron Diffraction (ED) patterns [3, 19, 20].

Sinterable silicon carbide powders have been prepared by the attrition grinding and chemical processing of an Acheson type α-SiC. The pressureless sintering of these powders was achieved by the addition of aluminium nitride and boron carbide together with carbon. Nearly 99% sintered density was obtained by this process due to the use of "nano particles". The mechanism of sintering was studied by SEM and TEM. This study shows that the mechanism of sintering is a solid-state sintering process.

2.4.1. Role of Dopants

Due to the high amount of covalent bonding (~ 90%) in silicon carbide, it is not possible to obtain high densities with a 'pure material' by using a normal sintering process. It has been observed that the powder compact of very fine pure crystalline solids does not densify without any applied pressure. The reason for this lack of sintering characteristic involves kinetic as well as thermodynamic considerations. The 'hot pressing' technique with the addition of small amounts of boron, aluminium, nickel among others, is well known, but it is limited to the production of rather simple shapes, as found out by Alliegro et al [21] and Kriegesman [22]. The 'hot isostatic' pressing technique may be useful for the fabrication of more complicated geometric shapes.

The pressureless sintering of silicon carbide has been reported by Prochazka [23] who synthesized β-silicon carbide in the submicron range, which was sintered after the addition of 0.5 wt% boron and 1 wt% carbon at 2050°C to 2150°C in order to obtain densities of 85 to 96% of the theoretical density in the argon atmosphere. Three years later, Coppola and Mcmurtry [24] achieved the sintering of more readily accessible α-silicon carbide powder in a mixture containing boron and carbon additives.

However, in both boron doped sintered α- and β-silicon carbides, the exaggerated grain growth was observed by Johnson and Prochazka [25], if the sintering temperature exceeded a critical limit. The micro structure of boron doped SiC, sintered at higher than 2075°C, usually contains large tabular grains, which may act as 'stress concentrators', which seriously degrade the strength properties.

In its pure form, SiC powder will not sinter to a fully dense state. By heating at the temperature range of 1900–2300°C with pressures ranging from 100 - 400 MPa, Kriegsman et al [22] were able to make 'pore-free' dense particles of polycrystalline SiC, starting from green bodies of 50% green density without the use of any sintering aids. Prochazka [23] proposed that during the sintering of pure submicron powders of covalently bonded solids, the densification is prevented by a hypothetically high ratio of 'grain-boundary' to 'solid-vapour' surface energies, *i.e.* γ_{GB}/γ_{SV} (see the section 2.2.3). In order to get a pore 'surrounded by three grains' shrink to closure, the equilibrium dihedral angle (θ) must be > 60° or $\gamma_{GB}/\gamma_{SV} < 3$. The equation relating these 'important sintering' parameters that is applicable at solid-vapour interface is written as

$$\gamma_{GB} = 2\gamma_{SV} \cos\left(\frac{\theta}{2}\right)$$

The following argument was advanced by Prochazka [23] in order to explain the role of boron and carbon :

1. As γ_{GB} for SiC is expected to be high due to strong 'directionality' of Si-C bonds, γ_{GB}/γ_{SV} ratio is also too high, and the sintering is thus inhibited,
2. Boron segregates selectively along the grain-boundaries decreasing γ_{GB} and consequently γ_{GB}/γ_{SV} ratio, and
3. Since γ_{SV} for SiC is high, its surface has a tendency to adsorb impurities and eventually lower γ_{SV}, thus increasing γ_{GB}/γ_{SV} ratio. Both silicon and silica which lower γ_{SV} for SiC by this process should be removed. The carbon removes silica and silicon by forming SiC and CO.

On the other hand, Lange and Gupta [26] used mixtures containing β-SiC with different amounts of B and C or B_4C. After sintering of these mixtures, their microstructural observations strongly suggested that a 'boron-rich' liquid was present during sintering. From these observations, they concluded that either 'reaction' sintering or 'liquid-phase' sintering is responsible for the densification of submicron silicon carbide with boron and carbon additions.

But, the solubility and lattice position of B in SiC are contradictory. Shaffer [27] reported on the solubility of 0.1 wt% B in SiC at 2500°C. By contrast, Murata and Smoak [28] found a solubility of 0.5 wt% at 2200°C. The lattice parameter measurements upon incorporation of B suggests that boron substitutes for Si or C. A comparison of these data with the theoretical calculations by Tajima and Kingery [29] suggested that the aforementioned lattice contraction is less than that would be expected if all the boron atoms replaced Si. The considerations of 'strain energy' and the 'bonding stability' favour the incorporation at C and Si sites simultaneously. Thus, it seems that boron may occupy both sites simultaneously.

The Scanning Transmission Electron Microscope (STEM), Auger Electron Spectroscopy (AES), Wavelength-Dispersive X-ray Analysis (WDXRA) and Micro-Radiography (MRG) studies were conducted by Hamminger et al [30] on several α-SiC separately doped with Al and C or B and C, which showed a homogeneous region of ≈ 200 μm diameter containing B, O, N, Al, and/or C-enrichments.

Let us take a critical view on these findings. The pressureless sintering of silicon carbide has been reported by Prochazka [23] as explained above, and also by Prochazka and Scanlan [31], based on the grain-boundary and solid-vapour surface energies. This theory is inconsistent with the experimental measurements of 'dihedral angles' in partially sintered silicon carbide powder compacts, as reported by Greskovich et. al. [32]. Moreover, Prochazka and Scanlan [31] did not find the presence of any boron in the grain-boundaries through their experiment of Neutron Activation Auto-Radiography (NAAR), which contradicted Prochazka's earlier theory that boron reduces the grain-boundary energies through absorption at the grain-boundaries, and enhances the diffusion process in silicon carbide.

As also mentioned above that Lange and Gupta [26] predicted a liquid-phase sintering. The extensive study of Si-B-C phase equilibrium by Kieffer et. al. [33] shows that only the SiC - SiB_6 - Si compatibility triangle contains liquidus at temperatures < 2100°C. The lowest eutectic temperature in this compatibility triangle is 1380°C. Lange and Gupta [26] also found that Si + B_4C powder compositions react to form SiC with traces of SiB_6, as observed by the microscopic examination.

The 'free silicon', which is a necessary ingredient for both sintering mechanisms, is available at high temperature as a result of the decomposition of SiC. Coppola and Smoak [34] achieved the sintering of the more readily accessible α-silicon carbide powder in a mixture containing boron and carbon additives. An explanation of the success of the densification, with the addition of boron nitride, boron phosphate and boron carbide, was provided by Murata and Smoak [28].

Contrary to the view of Prochazka [23] as mentioned above, Suzuki and Hase [35] observed that both boron and carbon effectively inhibited the grain growth, but they did not cause sufficient densification, when used individually, indicating that the surface diffusion was not completed. By using high-resolution microscopy, these workers found that at above 1900°C, a second phase with no more than **80 nm** in size, that influences the material movement was formed. It was considered that boron dissolves very little into silicon carbide and that a large amount of silicon dissolves in this inter-granular phase. They did not observe a B-Si-C liquid phase formation. Above 1950°C, this inter-granular secondary phase dissolves in the matrix up to 0.2 to 0.3 %, and then distributed widely in the material.

Quite interestingly, these observations have been confirmed by Davis et al. [36]. Subsequently, Hamminger et al. [30] also investigated the fracture surfaces of B-C and Al-C doped silicon carbide to show the formation of solid solution, which also showed a heterogeneous region of @ 200 mm diameter containing B, O, N, Al, and/or C enrichments. Tajima and Kingery [29] found by using lattice constant evaluations that aluminium entered into the 'solid solution' in β-silicon carbide, occupying the Si sites. Tajima and Kingery [37] also observed that there is some segregation of aluminium in the grain-boundaries.

The above reports show that the sintering of silicon carbide is not an easy affaire. Moreover, there is quite a difference of opinion among various important workers on the mechanism of sintering in the atomistic level. The sintering of SiC needs to be properly understood so that its various applications can be tailored at ease. It has already been mentioned above that the route to 'nano particles sintering' is a better way to have an adequate knowledge on the sintering mechanism, after dealing properly with the microstructure of the sintered products with 'nano-crystalline' SiC. This amply justifies our intention of going to the "nano route" of material processing for an important material like SiC for various useful applications.

The studies on the sinterability of nano-crystalline α-silicon carbide with the addition of aluminium nitride and boron carbide are presented here with their effect on the microstructure and on the mechanism of sintering (see later in the **section 2.7.3**). The next step is to understand the precise 'role of carbon' in the sintering process.

2.4.2. Role of Carbon

The effect of carbon has been studied by different workers. The experimental evidence that the excess carbon may control the grain growth of silicon carbide, has been reported by Mirzah et. al. [38] and Hollenberg and Crane [39]. The work of Hojo et. al. [40] has shown that the highest sintered densities are achieved when the amount of free carbon is just enough to completely remove the surface silica layer according to the equation

$$SiO_2 + 3C = SiC + 2CO$$

The final densities are lower than this maximum, if the amount of free carbon is either higher or lower than this value, suggesting that the effect of carbon is associated with the reduction of the surface silica. According to Clegg [41], the role of carbon is to remove the silica layer preventing the deleterious reaction of silicon carbide with silica. Based on the studies of different diffusion data, Rijswijk and Shanefield [42] reported that carbon helps in the sintering process in different ways, such as

(*a*) It reduces the silica layer,

(*b*) It lowers SiO_2 vapour pressure, and

(*c*) It prevents the transport of elemental silicon along the grain-boundary.

The self-diffusion of SiC has been studied in both α- and β-silicon carbides, in pure and doped samples, in the single crystals and also in the polycrystalline materials. It was first observed by Friederich and Coble [43] that a large discrepancy exists between the diffusion data measured in silicon vapour, as done by Hong et al [44] and those measured in a high carbon vapour pressure by Ghostagore and Coble [45]. They provided an explanation based on the difference in the defect structure in the silicon carbide with the changes in atmospheric condition.

Hence, a study was carried out with the aim of achieving full densification of both nano-crystalline α- and β-silicon carbides doped with 0.5 wt% boron carbide + 1 wt% carbon and 2 wt% aluminium nitride + 1 wt% carbon. The effect of carbon on sintering was studied by varying the carbon content keeping all other parameters unchanged. Here, the effect of carbon on the sintering behaviour of both the silicon carbides is reported in order to find out whether the sintering mechanism is through bulk diffusion (see later in the **section - 2.7.3**). The next step is then to explore the studies on the effect the 'effect of atmosphere' on the sintering mechanism of silicon carbide.

2.4.3. Role of Sintering Atmosphere

The effect of atmosphere on the sintering of nano-crystalline α-silicon carbide, prepared by attrition milling, has been studied under vacuum, argon and nitrogen atmospheres between 2000° - 2100°C after doping with boron carbide and carbon. It has been found that the sintering atmosphere has a very critical influence on the sintering of silicon carbide. The vacuum atmosphere helps in the sintering of nano-crystalline silicon carbide doped with boron carbide and carbon to about maximum theoretical density, whereas the nitrogen atmosphere has a retarding effect and does not yield full densification, while the effect of argon atmosphere takes an intermediate role in the sintering process [16].

Silicon carbide, when sintered to about theoretical density, shows a remarkable behaviour in terms of many useful technical applications like ceramic engines, components in aerospace etc. This is due to some interesting and favourable thermo-mechanical properties at high temperatures. Therefore, the sintering behaviour of silicon carbide, particularly of nano particles, assumes great importance [2, 3]. The sintering of such nano-crystalline silicon carbide should not only be studied by using various dopants to enhance the diffusion of both silicon and carbon at different temperatures, but it should be also studied under different atmospheres to see the effect of sintering atmosphere on the overall sintering behaviour [16].

The effect of sintering atmosphere was found to be very important in addition to the sintering additives and temperatures in the microstructural evolution of sintered silicon carbides. The atmospheric effects have been studied thoroughly by Prochazka et. al. [46] for sintering of silicon carbides. The work was later supplemented by that of Murata and Smoak [28] who confirmed by the use of the 'compound' additives that the diffusion of not only boron, but also of nitrogen and phosphorous may take place simultaneously during the sintering process. One of their significant findings is that the maximum density of sintered silicon carbide was achieved at the concentrations of the maximum 'solid solubility' of the additives.

Mirzah et. al. [38] further continued the studies on pressureless sintering in vacuum, Ar, He and N2 atmospheres by using α-silicon carbide. No differences in the densities or microstructures could be detected between vacuum, Ar and He as sintering atmospheres. They found that the samples of sintered silicon carbide in vacuum exhibited a carbon-rich surface layer, which was due to the decomposition of silicon carbide. These workers also observed that the sintering in nitrogen required approximately 150°C higher sintering temperature to achieve a full densification. A thermodynamic analysis was carried out by Venkateswaran and Kim [47] who reported by XPS studies about the formation of a thin layer of boron nitride (BN) when α-silicon carbide was sintered in nitrogen atmosphere. They observed that higher temperature is required for the complete densification of silicon carbide in nitrogen atmosphere.

In the investigation of Datta et al. [16, 18], nano-crystalline SiC particles, meticulously prepared by controlling and eventually optimizing various grinding parameters, have been sintered between 2000° - 2100°C under three different atmospheres, like vacuum, argon and nitrogen, in order to see the effect of atmosphere on the sintering behaviour. The results are explained in terms of a diffusion model (see later in the **section - 2.7.3**).

2.5. X-RAY DIFFRACTION DATA

As explained above, it is difficult to sinter silicon carbide under normal conditions. However, the nano particles of SiC prepared by attrition milling could be sintered under suitable conditions for a variety of novel applications. The silicon carbide is known to develop different polytypes during and after proper sintering. This is developed through a dislocation mechanism, which is discussed here along with the quantitative aspects of determining the contents of different polytypes by X-ray diffraction analysis.

Different polytypes are formed during and after sintering (4H, 6H, etc.), through the nucleation of stacking faults by a dislocation mechanism in these different polytypes [48]. In order to throw light on the formation of these polytypes, it is important to make quantitative determination of these polytypes in the sintered masses, which has been attempted here by following the standard techniques [49].

The identification and quantitative analysis of different polytypes present in the sintered silicon carbides, doped with different additives and fired at different temperatures and time, were done by the X-ray diffraction. The X-ray diffraction pattern was taken by using a X-ray Diffractometer (Model : PW 1840 of M/s Philips NV, Holland). All X-ray diffractograms were taken by using Cu-K_α radiation (wavelength, $\lambda = 1.54178$ Å) at a scanning speed of 0.10 per second in 2θ. A tube voltage of 40 KV, a tube current of 20 mA, and a time constant of 10 seconds were used. The lines were drawn by one line recorder (Model : PW 1879 of M/s Philips NV, Holland) with a speed of 10 mm per degree 2θ. The diffraction patterns are standard patterns with the intensity (I) vs. 2θ curves, and are not shown here.

2.5.1. X-ray Data Analysis

The quantitative analysis by X-ray diffraction is based on the fact that the intensity of the diffraction pattern of a particular mixture depends on the concentration as

$$I = \left(\frac{I_0 A \lambda^3}{32\pi r}\right)\left[\left(\frac{\mu_0}{4\pi}\right)^2 \frac{e^4}{m^2}\right]\left(\frac{1}{V^2}\right)\left[|F|^2 P\left(\frac{1+\cos^2 2\theta}{\sin^2\theta\cos\theta}\right)\right]\left(\frac{e^{-2M}}{2\mu}\right)$$

where, I = Integrated intensity per unit length of diffraction line, I_0 = Intensity of incident beam (Joules $sec^{-1}\ m^{-2}$), A = Cross sectional area of incident beam (m^2), λ = Wavelength of incident beam (*m*), *r* = Radius of diffractometer circle (m), V = Volume of the unit cell (m^3), F = Structure Factor, P = Multiplicity Factor, θ = Bragg Angle, m = Linear Absorption Coefficient (which enters ½, *i.e.* the absorption factor).

Let us write different important parameters as

$$R = \left(\frac{1}{V^2}\right)\left[|F|^2 P\left(\frac{1+\cos^2 2\theta}{\sin^2\theta\cos\theta}\right)\right] e^{-2M}$$

$$K = \left(\frac{I_0 A \lambda^3}{32\pi r}\right)\left[\left(\frac{\mu_0}{4\pi}\right)^2 \frac{e^4}{m^2}\right]$$

which is a constant for a fixed incident beam and particular diffractometer, and is represented by K. The temperature factor is not important and thus, it can be neglected.

Then, the expression for the intensity can be represented by

$$I = \frac{KR}{2\mu}$$

It applies only to a pure substance for a particular phase in a mixture. Thus, the intensity of a particular line of the phase becomes the intensity of the amount of the phase present in the mixture. Moreover, there are several kinds of polytypes of silicon carbide, which contain mainly 6H, 4H, 3C (Cubic) and 15R peaks. Therefore, the equations have been developed to calculate different peaks at the same '*d*' value, although their intensities are different for different polytypes at different '*d*' values. Different phases of polytypes present in the sintered silicon carbides are calculated from these equations.

Table 2.2 : X-ray Data for various Polytypes of SiC.

3C Polytype :

h k l	*d*-Spacing (nm)	R
1 1 0	0.25109	3.813
2 0 0	0.21745	5.728

4H Polytype :

h k l	*d*-Spacing (nm)	R
1 0 0	0.26613	3.766
1 0 1	0.25727	14.834
0 0 4	0.25133	9.568
1 0 2	0.23520	12.988

6H Polytype :

h k l	*d*-Spacing (nm)	R
1 0 1	0.26208	8.435
0 0 6	0.25132	9.528
1 0 2	0.25095	14.292
1 0 3	0.23520	7.710
1 0 4	0.21741	2.786

15R Polytype :

h k l	d-Spacing (nm)	R
1 0 1	0.266	1.215
0 1 2	0.263	4.261
1 0 4	0.257	9.904
0 0 1 5	0.251	9.564
0 1 5	0.235	4.564
1 0 1 0	0.217	0.923

As stated above, the X-ray intensities are proportional to the volume fraction of a phase in a mixture of phases. The calculated intensities, as shown in Table - 2.2, were normalized setting 1 (111) of 3C polytype to 100, and were then used to derive the equation system given in this table for the volume fraction of different polytypes A, B, C and D, which are the volume fractions of polytypes, such as 15R, 6H, 4H and 3C respectively.

Thus, the following equations were derived :

	Peak	Peak at d (nm)
$K(3.2a + 9.9c) =$	A	0.266
$K(11.2a + 19.4b) =$	B	0.263
$K(26a + 38.9c) =$	C	0.257
$K(31.1a + 59.2b + 25.1c + 100d) =$	D	0.251
$K(18.1b + 34.1c) =$	E	0.235
$K(2.4a + 6.5b + 13.1d) =$	F	0.217

By solving these equations, different polytypes present in the sintered silicon carbide were determined from the observed X-ray diffraction intensities.

The above results can be summatized as follows :

A. For α-SiC doped with boron carbide, the X-ray diffraction analysis shows 6H to 4H polytype transformation with about 72% 4H polytype.

B. For β-SiC doped with boron carbide, it is 72% 6H transformed from 3C polytype.

C. For α-SiC doped with AlN, it is 72% 4H transformed from 6H.

D. For β-SiC doped with AlN, it shows a polytype transformation to 4H, through 15R polytype route, from the 3C-cubic polytype, with 50% 4H and 50% 6H.

This is the result from the analysis of the X-ray diffraction patterns through the above standard formalism. From the dislocation behaviour of SiC crystals, this transformation seems to result from a suitable layer displacement mechanism, which is caused by the nucleation and expansion of the stacking faults within the silicon carbide crystal structure, as detailed below.

2.5.2. The Dislocation Mechanism

As shown later in this section, both TEM and SEM micrographs show that both boron and aluminium have entered within the structure of silicon carbide, but did not show the presence of any dopant-related phases or any liquid/glassy phase in the grain-boundaries. The presence of boron or aluminum within the structure was further confirmed by the electron diffraction patterns, wherein double spots were observed. It can be said that the sintering mechanism of silicon carbides doped with boron carbide and aluminium nitride is a solid state sintering process through the creation of vacancies and solid state diffusion through the increase of the diffusion coefficients by many orders of magnitude [2, 3]. It should be pointed out that the fracture mechanical behaviour of these materials also confirm the absence of any liquid/glassy phase, where both the flexural strength and the fracture toughness, *i.e.* the critical stress intensity factor or K_{IC}, remain constant over a wide temperature range upto 1400°C showing no degradation of strength [19, 20].

2.5.2.1. Dislocation Behaviour

At this point, some discussion is necessary on the dislocation behaviour of the polytype formation in α-silicon carbide. The formation of SiB_4 increases the cell dimension of 6H α-silicon carbide, whereas the formation of B_4C decreases the cell dimension. During sintering, the structure of SiB_4 and B_4C may begin to coalesce in different cells in layer displacement mechanism. The cells are likely to become disordered and strained.

The cells containing SiB_4 structure begin to recrystallize as 4H polytype, whereas the cells containing B_4C begin to precipitate from the cells having B_4C structure. Thus, it may be the reason for obtaining 72% 4H polytype from the grains containing mainly 6H polytype, which was determined by X-ray diffraction analysis. The quantitative measurement of silicon carbide polytypes from X-ray diffraction peak intensities developed by Ruska et al. [49] has been utilized for this calculation.

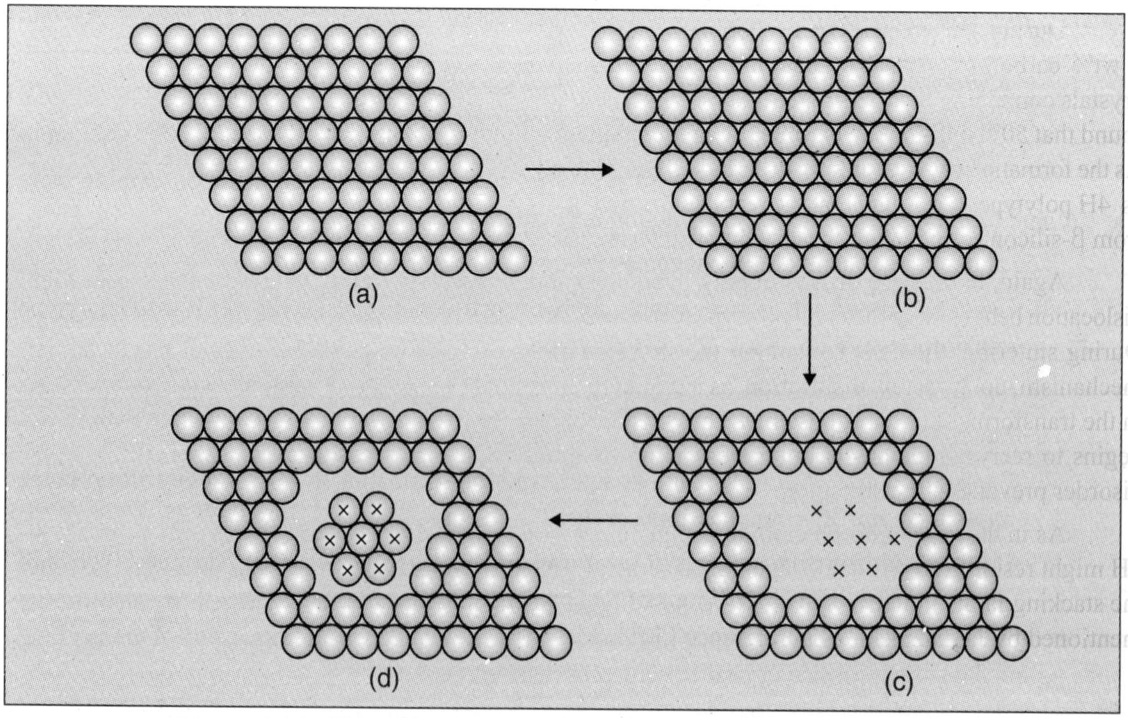

Figure 2.14: Dislocation behaviour of different polytypes of SiC [50].

Figure 2.14 shows the dislocation of various atomic sites in silicon carbide crystals. The polytype transformation from 6H to 4H might result from suitable layer displacements, which is caused by the nucleation and expansion of the stacking faults in individual close-packed double-layers of Si and C. This process is governed by the thermal diffusion, such as the grain boundary diffusion, since the nucleation of a stacking fault would require the migration of atoms inside the crystal.

This layer displacement is likely to occur in such a manner so as to minimize the free energy, and take the structure towards the stable state. If the stable state happens to be the one with a different order, such an order will tend to result. Such a mechanism was also suggested by Jagadzinski, as mentioned in reference [50].

During sintering by the addition of boron and carbon, and the increase of temperature, the vacancies would be created, and the atoms would become free to migrate within the crystal and to the surface by diffusion. If the number of vacancies come closer together, it will become possible for the neighbouring atoms to move into "B" sites thereby nucleating a fault. Within the region, the atoms are in "B" sites, while the rest of the atoms are in 'A' sites causing a partial dislocation to bind the fault.

This partial dislocation glides, which causes the fault to expand, until the entire layer of atoms is displaced into 'B' sites, and the partial dislocation moves out of the crystal. The entire layer is then displaced from 'A' to 'B' orientation, as shown in **Figure - 2.14.** By the same mechanism, the displacements occur for other layers throughout the structure in such a manner so as to result in a new structure.

Moreover, by this mechanism, the transformation would proceed through one-dimensional disorder caused by the nucleation of stacking faults and it is expected to exhibit a considerable one-dimensional disorder, as observed by the needle-like crystals of 4H polytype. The growth of these needle-like crystals in the later stage of sintering may also be governed by the screw dislocation, which might be generated from the strains in the crystal structure due to the incorporation of boron atom in the α-silicon carbide crystal.

During recrystallization of β-silicon carbide during sintering with 2 wt% aluminium nitride and 1 wt% carbon, the crystals containing Al_4C_3 presumably recrystallize as 4H polytype, whereas the crystals containing Si_3N_4 recrystallize as 6H polytype. From the quantitative analysis of polytypes, it is found that 50% of β-silicon carbide (3C) recrystallizes as 6H polytype, whereas the rest as 4H polytype. As the formation of Al_4C_3 in silicon carbide creates 'strain' in the structure, this leads to the recrystallization as 4H polytype. It is found that the recrystallization of 4H polytype occurred through 15R transformation from β-silicon carbide with the change of sintering time, as shown in **Figure 2.14.**

Again, in the case of AlN doping in silicon carbide, similar discussion may be made about the dislocation behaviour of different polytype formation. The formation of Al_4C_3 increases the cell dimension. During sintering, the cells containing the structure of Al_4C_3 begin to coalesce by layer displacement mechanism, not by screw dislocation, as it would require clockwise as well as anti-clockwise movement in the transformation that is possible. The cells become disordered and strained, and thus the structure begins to recrystallize as 4H polytype leading to an increase of the entropy, *i.e.* a higher degree of disorder prevails in the structure.

As in the case of boron carbide doping, as described above, here the transformation from 6H to 4H might result from a suitable layer displacement, which is caused by the nucleation and expansion of the stacking faults in individual close-packed double layers of Si and C. This process is similar to that mentioned by Jagadzinski, as in reference [50].

2.6. ELECTRON MICROSCOPY

2.6.1. Scanning Electron Microscope

In 'Scanning Electron Microscopy' (SEM), a film beam of electron (10-40KeV) is caused to scan the sample in a series of parallel tracks. These electrons interact with the samples producing secondary emission of electron (SEE), the back scattered electrons (BSE), the light of cathode luminescence and the X-ray. Each of these signals can be detected and deployed on the screen of a cathode ray tube like a television picture. The examinations are generally made on the photographic records of the screen.

The SEM is considerably faster and gives more of three dimensional details than TEM. The samples as large as 25 mm × 25 mm can be accommodated and parts viewed at magnifications varying from 20 to 100,000 at a resolution of **15 nm - 20 nm** as compared to **0.3 nm - 0.5 nm** for transmission electron microscopy. Due to its great depth of focus of a particle and its surface morphology, its depth of focus is nearly 300 times that of an optical microscope. In both, the scanning electron microscope and back scattered electron modes, the particles appear to be viewed from the above.

In SEM mode, where the particles appear as diffusely illuminated, the particle size can be measured and the aggregation behaviour can be studied, but there is little indication of the height. The BSE mode in which the particles appear to be illuminated from the point source gives a good impression of the height due to the shadows. Several of the current methods of particle size analysis can be adopted for the quantitative measurement of images in SEM photographic records.

2.6.2. Sample Preparation for Microstructural Study

The micro-structural evolution for sintered α-silicon carbide, for various sintering atmospheres, was studied by Scanning Electron Microscope (SEM). The 'small chips' of the samples are mounted in resin for grinding and polishing. The mounted samples are initially ground by alpha silicon carbide of grit size of – 200 and – 400. These are then polished with a diamond paste upto 1 micron. They are finally polished by 0.5 micron alumina powder in a vibratory polisher. The optically smooth polished surfaces are etched to reveal the full details of the microstructure according to the requirement.

Alpha Etch : sintered α-silicon carbide samples are boiled in 100 mL of water containing Murakami reagent [10 g of Sodium Hydroxide (NaOH) and 10 g of Potassium Ferro Cyanide [$K_3Fe(CN)_6$] for 30 mins. This etching attacks α-phase preferentially. The microscopy of the samples was studied by scanning electron microscope (Model - JSM 5200, JEOL, Japan).

Beta Etch : A fused salt mixture of Potassium Hydroxide and Potassium Nitride at 480°C for 5 minutes is used. This mixture preferentially attacks β- Silicon carbide revealing α/β interfaces, and also etches the grain boundaries between the β-grains. The microscopy of the samples was also studied by scanning electron microscopy (Model - JSM 5200, JEOL, Japan).

2.6.3. Transmission Electron Microscope

In 'Transmission Electron Microscope' (TEM), a thin specimen is irradiated with an electron beam of uniform current density : the electron energy is in the range of 60 -150 KeV (usually, 100 keV), or 200 KeV-1 MeV in case of the 'high voltage electron microscope' (HVEM) or 'high resolution transmission electron microscope' (HRTEM).

The electrons are emitted in the electron gun by the 'thermionic emission' from tungsten cathodes or LaB_6 rods or by the field emission from the pointed tungsten filaments. The latter are used when high gun brightness is needed. A two-stage condenser-lens system permits the variation of the illuminated aperture, and the area of the specimen is imaged with a three- or four-stage lens system onto a fluorescent screen. The image can be recorded in emulsion inside the vacuum.

The lens aberrations of the objective lens are so great that it is necessary to work with very small objective apertures, of the order of 10-25 mrad, to achieve a resolution of the order of **0.2 nm -0.5 nm.** The bright-field contrast is produced either by the adsorption of the electrons scattered through the angles, which are larger than the objective aperture (*i.e.* scattering contrast), or by the interference between the scattered wave and the incident wave at the image point (*i.e.* phase contrast). The phase of the electron waves behind the specimen is modified by the wave aberration of the objective lens. This aberration, and the energy spread of the electron gun, which is of the order of 1-2 eV, limits the contrast transfer (*i.e.* Fourier transform) of high spatial frequencies.

The electrons interact strongly with the atoms by elastic and inelastic scattering. The specimen must therefore be very thin, typically of the order of **5 nm - 0.5 μm** for 100 KeV electrons, depending on the density and the elemental composition of the object, and the resolution desired. The special preparation techniques are needed for this purpose.

The TEM can provide high resolution, because the elastic scattering is an interaction process that is highly localized to the region occupied by the screened Coulomb potential of an atomic nucleus, whereas the inelastic scattering is more diffuse. It spreads out over about a nanometer.

A further capability of the modern TEM is the formation of very small electron probes, **2 nm - 5 nm** in diameter, by means of a three-stage 'condenser-lens' system, the last lens field of which is the objective pre-field in front of the specimen. This enables the instrument to operate in a scanning transmission mode with a resolution determined by the electron probe-diameter. This has the advantage for imaging thick or crystalline specimens, and for recording secondary electrons and back-scattered electrons, cathode-luminescence and electron-beam-induced currents.

The main advantage of equipping a TEM with a STEM attachment is the formation of a very small electron probe, with which the elemental analysis and micro-diffraction can be performed on extremely small areas. The X-ray production in thin foils is confined to small volumes excited by the electron probe, which is only slightly broadened by the multiple scattering. Therefore, a better 'spatial resolution' is obtainable for the 'segregation effects' at crystal interfaces or precipitates, for example, than in an X-ray micro-analyser with the bulk specimens, where the spatial resolution is limited to 0.1-1 mm by the diameter of the electron-diffusion cloud.

2.6.4. Sample Preparation for TEM Study

For TEM study, the cylindrical specimen of 3 mm diameter and 1 mm high, which is a suitable size for the fabrication of TEM specimens, is cut directly from the bulk sintered pellets of alpha silicon carbide. The specimens are prepared from these samples by mechanical thinning to ≅ 75 μm, which is followed by dimpling and subsequent low-energy (5 to 6 kV) and low angle (15°) Ar^+ ion beam milling. The films are then examined in a transmission electron microscope (Model - TEM-400CX, JEOL, Japan), which was operated at an accelerating voltage of 100 KeV.

2.7. SINTERING OF NANO PARTICLES

2.7.1. Preparation of Materials

Both aluminium nitride and boron carbide from M/s Starck AG, Germany, having a purity of 99.5% were used as the sintering additives along with phenolic resin and organic lubricant. The phenolic resin from M/s Allied Resin & Chemicals Ltd., India, was used as a source for carbon. The phenolic resin pyrolizes at high temperature with carbon yield of 47%. The carbon content of resin was measured by firing the phenolic resin at 700°C for 1h in argon atmosphere. This particular test is necessary in order to prepare the 'compacts' from the nano particles.

The milled nano crystalline silicon carbide powder and the additives were thoroughly mixed under acetone and 1 wt% of Oelic acid was added as the pressing aid in solution. After drying of the slurry at 80°C, the mixture was passed through a sieve of 60 mesh to granulate the particles. The sieved granules were compacted isostatically from a pressure of 50 to 250 MPa in order to study the effect of pressure on the green density. The optimum pressure, which was used for pressing the compacts, was found to be 250 MPa for an optimum green density of 62% of the theoretical density (TD) of silicon carbide [3, 4]. The importance of obtaining a higher green density in order to get a better sintered density is explained in the **section 2.2.3.**

2.7.2. Sintering of Nano Particles of SiC

The sintering was carried out on the bars of 50 mm x 10 mm x 20 mm, and on the pellets of size 10 mm in diameter × 10 mm in thickness in a high temperature Astro Furnace (Model : 1000-3600-FP-20-F8204025 of M/s Thermal Technology Inc., USA), which was fitted with a control arrangement for various atmospheres of sintering.

The effect of 'sintering time' on the sintering behaviour of silicon carbide compacts containing nano particles was studied from 15 to 30 minutes on the above two compositions with two different dopants, and the optimum time was chosen as 15 minutes [3, 4]. The effect of temperature on sintering was already studied between 2000°C–2100°C [3, 4], and the optimum temperature was between 2050°C and 2100°C The effect of atmospheres on sintering was studied under vacuum, argon and nitrogen gas atmosphere respectively, and the optimum atmosphere for sintering for the present study was chosen as vacuum (3 mbar) at the above temperatures for a sintering time of 15 minutes.

2.7.3. Analysis of the Sintering Data

2.7.3.1. Role of Aluminum Nitride

The change in bulk density of the sintered body as a function of aluminium nitride concentration was measured at different sintering temperatures. The results are shown in **Figure 2.15,** where it is seen that with an increasing AlN content, the density increases up to a maximum of 3.16 gm/cc, or 98.4% of TD, *i.e.* a maximum sintered density is obtained at a concentration of 2 wt% AlN at each temperature from 2000° – 2100°C. Above 2 wt% AlN concentration, the density decreases from its peak value. As shown later, on a molar basis, this almost corresponds to the same amount of boron, which was found to be an 'optimum concentration' of dopant. In order to achieve high sintered density, the simultaneous addition of carbon is necessary like in the case of boron addition. These results were obtained at a constant amount of carbon addition, *i.e.* 1 wt%.

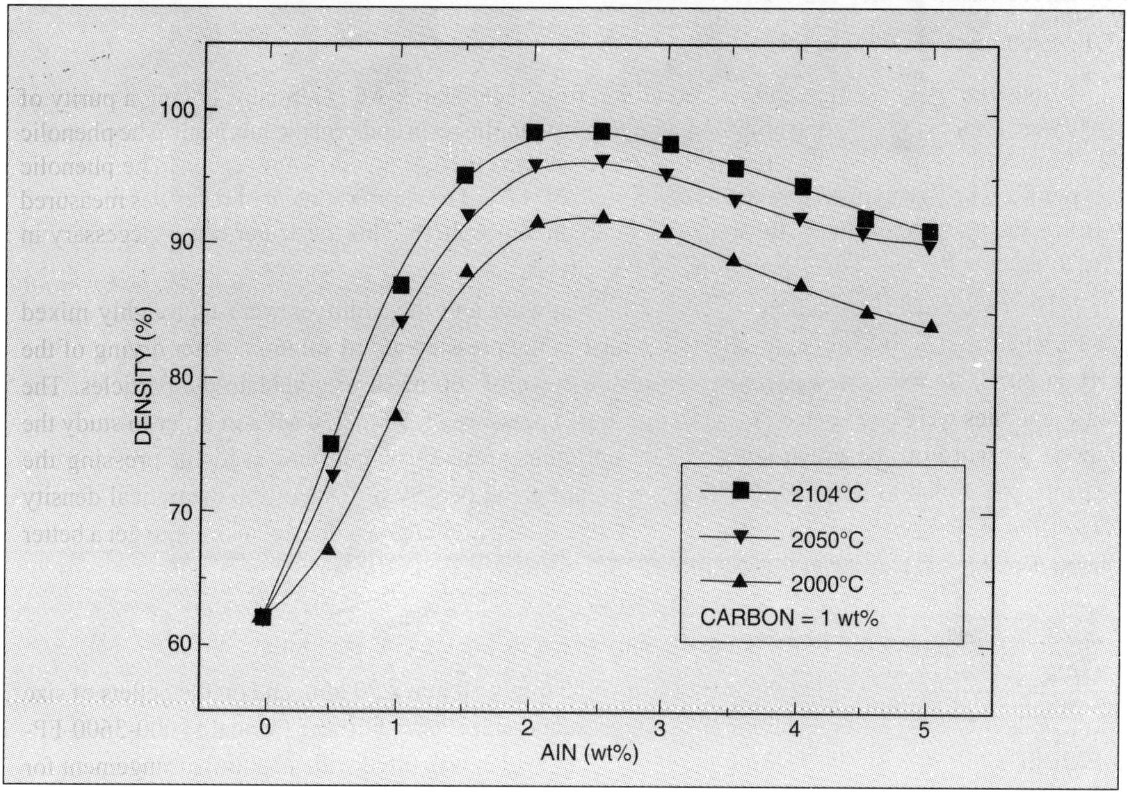

Figure 2.15: Bulk density of sintered α-SiC against AlN content.

It is a significant result that the sintered density changes from 62% of TD to 99% of TD at 2 wt% AlN at a temperature of 2100°C for a time of 15 mins under vacuum (3 mbar). These results suggest that the solid solubility of Al in SiC may be 1 wt% from 2000°–2100°C. As a matter of fact, the body having 99% TD exhibits a dense microstructure with relatively fine grains, the average size of which is around 3.7 μm. Two types of porosity could be identified in the microstructure :

(*a*) Between the grain boundaries, and

(*b*) Within the bulk crystal

The latter were developed during the grain growth of the original nano-sized SiC grains. The needle like growth of the grains of SiC points out to the fact that the rate of crystallization from the original (rather spherical) particles, produced by grinding and leaching etc., has been very fast favouring the growth in a particular direction, according to the energy consideration. Due to the growth from **37 nm to 5700 nm (Figure 2.16),** it is suggested that a group of grains after sintering should have the grain boundaries with a relatively low energy.

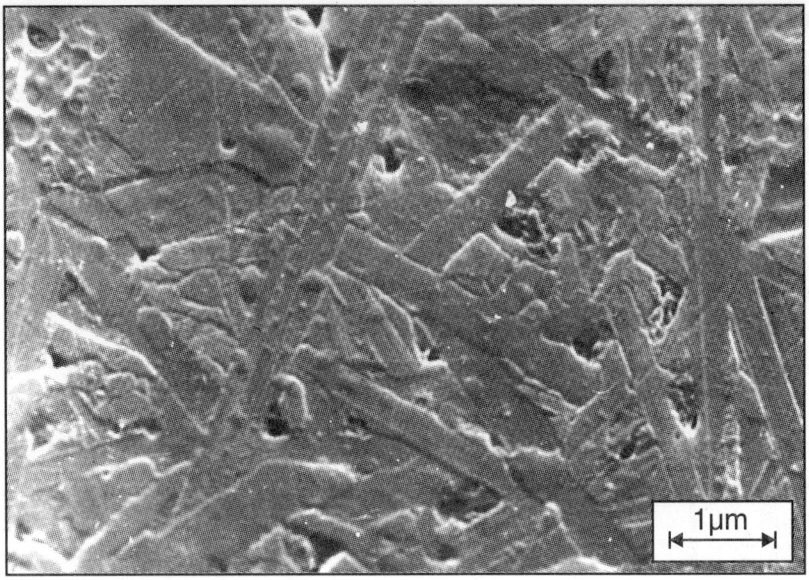

Figure 2.16: TEM photo of sintered α-SiC doped with aluminium nitride.

Since the sintering takes place through a 'diffusion', the following mechanism for the enhancement of diffusion is proposed as [2] :

$$SiC \longrightarrow Si + C$$

$$SiC \xrightarrow{AlN} Al_{Si}''' + N_C''''$$

$$Si_3N_4 \xrightarrow{SiC} 3Si''' + 4N''' + V_{Si}$$

$$Al_4C_3 \longrightarrow 4Al''' + 3C''' + V_C$$

AlN and SiC form solid solution for a limited range. When aluminium enters into Si site, Al_4C is likely to be formed leading to the creation of carbon vacancy, whereas when nitrogen enters into the carbon site, Si_3N_4 is likely to be formed, which leads to the formation of a silicon vacancy. Thus, the vacancies are created, which lead to the increase of the diffusion coefficient of both silicon and carbon. In argon atmosphere, the carbon diffusion coefficient is two orders of magnitude larger than that of silicon. However, in carbon rich atmosphere, the carbon diffusion coefficient is less than the diffusion coefficient of silicon, and also the enhancement of the diffusion coefficient of silicon occurs, as reported by Rijswijk and Shanefield [42]. The details of this mechanism working for the better densification of nano-particles of SiC is given in some more details in the **section - 2.7.3.3.**

2.7.3.2. Role of Boron Carbide

The change in the bulk density of the sintered SiC, made from nano-crystalline α-SiC doped with B_4C, is shown in **Figure 2.17** as a function of boron carbide content. It is seen that the maximum sintered density is obtained at 0.5 wt% B_4C at each temperature between 2000° and 2100°C. These data were obtained at constant concentration of carbon, *i.e.* 1 wt%. Above 0.5 wt% B_4C content, the density decreases from the peak value with increasing B_4C. As in the previous case with AlN doping, 99% TD is obtained after sintering. The results suggest that the solid solubility of B_4C in SiC may be up to 0.5 wt% in the above temperature range, as also reported by Murata and Smoak [28].

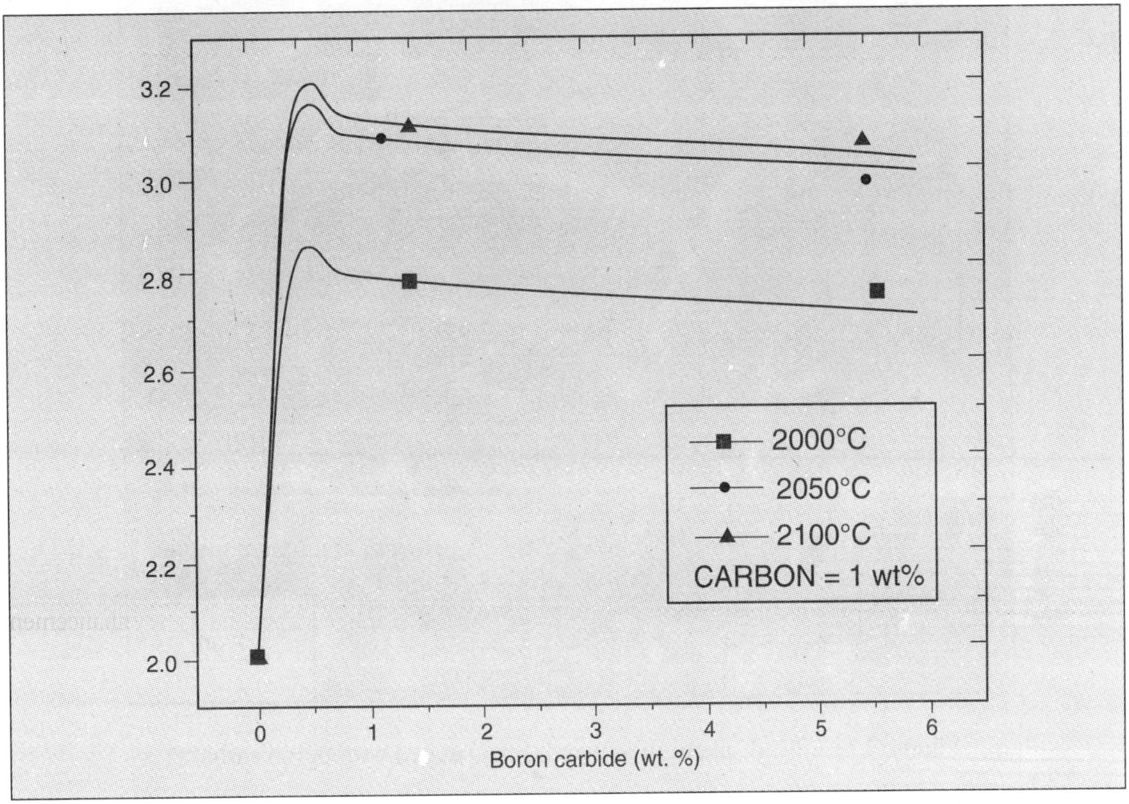

Figure 2.17: Bulk density of sintered α-SiC against boron carbide content.

The sintered SiC shows a dense microstructure with relatively fine grains having an average size of about 5.7 μm, measured from the micrographs (see **Figure - 2.18**). Since the sintering process of nano-crystalline SiC with boron carbide doping is very similar to that doped with AlN, the discussion is not repeated here, except that the following mechanism for the enhancement of diffusion is proposed as [3] :

$$SiC \longrightarrow Si + C$$

$$B_4C \xrightarrow{SiC} 4B'_{Si} + C'''' + 3V_C$$

$$SiB_4 \xrightarrow{SiC} Si'''' + 4B' + 3V_{Si}$$

When boron enters into C site, SiB_4 is likely to form. Thus, the silicon vacancies are created. Whereas, when boron enters into silicon site, the boron carbide is likely to form. Hence, the carbon vacancies are created. The difference of covalent radius of B and C is 0.011 nm, and that between B and Si is 0.029 nm. The amount of C and Si replaced is calculated from the change in lattice parameter (*c*), and these are 71.85% and 28.15% respectively. Thus, SiB_4 is likely to be formed 2.55 times more than B_4C. Hence, the formation of silicon vacancy is 2.55 times more than carbon vacancies, and the diffusion coefficient of silicon increases more than twice that of carbon, which is further increased by carbon-rich atmosphere. Some more details are given in the **section - 2.7.3.3.**

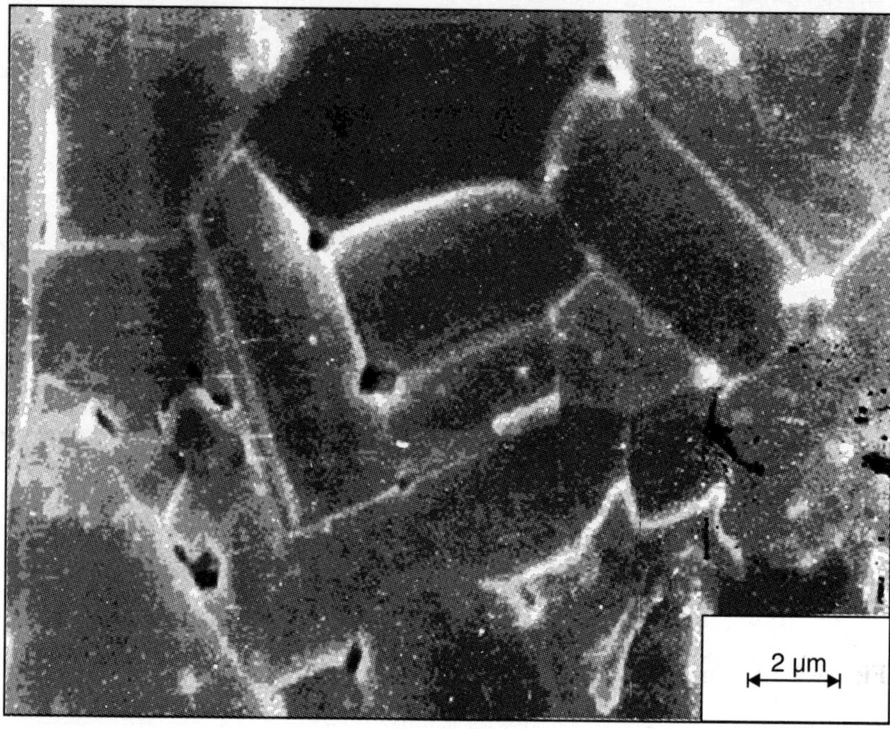

Figure 2.18 : TEM photo of sintered α-SiC doped with boron carbide.

2.7.3.3. Effect of Carbon

The doping level, temperature and other parameters being constant, the role of carbon seems to be important from the thermodynamic point of view and for generating vacancies, eventually to drive a bulk diffusion mechanism for both silicon and carbon to better densification of silicon carbides. The effect of carbon has been studied as a function of carbon content to ascertain the role of carbon in the sintering process in both the silicon carbides.

Figures 2.19 and 2.20 show the sintering curve as a function of temperature of α-silicon carbide doped with both boron carbide and aluminium nitride. It is seen that in both cases, 99% of TD is achieved at 2050°C and 2100°C for boron carbide and aluminium nitride respectively for a sintering time of 15 minutes under vacuum (3 mbar). A study of varying the concentration of the dopants showed that the optimum densification occurs at 0.5 wt% boron carbide and 2 wt% aluminium nitride with 1 wt% carbon, as already shown above. In order to see the effect of carbon on sintering, the carbon content was varied up to 5 wt%, and the results are shown in **Figures 2.21 and 2.22** for α-silicon carbide. **Figures 2.23 and 2.24** show the microstructure with optimum carbon content for β-silicon carbide, for both these dopants, keeping the doping level constant as above.

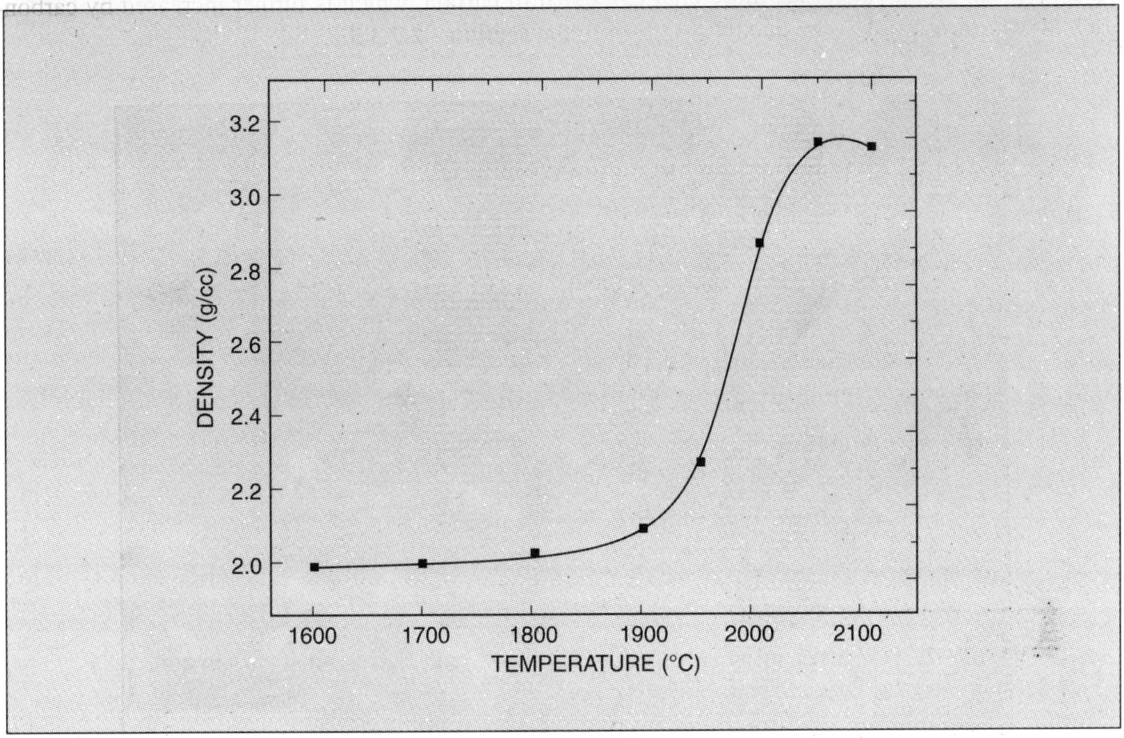

Figure 2.19: Densification curve of α-SiC doped with B_4C against temperature.

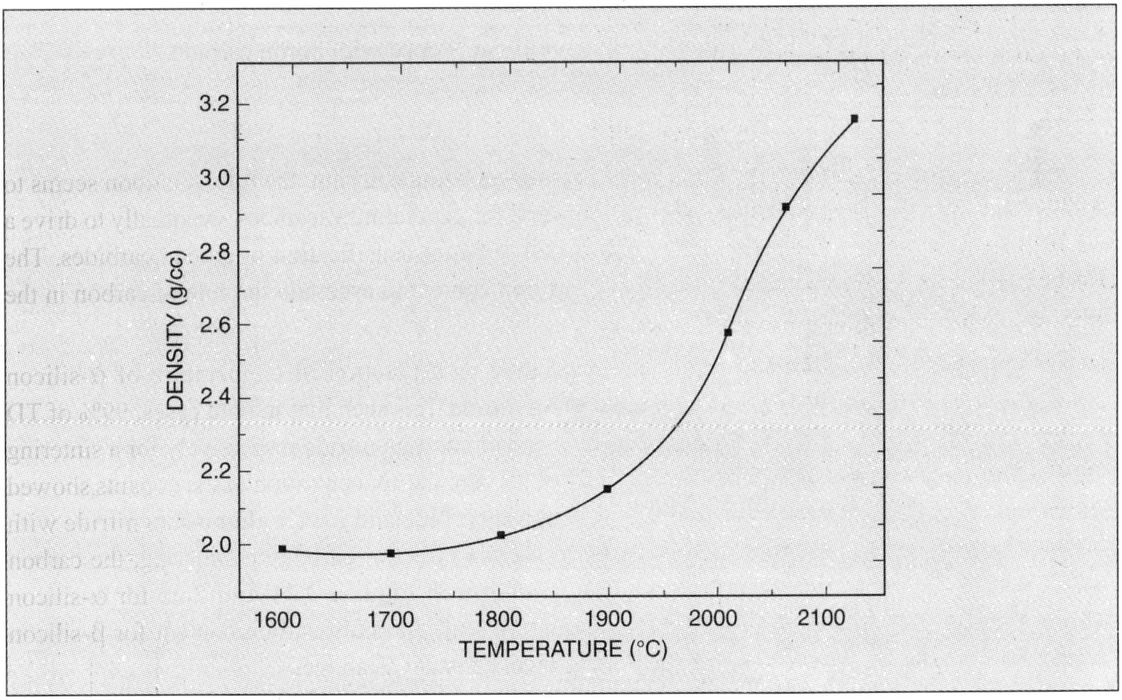

Figure 2.20: Densification curve of α-SiC doped with AlN against temperature.

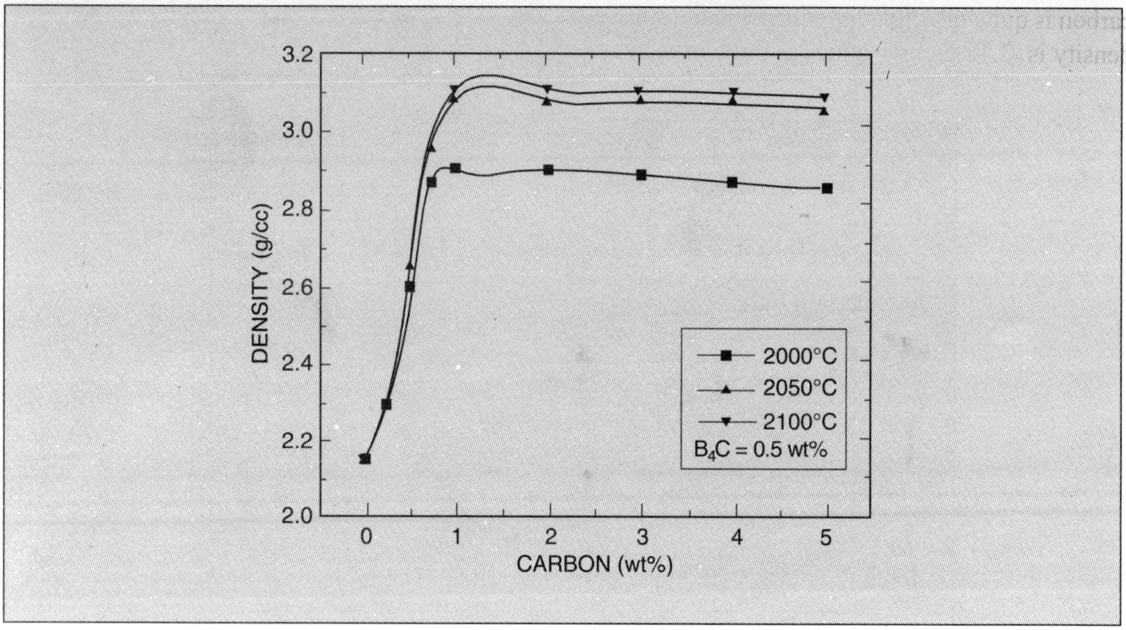

Figure 2.21: Densification curve of α-SiC doped with B_4C against carbon content.

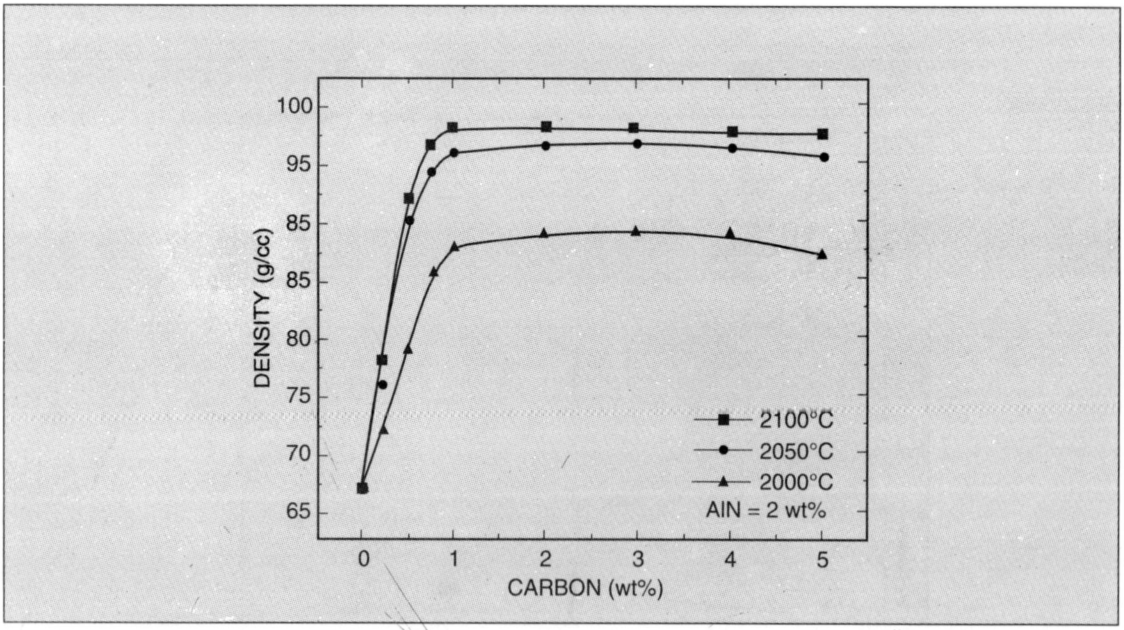

Figure 2.22: Densification curve of α-SiC doped with AlN against carbon content.

It is seen that in both the cases, the maximum density was obtained at 1 wt% carbon for both the dopants at each temperature from 2050° – 2100°C. These curves also demonstrate that the addition of carbon is quite effective for enhancing the sintered density. For instance, at zero % carbon, the sintered density is 2.15 g/cm^3, which is only an increase of 7.5% from the green density of 2.0 g/cm^3.

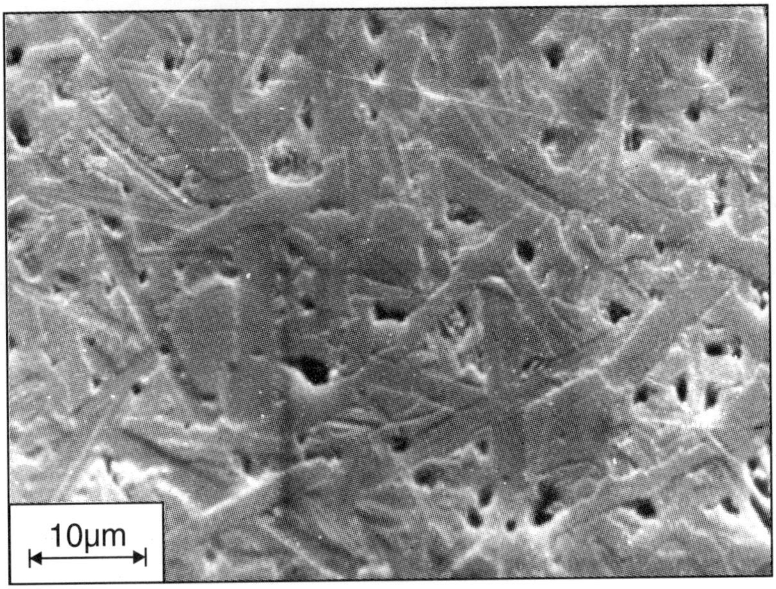

Figure 2.23: Microstructure of sintered β-SiC doped with B_4C and carbon.

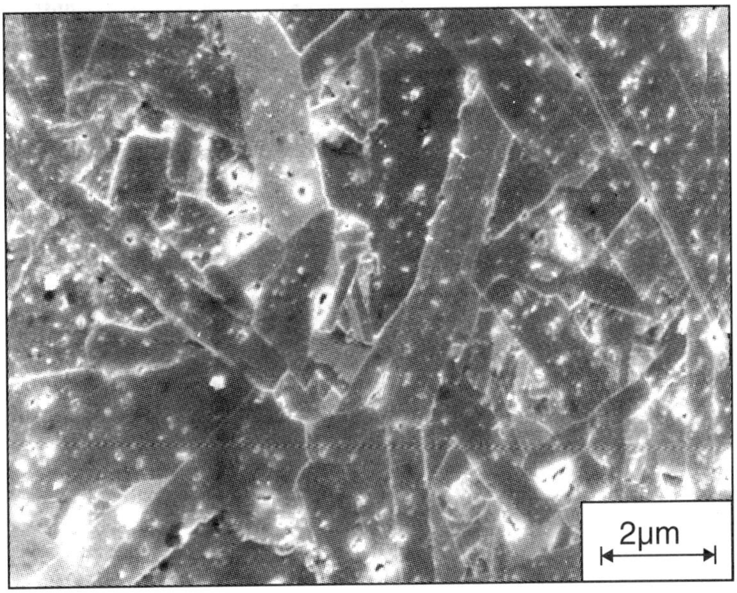

Figure 2.24: Microstructure of sintered β-SiC doped with AlN and carbon.

Now, a detailed analysis is necessary on the role of carbon during sintering of silicon carbide, which is important primarily due to the following reasons :

A. De-oxidation of SiO_2 films originally developed on the silicon carbide grains,

B. Coating of the surface of silicon carbide to lower the free energy of the surface and consequently lowering the surface vapour energy, and

C. Decreasing the diffusivity of C atoms to match with the diffusivity of Si atoms, which is essential for the effective sintering.

One of the roles of carbon during sintering of SiC was examined by the measurement of the grain size of the sintered body. It was found that the grain size decreases with increasing content of carbon. This means that carbon decreases the surface vapour energy, and thus prevents the surface vapour transport responsible for the grain growth.

In **Figure 2.25,** the microstructure of a sintered α-SiC specimen is shown with a starting composition of 97 wt% α-SiC, 2 wt% AlN and 1 wt% C. The grain size is in the range of 5-10 μm. The grain size is smaller compared to the sample doped with boron as a sintering aid. In order to study the sintering mechanism, the grain boundary regions have been thoroughly studied by using transmission electron microscope, as shown in **Figure 2.26.** No liquid phase is found to be present in the micrographs. From the "electron diffraction" patterns, as shown in **Figure 2.27,** it is also found that aluminium has entered into the structure of silicon carbide. Thus, it may be concluded that the sintering of silicon carbide is a solid state process involving 6H to 4H polytype transformation for which some evidences have been presented in the **section 2.5.1** on the quantitative analysis of the X-ray data.. The results are very similar for boron carbide doping.

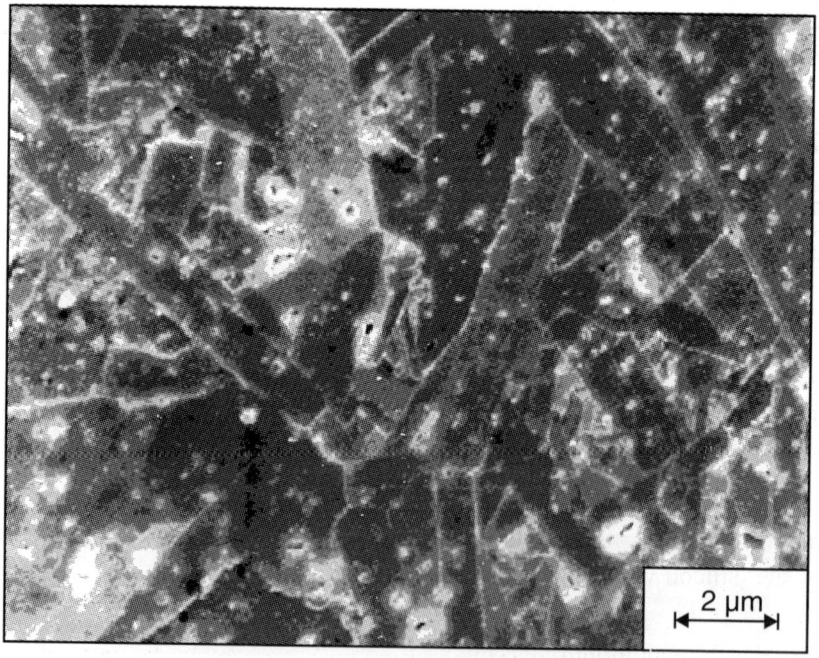

Figure 2.25: TEM photo of α-SiC doped with 2 wt% AlN

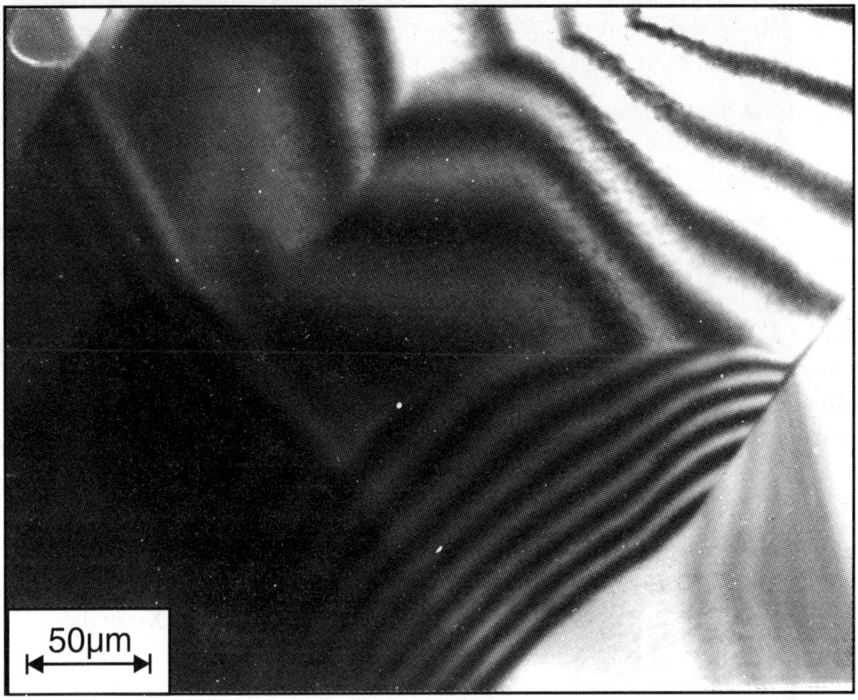

Figure 2.26: TEM photo of α-SiC doped with AlN showing no liquid phase present in the grain boundary region.

The sintered density increases almost linearly with increasing carbon content up to 1 wt%, and above 1wt%, it becomes almost constant. The role of carbon during sintering of silicon carbide has several effects. The silicon carbide always contains a very small percentage of oxygen, which is present as a very thin layer of silica, which is usually present on the surface of silicon carbide powder particles [23, 30]. A detrimental side reaction can take place in which the silica layer is reduced by the silicon carbide itself by the reaction as

$$2SiO_2 + SiC \rightarrow 3SiO + CO \quad ...(2.1)$$

for which $\Delta G = 0$ at 1870°C. This volatile SiO is further reduced at higher temperature through the reaction as

$$SiC + SiO \rightarrow 2Si + CO \quad ...(2.2)$$

This reaction with $\Delta G = 0$ at 1950°C is very efficient at usual sintering temperatures (2050° - 2150°C). Thus, the silicon vapour is usually present at sintering temperatures, when carbon is not added.

It is known that the equilibrium concentration of point imperfections, *i.e.* the vacancies or Schottky defects for the minimum free energy at a particular temperature is given by [51]

$$n = N \exp(-\Delta H_f / kT) \quad ...(2.3)$$

where N is the number of sites, and n the number of vacancies.

Figure 2.27: Electron diffraction pattern α-SiC doped with AlN.

The silicon carbide is a covalent compound. Thus, the carbon vacancies and silicon vacancies can be written as

$$n_C = N \exp(-\Delta H_C/kT) \quad ...(2.4)$$

$$n_{Si} = N \exp(-\Delta H_{Si}/kT) \quad ...(2.5)$$

$$n_C \cdot n_{Si} = N_2 \exp[-(\Delta H_C + \Delta H_{Si})/kT] \quad ...(2.6)$$

This product of $n_C \cdot n_{Si}$ is constant for a particular temperature and for a particular atmosphere for silicon carbide. Thus, the increase of carbon vacancies must be followed by a decrease of silicon vacancies and vice-verse. The diffusion coefficient is simply the product of vacancy concentration, the jump frequency and one sixth of the jump distance squared. As there are 12 possible adjacent carbon and silicon sites, the diffusion coefficient of carbon and silicon can be represented by

$$D_{Si} = 12\,[V_{Si}]\,\omega_0\,\frac{\alpha^2}{6} \quad ...(2.7)$$

$$D_C = 12\,[V_C]\,\omega_0\,\frac{\alpha^2}{6} \quad ...(2.8)$$

where ω_0 is the jump frequency, and α is the jump distance [52].

Thus, the increase of diffusion coefficient of carbon will increase the carbon vacancies, which will be followed by a decrease of the diffusion coefficient of silicon due to the decrease of silicon vacancies, and vice-verse. The chemical analyses of both pure 6H α- and β-silicon carbide single crystals have shown that the crystals are silicon rich. The ratio of silicon to carbon of 6H α-SiC and β-SiC were reported to be 1.032 and 1.049 respectively. Thus, the excess of silicon must result in the carbon vacancies.

From the equation (2.6), it can be concluded that the higher carbon vacancies with respect to silicon would result in the higher diffusion coefficient of carbon. This was exactly found by Hong et al [53-55] that the carbon diffusion coefficient is two orders of magnitude larger than that of silicon in argon atmosphere, as shown in **Figure 2.28.**

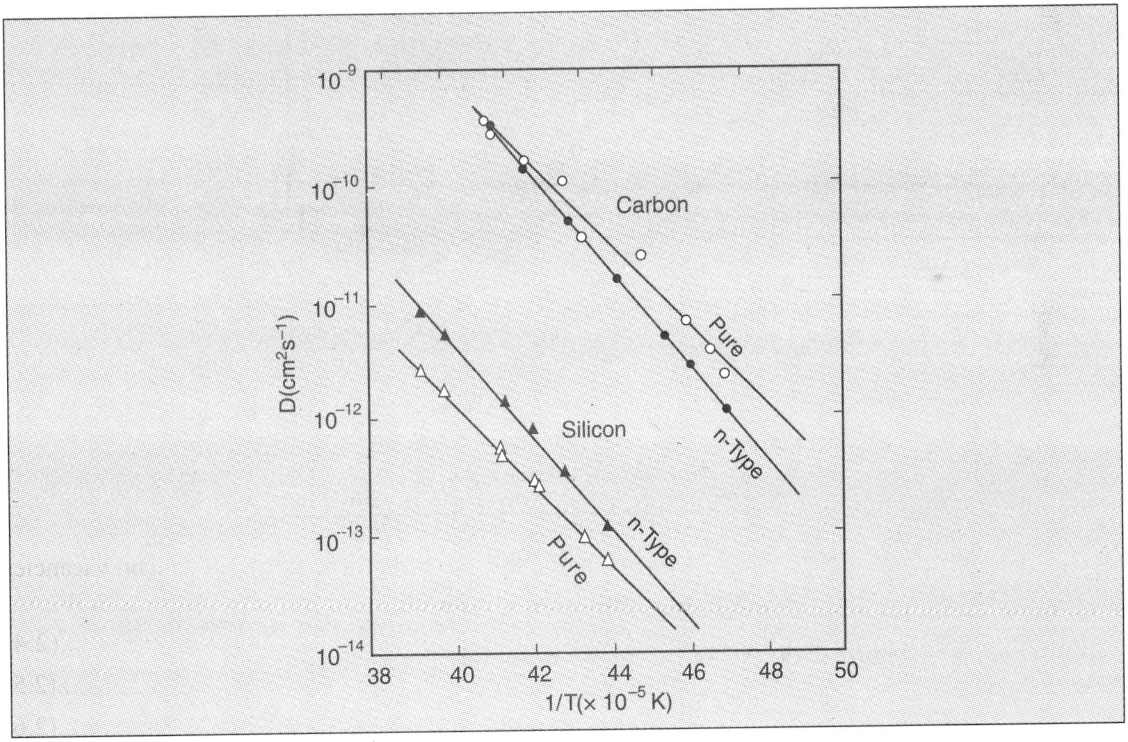

Figure 2.28: Carbon and silicon self-diffusion rates in SiC single crystals. [53]

Also carbon and silicon vacancies are related to partial pressure of carbon and silicon. The higher carbon partial pressure would result in a decrease in the carbon vacancies, but would promote an increase in the silicon vacancies, according to the equation (2.6), as already explained.

An atmosphere rich in silicon would result in the higher carbon vacancies, thereby showing an increase in the carbon diffusion coefficient, which is followed by a very high decrease in the silicon diffusion coefficient. This was also found by Hong et al. [53]. that the carbon diffusion co-efficient is three orders of magnitude larger than that of silicon and the diffusion coefficient of silicon is also low, as shown in **Figure 2.29.**

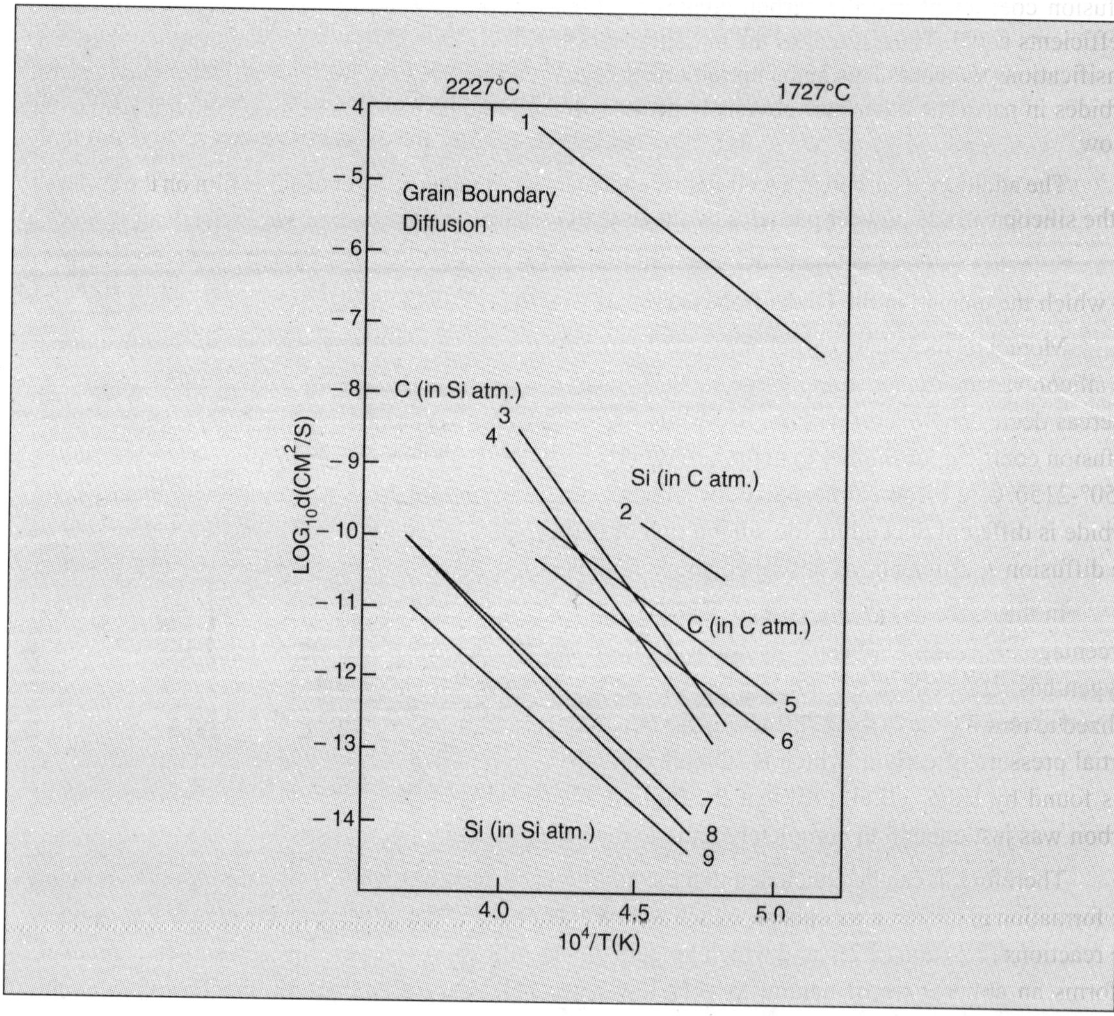

Figure 2.29: SiC self-diffusion in various atmospheres. [53]

This large discrepancy between the diffusion coefficients of carbon and silicon in silicon rich atmosphere, and generally the very low value of diffusion coefficient of both silicon and carbon, prevent the sintering of silicon carbide in silicon rich atmosphere, which is generally present in the sintering range of 2050° - 2150°C. This role of the diffusion coefficients was found to be effective in the case of doping with boron carbide in α-silicon carbide [3].

It is found that the self-diffusion rate of each element is enhanced by the presence of the other element, due to the creation of the vacancies. Thus, the carbon atmosphere would equalize the ratio of Si to C to unity or less than unity in silicon carbide, thereby creating silicon vacancies, and thus increasing the bulk diffusion of silicon, which is further activated by the addition of boron carbide [3]. For the densification to occur, the mass transport of silicon and carbon should be more or less equal.

Without the addition of boron and carbon, the mass transport of silicon and carbon is not equal, and it is also low, and the densification does not occur. The addition of boron carbide increases the diffusion coefficient and the carbon creates a partial carbon atmosphere, which make the diffusion coefficients equal. Thus, it makes the mass transport of silicon and carbon equal in order to make the densification to occur. This sums up the effect of carbon along with boron in the sintering of silicon carbides in part. The other part obviously deals with the removal of surface silica, which is discussed below.

The addition of carbon in a well-distributed manner reduces the layer of silica film on the surfaces of the silicon carbide powder particles by the overall reaction

$$SiO_2 + 3C \rightarrow SiC + 2CO \qquad ...(2.9)$$

for which the change in the Gibbs free energy $\Delta G = 0$ at 1520°C.

Moreover, the excess partial pressure of carbon decreases the carbon vacancies, thereby promoting the silicon vacancies, according to the equation (2.6) by increasing the diffusion coefficient of silicon, whereas decreasing that of carbon. It was also found by Ghostagore and Coble [45] that the silicon diffusion coefficient is higher than the carbon diffusion coefficient in a sintering temperature range of 2050°-2150°C in carbon atmosphere for silicon carbide. Therefore, the sintering behaviour of silicon carbide is different depending on silicon rich or carbon rich atmosphere, due to the respective role of the diffusion coefficients, as described above in the entire discussion.

In this work, it has been found that the highest sintered densities have been obtained when the percentage of carbon has been 1 wt%. The oxygen content of the processed powder is 0.33 wt%. This oxygen has remained as silica on the surface of the processed powder. The most of carbon has been utilized to remove the surface silica according to the equation (2.9), and the rest of carbon creates a low partial pressure of carbon, which is suitable enough to increase the diffusion coefficient of silicon. It was found by Hojo et. al. [40] that the highest sintered density was achieved when the amount of carbon was just enough to completely remove the surface silica.

Therefore, it can be concluded that the addition of 1 wt% of carbon is required for not allowing the formation of silicon atmosphere, which is normally present when carbon is not added, according to the reactions (2.1) and (2.2), and which hinders the densification process, as stated earlier. Moreover, it forms an atmosphere of neutral to very low partial pressure of carbon, and hence enhances the densification by increasing the diffusion coefficient of silicon, which is very low at silicon rich atmosphere.

As shown in **Figures 2.21** to **2.22** that the sintered density actually tends to decrease as the carbon content increases beyond 1 wt%. The addition of excess carbon increases the partial pressure of carbon. As carbon vacancies are inversely related to the partial pressure of carbon, the number of carbon vacancies would, therefore, decrease thereby decreasing the diffusion coefficient of carbon resulting in a decrease of the sintering rate, as observed.

It was found by Suzuki and Hase [35] that the grain growth also becomes very prominent above 1900°C. The observed 'structure development' could be described as (*a*) The larger grains dissolve the surrounding smaller grains, and

(*b*) The mass of dissolving grains engulf the pores, which are located at the surface of the grains.

Thus, when the sintering rate decreases due to the decrease in the diffusion coefficient of carbon in the presence of excess carbon partial pressure, two possibilities are as follows

1. The exaggerated grain growth which is very prominent at this sintering temperature becomes operative at a very fast rate over the sintering rate, engulfing the pores and causing entrapment of these pores, thereby making a decrease in the sintered density.
2. To coat the surface of silicon carbide to lower the free energy of the surface and consequently lowering the surface vapour energy and thus prevent vapour surface transport of materials, thereby preventing the exaggerated grain growth.

For a study on the effect of atmosphere on the sintering of silicon carbides, it is important to reconcile with the concept of vacancy formation. As already mentioned above, the product of the number of Schottky defects ($n_C \cdot n_{Si}$) is constant for a particular temperature and for a particular atmosphere. Therefore, the effect of atmosphere needs to be discussed later in the light of the above diffusion model as a confirmation of the model [16, 18].

In summary, it can be stated that even for nano particles of SiC, it is important to study the sintering behaviour of SiC with different "dopants mixed with 1 wt% carbon", which is an optimum quantity for a maximum densification. The effect of carbon is to remove the surface silica from the 'nano particles' of silicon carbide to help these particles to sinter better. The effect of addition of carbon up to 1 wt% also helps in the formation of vacancy of both silicon and carbon, and thereby increasing the bulk diffusion coefficients in silicon carbide, which increase the densification of silicon carbides to nearly theoretical density. The diffusion model shows that the relation between number of vacancies and the diffusion coefficients of both silicon and carbon explains the densification behaviour of both α- and β-silicon carbides.

2.7.3.4. Effect of Atmosphere

In order to understand the sintering process of silicon carbide under various atmospheres, the general sintering behaviour should be made clear. It has been shown by SEM, TEM and electron diffraction experimental data [3] that the sintering mechanism of nano particles of SiC, through doping of boron carbide and carbon, is a solid state process. This might have occurred by the creation of vacancies formed by the dopants and by an increase of the bulk diffusion by many orders of magnitude.

This solid-state sintering mechanism was also evident from an extensive analysis of fracture toughness data as a function of temperature [19, 20]. The sintering process does not appear to be either a surface phenomena, as proposed by Prochazka [23] , or a liquid phase sintering, as proposed by Lange and Gupta [26]. The chemical analysis of both β-silicon carbide and 6H α-silicon carbide single crystals by three different laboratories has shown the crystals to be silicon rich. The ratio of silicon to carbon of both β-silicon carbide and 6H α-silicon carbide was reported to be 1.049 and 1.032 respectively [56, 57], as said earlier. The excess silicon must result from the carbon vacancies. A high concentration of carbon vacancies is precluded by the following reasons

1. The low diffusion coefficient,
2. A high activation energy, and
3. A large pre-exponential term

This is a conclusion which is supported by a high value of the calculated Schottky energy by Vechten [51]. It was found that the self-diffusion rate of each element is enhanced by the presence of the other element, due to the creation of vacancies. Thus, the carbon atmosphere would equalize the ratio of Si to C to unity or less than unity in silicon carbide, thereby creating the silicon vacancies and thus increasing the bulk diffusion of silicon. A further enhancement of silicon diffusion coefficient occurs due to carbon rich atmosphere. In 'argon atmosphere', the carbon diffusion coefficient is two orders of magnitude larger than that of silicon.

However, in 'carbon rich atmosphere', the carbon diffusion coefficient is less than the diffusion coefficient of silicon, and also the enhancement of diffusion coefficient of silicon occurs, as reported by Rijswijk and Shanefield [42]. Here, carbon also prevents the formation of B_4C structure, and also decreases the vacancies of carbon. Thus, it lowers the diffusion coefficient of carbon as reported in carbon-rich environment.

By this process, the addition of boron increases the bulk diffusion of silicon and carbon by many orders of magnitude, and they eventually become equal to grain boundary diffusion coefficient of both silicon and carbon vacancies. For the densification to occur, the mass transport of silicon and carbon should be more or less equal. Without the addition of boron carbide and carbon, the mass transport of silicon and carbon is not equal, and it is also low, and the densification does not occur [3]. The addition of boron carbide increases the diffusion coefficient and the carbon creates a partial carbon atmosphere, which make both the diffusion coefficients equal. Thus, the addition of dopants makes the mass transport of silicon and carbon equal in order to make the densification to occur [3].

It is already noted from the above description on the creation of vacancies and consequent diffusion for sintering of nano particles of SiC in terms of the vacancy formation and consequent diffusion should be helpful in order to understand the overall densification behaviour under different atmospheres, as detailed below.

The evolution of the density for the α-silicon carbide samples, sintered under three different atmospheres like vacuum, argon and nitrogen, is shown as a function of sintering temperature in **Figure 2.30.** It is seen that the sintered density is higher for vacuum atmosphere compared to those of the other two atmospheres at all temperatures. It is also seen that the maximum densification occurs in vacuum, while the minimum densification is observed in nitrogen atmosphere.

The densification behaviour of α-silicon carbide in vacuum and argon atmosphere is quite similar, but they are distinctly different than that in nitrogen atmosphere, although the densification level in case of argon is lower than that in vacuum. The sintered density initially increases linearly up to 2050°C for all the samples and then almost saturates towards higher temperatures, or rather the sintering rate is considerably slowed down. This shows that the optimum sintering temperature of nano-crystalline α-silicon carbide doped with boron carbide and carbon is 2050°C.

Moreover, it is noteworthy that the initial rate of increase of density is somewhat higher for vacuum atmosphere compared to those of the other two atmospheres. This can be ascribed to higher diffusion of carbon and silicon under vacuum atmosphere. The number of vacancies of carbon and silicon are high in vacuum, compared to that in argon atmosphere. The creation of a higher vacancy leads to a higher diffusion as a function of temperature, which is the main driving force for densification. Therefore, the higher sintered density of α-silicon carbide has been obtained in vacuum than in argon atmosphere as mentioned earlier, as shown in **Figure 2.30.**

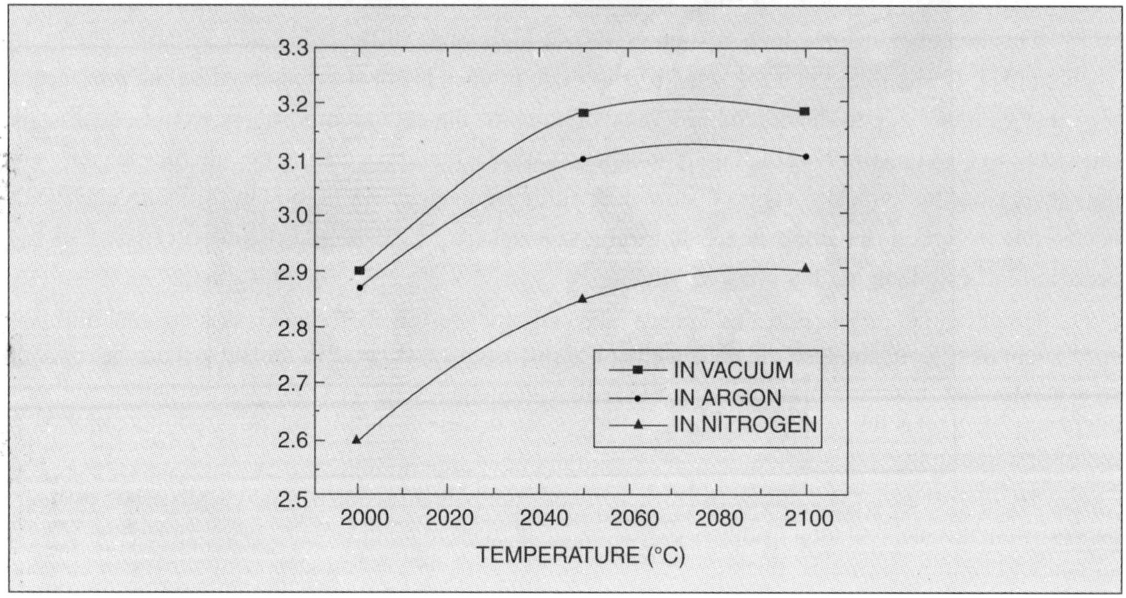

Figure 2.30: Sintered density of α-SiC in three different atmospheres.

The microstructures of α-silicon carbide, doped with 0.5 wt% boron carbide and 1 wt% carbon sintered at 2050°C in three different atmospheres are shown in **Figures 2.31, 2.32 and 2.33** respectively, which show the different levels of densification.

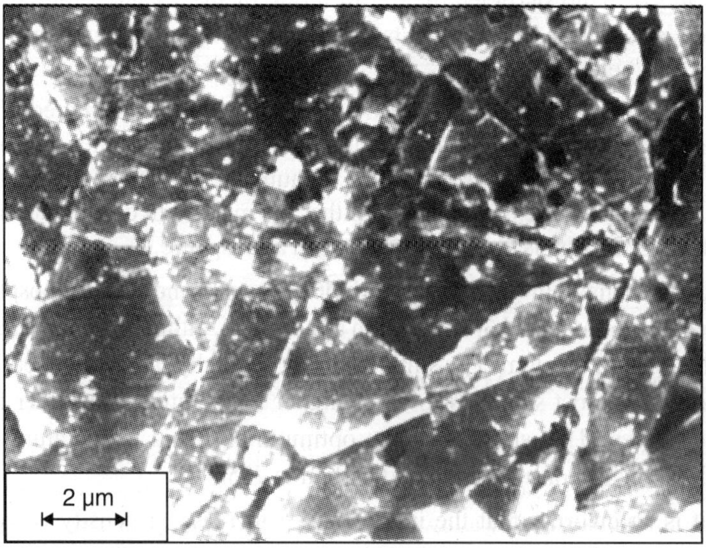

Figure 2.31. TEM photo of α-SiC sintered in vacuum.

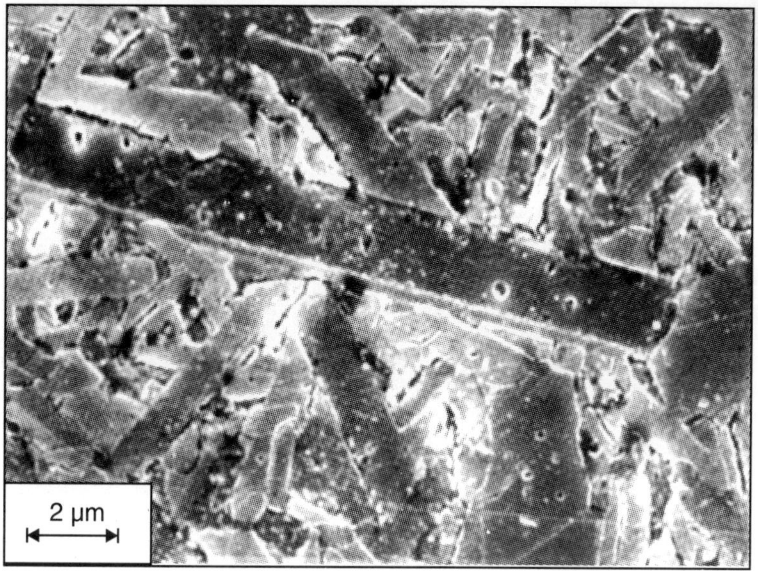

Figure 2.32: TEM photo of α-SiC sintered in argon.

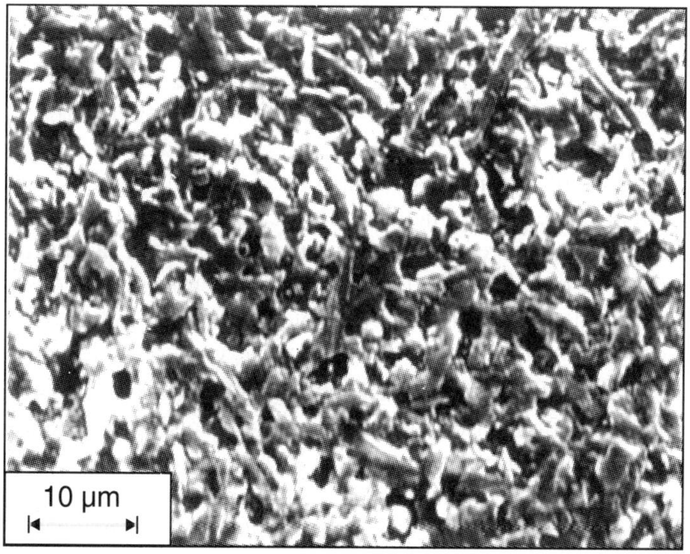

Figure 2.33: TEM photo of α-SiC sintered in nitrogen.

The intermediate behaviour of densification in argon atmosphere can be ascribed to that during later stage of sintering, the argon gas could have remained entrapped in the disconnected pores, which prevent full densification, as observed by a lower level of densification in argon atmosphere compared to that in vacuum. From the SEM micrograph of the sample sintered under argon atmosphere, some evidence of such disconnected pores was observed. The disconnected pores do play a role in sintering, but here the entrapment of argon gas in these pores may prevent mass transport into these pores, thereby causing a lower level of densification, compared to that in vacuum.

In case of the densification in nitrogen atmosphere, the Schottky defects, *i.e.,* the vacancy formation of carbon and silicon will be lower compared to the atmosphere of vacuum or argon. Moreover, in boron carbide doped α-silicon carbide samples, nitrogen may form BN with boron as observed by Venkateswaran and Kim [47], as mentioned earlier, preventing the formation of SiB_4 and B_4C, which actually increases the diffusion coefficients of silicon carbide. There is no experimental proof in this case, since small amounts of BN formed on the surface of the SiC grains is difficult to detect by the usual technique of SEM. But under the thermodynamic situation discussed above in the **section - 2.7.3.3,** it is most likely that BN forms under nitrogen atmosphere. Hence, the possible formation of boron nitride, which is not helpful in increasing the diffusion as an aid to sintering, might be responsible for the lower level of densification for sintering under nitrogen atmosphere.

It is quite plausible to mention here that in the case of nitrogen atmosphere, the onset of densification has been observed to be retarded by 150°C in the present investigation, which was also observed by Mirzah et al [38]. Therefore, the slower densification can be considered to be due to the decrease of diffusion coefficient, as opposed to the situation under vacuum. The level of densification also becomes lower in case of nitrogen atmosphere compared to vacuum or argon atmosphere, where the densification has been increased by the addition of dopants like boron carbide and carbon to speed up the required vacancy formation with a consequent increase in diffusion, as stated above through a diffusion model proposed earlier to explain the sintering behaviour of α-silicon carbide [3]. The importance of the diffusion mechanism through vacancy formation, as detailed above, may be considered relevant for explaining the maximum sintering under vacuum, and for the lowest level of densification in case of nitrogen atmosphere.

In summary, it can be stated that the diffusion model proposed here predicts the formation of SiB_4 and B_4C, which create vacancies and thereby increase the diffusion coefficients. This situation prevails in case of sintering of α-silicon carbide in vacuum and a maximum densification has been observed. The scenario is quite similar, but to a lesser extent in argon atmosphere. In nitrogen atmosphere, the possible formation of BN with boron prevents the formation of SiB_4 and B_4C, thereby decreasing the formation of vacancies and consequently their diffusion coefficients. Moreover, the formation of the Schottky defects in nitrogen atmosphere is also lower compared to that in vacuum and argon atmosphere, thereby a further decrease in diffusion coefficient occurs. Therefore, the onset of densification of nano-crystalline silicon carbide doped with boron carbide (0.5 wt%) and carbon (1 wt%) retarded by 150°C is due, presumably, to the formation of BN and to the creation of lower amount of defect sites as compared to that occurred in vacuum and argon atmospheres.

Finally, it should be mentioned from the recent studies that silicon carbide can be sintered up to high densities by means of 'liquid phase sintering' under low gas pressure with a combination of 'sintering aids' such as AlN-Y_2O_3, AlN-Yb_2O_3, and AlN-La_2O_3. As-sintered materials of SiC exhibited fine-grained homogeneous microstructure, and this has the advantage of sintering SiC at comparatively low temperature with improved mechanical properties [58 - 60].

The densification behaviour of materials with Y_2O_3 and La_2O_3 additions has been shown to be opposite. While the former materials densify better in Ar, the latter show better sinterability in N_2. Moreover, the densification of La_2O_3 containing materials has a complex nature most probably due to a complex phase formation sequence on heating, which has to be confirmed yet. The Yb_2O_3 containing materials exhibited sintering behaviour more similar to the behaviour of Y_2O_3 containing ones [58].

The transformation-controlled 'grain growth' during post-sintering heat-treatment of the developed materials was established. A high degree or even a complete β-SiC to α-SiC transformation was achieved by such techniques in reasonably short time intervals. The kinetics of the phase transformation has been shown to depend on the composition of the sintering additives, the use of less refractory rare earth oxides accelerating the process. Thus, the possibility of in-situ platelet reinforced SiC ceramics was shown, which might be the reason of higher flexural strength and better fracture toughness [58]. However, an optimization process of several factors involving these different 'sintering additives' vis-à-vis the processing techniques has to be undertaken to get better information for SiC for high performance applications.

REFERENCES

1. M. S. Datta, A. K. Bandyopadhyay and S. Maitra, 'Nano crystalline sintered α-silicon carbide for special applications', Indoceram, 39 (2002) 19.
2. M. S. Datta, A. K. Bandyopadhyay, B. Chaudhuri, 'Sintering of nano crystalline α-silicon carbide doping with aluminium nitride', Bull. Mater. Sci., 25 (2002) 121.
3. M. S. Datta, A. K. Bandyopadhyay and B. Chaudhuri, 'Sintering of nano crystalline α-silicon carbide by doping with boron carbide', Bull. Mater. Sci., 25 (2002) 181.
4. M. S. Datta, A. K. Bandyopadhyay and B. Chaudhuri, 'Preparation of nano α-silicon carbide crystalline particles by attrition grinding', Ind. J. Phys.-A, 78 (2004) 257.
5. M. S. Datta, A. K. Bandyopadhyay and B. Chaudhuri, Interceram, 53 (2004) 242.
6. H. J. Hwang, K. Nihara, 'Fabrication of surface barrier layer capacitor on $BaTiO_3$-based composite containing particulate SiC', J. Mater. Res., 33 (1998) 2866.
7. J. Zhao, L. C. Stearns, M. P. Harmer, H. M. Chan, G. A. Miller, R. F. Cook, 'Mechanical behaviour of alumina-silicon carbide nano-composites', J. Am. Ceram Soc., 76 (1993) 503.
8. H. J. Hwang, T. Sekino, K. Ota, K. Nihara, 'Perovskite-type $BaTiO_3$ ceramics containing particulate SiC', J. Mater. Sci., 31 (1996) 4617.
9. X. Pan, J. Mayar, M. Ruhle, K. Nihara, 'Silicon nitride-based nano-composites', J. Am. Ceram. Soc., 79 (1996) 585.
10. K. Singh, L. Stachowicz, S. L. Girskick, E. Pfender, 'Thermal plasma synthesis of SiC from rice husk', J. Mat. Sci. Let., 12 (1993) 659.
11. L. Chen, T. Goto and T. Hirai, 'Preparation and pressure-less sintering of chemical vapour deposition SiC-B composite powder', J. Mater. Sci.., 31 (1996) 679.
12. H. P. Martin, R. Ecke, E. Muller, 'Synthesis of nano crystalline silicon carbide powder by carbothermal reduction', J. Eur. Ceram. Soc., 18 (1998) 1739.
13. H. P. Martin, G. Irmer, E. Muller, 'Submicron structure of silicon carbide derived from poly (methylchlorosilane)', J. Eur. Ceram. Soc., 18 (1998) 193.
14. H. P. Martin, E. Muller, R. Richter, G. Roewer, E. Brendler, 'Conversion process of chlorine-chlorine containing polysilanes into silicon carbide', J. Mater. Sci., 32 (1997) 1381.
15. B. S. Mitchell, H. Zhang, N. Maljkovic, M. Ade, D. Kurtenbach, E. Muller, 'Formation of nano crystalline silicon carbide powder from chlorine-containing polycarbosilane precursors', J. Am. Ceram. Soc., 82 (1999) 2249.
16. M. S. Datta, A. K. Bandyopadhyay and B. Chaudhuri, 'The effect of Atmosphere on the sintering of nano crystalline α-silicon carbide doped with boron carbide', Trans. Ind. Ceram. Soc. 63 (2004) 105.

17. M. S. Datta and A. K. Bandyopadhyay, 'Effect of carbon on the sintering of nano particles of silicon carbide', Bull. Mater. Sci., (2005), In Press.

18. M. S. Datta, A. K. Bandyopadhyay and B. Chaudhuri, Studies on sintering atmosphere for nano-crystalline particles of α-silicon carbide doped with boron', Interceram, 54 (2005) 99.

19. M. S. Datta, A. K. Bandyopadhyay and B. Chaudhuri, 'Strength and microstructure of sintered α-silicon carbide prepared from nano crystalline particles by doped with boron carbide', Trans. Ind. Ceram. Soc., 61 (2002) 75.

20. M. S. Datta, A. K. Bandyopadhyay and B. Chaudhuri, 'Strength and microstructure of sintered α-silicon carbide prepared from nano crystalline particles doped with aluminium nitride', Interceram, 54 (2005) 22.

21. R. A. Alliegio, L. B. Coffin and J. R. Tinklepaugh, 'Pressure-less-sintered silicon carbide', J. Am Ceram. Soc., 39 (1956) 386.

22. J. Kriegesman, K. Hunold, A Lipp, K. Reinmuth, and K. Schwetz, "Shaped polycrystalline silicon carbide articles and isostatic hot-pressing', U.S. Patent, 4,564,601 (1986).

23. S. Prochazka, 'The role of boron and carbon in sintering of silicon carbide', Special Ceramics (Vol. 6), Ed. P. Popper, Academic Press, New York, (1975) pp. 171.

24. J. A. Coppola and C. H. McMurtry, 'Substitution of ceramics for ductile materials design', National Symposium on Ceramics in the Service of Man, Carnegie Institute, Washington D. C., (1976).

25. C. A. Johnson and S. Prochazka, 'Microstructures of sintered SiC', Ceramic Microstructures, Eds. R. M. Fulrath and J. A. Pask, (1977) pp. 76.

26. F. F. Lange and T. K. Gupta, 'Sintering of silicon carbide with boron compounds', J. Am. Ceram. Soc., 59 (1976) 539.

27. P. T. B. Shaffer, 'The phases in the system SiC-B_4C-C', Mater. Res. Bull., S4 (1969) 97.

28. Y. Murata and R. H. Smoak, 'Densification of silicon carbide by the addition of BN, BP and B_4C and correlation to their solid solubilities', Intl. Symposium of Factors in Densification and Sintering of Oxide and Non-oxide Ceramics, Hakona, Japan, (1978).

29. Y. Tajima and W. D. Kingery, 'Solid solubility of aluminium and boron in silicon carbide', J. Am. Ceram. Soc., 65 (1982) C-27.

30. R. Hamminger, G. Grathwohl and F. Thummler, 'Micro-analytical investigation of sintered SiC', J. Mater. Sci., 18 (1983) 353.

31. S. Prochazka and R. M. Scanlan, 'Effect of boron and carbon on sintering of SiC', J. Am.. Ceram. Soc., 58 (1975) 72.

32. C. Greskovich and J. H. Rosolowski, 'Sintering of Covalent Bonds', J. Am. Ceram. Soc., 59 (1976) 336.

33. R. Kieffer, E. Gugel, G. Lemer and P. E. Ettmayer, 'Investigations of the system B - C -Si', Dtsch. Keram. Ges., 49 (1972) 41.

34. J. A. Coppola and R. H. Smoak, 'Method of producing high density silicon carbide product', U.S. Patent, 4,230,286 (1980).

35. H. Suzuki and T. Hase, 'Some experimental consideration on the mechanism of pressure-less sintering of silicon carbide', International Symposium of Factors in Densification and Sintering of Oxide and Non-oxide Ceramics, Hakona, Japan, (1978).

36. R. F. Davis, J. E. Lane, C. H. Carter Jr., J. Bentley, W. H. Waldin, D. P. Griffis, R. W. Linton and K. L. More, 'Micro-analytical and microstructural analyses of boron and aluminium regions in sintered alpha silicon carbide', Scanning Electron Microscopy III, (1984) pp. 1161.

37. Y. Tajima and W. D. Kingery, 'Grain- boundary segregation in aluminium doped silicon carbide', J. Mater. Sci., 19 (1984) 3051.

38. T. Mirzah, M. Hoffmann and L. Gaukler, 'Pressure-less sintering of α-SiC, Powder Metall. Int., 16 (1984) 217.

39. G. W. Hollenberg and R. L. Crane, 'Method of fabricating silicon carbide articles', U. S. Patent, 3,966,855 (1980).

40. J. Hojo, K. Miyachi, Y. Okabe and A. Kato, 'Effect of chemical composition on the sinterability of ultrafine SiC powders', J. Am. Ceram. Soc., 66 (1983) C114.

41. W. J. Clegg, 'Role of carbon in the sintering of boron - doped silicon carbide', J. Am. Ceram. Soc., 83 (2000) 1039.

42. W. Van Rijswijk and D. J. Shanefield, 'Effects of carbon as sintering aid in silicon carbide', J. Am. Ceram. Soc., 73 (1990) 148.

43. K. M. Friederich and R. L. Coble, 'Influence of boron on chemical inter-diffusion in SiC during conversion of silicon carbide fibers to SiC', J. Am. Ceram. Soc., 66 (1983) C141.

44. J. D. Hong, M. H. Hon and R. F. Davis, 'Self-diffusion in alpha and beta SiC', Materials Science Monographs (Vol. 6), American Elsevier, New York, (1980) pp. 409.

45. R. N. Ghostagore and R. L. Coble, 'Self-diffusion in silicon carbide', Phys. Rev., 143 (1966) 623.

46. S. Prochazka, C. A. Johnson and R. A. Giddings, 'Atmosphere effects in sintering of silicon carbide, International Symposium of Factors in Densification and Sintering of Oxide and Non-oxide ceramics, Hakona, Japan, (1978).

47. V. Venkateswaran and J. H. Kim, 'Influence of sintering aids and atmosphere on pressureless sintering of alpha silicon carbide, Extd. Abstracts, Am. Ceram. Soc., 87th Annual eeting of the American Ceramic Society, (1985) pp. 56.

48. V. Raghavan, 'A First Course in Materials Science', Prentice-Hall of India (P) Ltd., New Delhi, (1974) pp.30

49. J. Ruska, L J Gauckler, J Lorenz and H U Rexer, 'Quantitative easurement of silicon carbide polytypes by X-ray diffraction peak intensities', J. Mater. Sci., 14 (1979) 2013.

50. A. R. Verma and P. Krishna, 'Polymorphism and Polytypism in Crystal', John Wiley and Sons Inc., New York, (1966) pp. 261.

51. J. A. Van Vechten, 'Simple theoretical estimate of the Schottky constant and virtual enthalpies of single vacancy formation in zinc blende and wurtzite type semiconductors', J. Electrochem. Soc., 122 (1975) 419.

52. D. P. Birnie III and P. Dunbar, 'A model for silicon self-diffusion in silicon carbide : anti-site defect otion', J. Am. Ceram. Soc., 69 (1986) C33.

53. J. D. Hong, M. H. Hon and R. F. Davis, Ceram. Intl., 5 (1979) 155.

54. J. D. Hong and R. F. Davis, 'Self-diffusion of carbon -14 in high-purity and N-doped α- SiC single crystals', J. Am. Ceram. Soc., 63 (1980) 546.

55. J. D. Hong, R. F. Davis and D.E. Newbury, 'Self-diffusion of silicon-30 in alpha-SiC single crystals', J. Mater. Sci., 16 (1981) 2485.

56. M. Nagatomo, H. Ishiwara and S. Furukawa, 'Studies of formation characteristics and mechanism of SiC on Si and metal-silicides by using ion back scattering techniques', Japan J. Appl. Phys., 18 (1979) 765.

57. P. T. B. Shaffer, Mater. Res. Bull., 16 (1980) 699.

58. V. A. Izhevskyi, L. A. Genova, A. H. A. Bressiani and J. C. Bressiani, 'Liquid phase sintered SiC : Processing and transformation controlled microstructure tailoring', Mat. Res., 3 & 4 (2000).

59. M. A. Mulla and V. D. Krastic, 'Pressureless sintering of β-SiC with Al2O3 additions', J. Mater. Sci., 29 (1994) 934.

60. K. Y. Chia, W. D. G. Boecker and R. S. Storm, 'SiC bodies having high toughness and fracture resistance and method of making same', U. S. Patent, 5,298,470 (1994).

Chapter 3

Nano Particles of Alumina and Zirconia

PREAMBLE

As already emphasized in **chapter 2,** the synthesis of nano-crystalline powder has drawn a considerable attention, as reported in the literature [1 - 8]. This is due to the fact that these powders, when sintered, have physical and chemical properties, which are superior to those made from bulk specimens containing submicron to micron particles. In particular, the emphasis is on the preparation of nano crystalline ceramic particles in the development of sintered products having higher density and lower sintering temperature [7]. Such nano-crystalline materials are important for a variety of applications including the fabrication of metal-ceramic laminate composites and also as a 'reinforcement' phase in polymers and brittle matrix composites.

One such industrially important ceramic material is 'alumina'. Recently, nano sized alumina has received more attention because of a significant improvement in mechanical properties of ceramic nano-composite **(see later in sections 3.5 on wear materials),** compared to the micrometer-sized particulates [8]. It also has many more potential applications including high-strength materials, electronic ceramics, and as catalysts. In particular, high quality nano-crystalline alumina is used as electronic substrates, bearing in expensive watches, and other fine precision equipment.

Although some work has been done on the preparation of nano sized alumina particles for better sintering and for better sintered properties, not much work has been done on its synthesis by high power Attrition milling. Dutta et al. [1-2] made a thorough study of Attrition milling taking different important grinding variables like the fluid medium for grinding, the grinding time and the weight ratio of the grinding balls to alumina. In this chapter, the results which have been optimized in terms of these variables on the synthesis of nano-sized alumina particles are presented to show the importance of these 'processing variables' on their synthesis. Some information on the processing of nano sized zirconia particles through sol-gel route are also presented, which have a variety of high performance applications, e.g., micro-electronics.

A number of other techniques have also been developed to synthesize nano particles. The inert gas phase condensation [9], laser ablation [10], molecular beam epitaxy [11], sputtering technique [12] are a few among the different physical methods. Various chemical methods such as sol-gel technique (to be discussed later in details in the **chapter 8)** [13], reduction of melt quenched glasses and electro-deposition method have also been employed. For the synthesis of nano-sized particles of alumina, there has been some efforts by various techniques like dc arc plasma [14], by sono-hydrolysis of alkoxide precursor [15], ultra-sonic flame pyrolysis [16] and by different novel chemical routes [17-23]. Most of these techniques are of some importance, and hence some of them will be described here in the **section - 3.2,** which are directly relevant to alumina, and some others in some more details in the **chapter 8.**

3.1. INTRODUCTION

Compared to these 'other methods', the Attrition milling has certain advantages. Moreover, the material processing by the 'other methods' has the following shortcomings :

(*a*) In some cases, the methods are not cost effective,

(*b*) The yield of materials is very small in some cases, and

(*c*) In sol-gel route in particular, there are problems of segregation, contamination and pore formation [13].

Hence, an alternative route has been adopted by Dutta et al. [1-2], *i.e.* high energy 'Attrition Grinding'. This is a very powerful method for the production of nano sized ceramic particles, and the yield of materials is quite high, depending on the size of the Attrition Mill [1-5], as also explained in **section 2.2.1.**

As discussed in **chapter - 2,** another important ceramic material like SiC has been prepared in the nano sized range of **37 nm** by Attrition Grinding [1, 2], which showed improved sintering behaviour through a diffusion mechanism [4, 5]. One of the most important work on high energy grinding of alumina has been done by Zhan et al [6] on a nano-composite containing polycrystalline zirconia in a toughened alumina matrix of **< 100 nm.**

The sintering behaviour of nano sized alumina particles has been already found to be very interesting by the microwave route [3] as shown later, and hence the grinding behaviour of these particles should be investigated. A comprehensive study is necessary to find out the effect of different variables like the fluid medium of grinding, the grinding time and the ratio of the weight of grinding balls to alumina on the preparation of nano sized particles of alumina by Attrition milling route. The idea is to optimize different grinding parameters in order to obtain suitable nano sized alumina particles eventually for a better sintered product.

As mentioned above, Zhan et al [6] used high energy attrition milling for making a dense γ-alumina matrix nano-composites with particle size **< 100 nm,** toughened by 3 mole%-yttria-stabilized zirconia, by using a novel technique of sintering. A combination of rapid sintering at the rate of 500°C/min and at a sintering temperature as low as 1100°C for only 3 min by the "spark-plasma-sintering" technique was used. The particle sizes for the alumina matrix and the toughening phase were **96 nm** and **265 nm** respectively. The toughness increased by almost 3 times that for pure nano-crystalline alumina. This work has clearly shown that our ability to obtain nano powders of alumina is not a great problem, but to manipulate the nano powder into a dense nano composite is definitely a challenging issue with nano-crystalline grain sizes **< 100 nm.**

From the above description, it is quite evident that the synthesis of 'nano powders' has become extremely important without any doubt. Now, before going to the high energy 'attrition milling' route, let us describe some of these processes of synthesis of 'nano powders', particularly nano-sized particles of alumina by different routes (collectively called 'Other Methods') in the next section.

3.2. OTHER METHODS FOR NANO MATERIALS

A preliminary account has been given in the previous sections on how the nano particles of alumina can be prepared by various novel routes. Although these processes have their shortcomings to a certain extent, as mentioned above, they also have many plus points, which need to be elaborated. A brief account of each of these processes is given here in this section to elucidate various points of interest about these methods of preparation of one of the most important ceramic materials like alumina, which has several important hi-tech application including the latest field of electronics. Although there are various methods of preparation of nano particles of alumina, only some of the 'novel' techniques will be explored here, which are arranged below not in any particular manner of importance, without any reflection on the other methods in terms of their importance in the nano world that are not described here.

3.2.1. Novel Techniques for Synthesis of Nano Particles

1. Gas Phase Condensation

The synthesis of nano-phase materials can be done by using 'Gas-Phase Condensation' method, *i.e.*, by evaporating a metal in an inert atmosphere and allowing it to condense on the surface of a cold finger, which is kept at liquid nitrogen temperature, *i.e.* 77°K. In order to eventually produce ceramic powders, the final metal clusters are oxidized and compacted in situ for the sintering process. The ultrafine metal particles are also produced by evaporating at a temperature-regulated oven, containing a reduced atmosphere of an inert gas, which is known as "inert gas phase condensation" [9]. Some details on these processes are given in the **chapter 8.**

2. DC Arc Plasma Method

The nano-crystalline alumina powders have been synthesized by 'DC Arc Plasma' by applying 300 Volts/15 Amp DC supply at atmospheric conditions by using aluminium electrodes, at an electrode separation of 3-5 mm in a closed chamber. In the beginning, the cathode and anode tips are positioned in order to facilitate ignition of the 'arc' by simply touching the anode to the cathode tip. The open circuit voltage (300 V) is applied to the electrode with a current limited to 7 A. The powder thus produced from the 'arc' is deposited onto the inner walls of the chamber. After sufficient cooling of the interior, this powder is gently scrapped only from the roof of the chamber and collected for the analysis.

This powder was characterized to be small nano particles by XRD and TEM techniques [14]. The distribution of such nano particles followed a log-normal behaviour. An interesting study was conducted by X-ray photoemission spectroscopy, which revealed that there is a presence of unreacted Al_2O_3 in the core of the nano particle, which is surrounded by Al^{3+} ions, *i.e.* a layer of alumina in the particulate structure [14].

3. Sonohydrolysis of Alkoxide Precursor

The nano particles of γ-Al_2O_3 with an average size of **5 nm** were synthesized by the 'Sono-Hydrolysis' of aluminium tri-isopropoxide under the influence of power ultra-sound, *i.e.* 100 W/cm^2, and in the presence of formic or oxalic acids as peptizers, which were followed by calcinations. The ultrasound-driven 'cavitation process' is shown to affect the agglomeration of inter-particle hydroxyls. The oxalate anions are found to be strongly adsorbed on the surface of the precursor nano particles, and thus have a retarding effect on the ultrasound-driven condensation process of inter-particle hydroxyls. The formic acid shows a lesser degree of adsorption on the surface of the precursor nano particles. The ultrasound-driven agglomeration of the primary particles and also the role of the organic modifiers on the microstructural properties of the precursor and the target alumina phases have been studied in details [15].

4. Ultra-Sonic Flame Pyrolysis

The nano-crystalline α-Al_2O_3 was synthesized in an 'Ultra-sonic Flame Pyrolysis' (UFP) set-up showing the formation of the nano particles in the flame. The aluminium nitrate can be 'ultra-sonically' dissolved in methanol-water mixture, which was pyrolysed in an oxy-propane flame in order to yield nano-crystalline α-Al_2O_3, which was confirmed by XRD. This technique, known as 'Flame Aerosol Technique' (FAT), is an emerging nano-crystalline synthesis method, which utilizes the 'power of the flame', *i.e.* the extreme spatial and temporal temperature gradients for bulk processing of nano materials. In a normal 'flame spray pyrolysis' (FSP), the liquid precursors are burnt. But, the latest variant, *i.e.* UFP, utilizes an ultra-sonic 'nebulizer' to atomize liquid into a spray having micron and sub-micron size droplets [16].

These droplets are then fed into an oxidizing stable pre-mixed burner that is used to produce an oxy-propane flame, which is 30 mm long in pre-mixed hot zone surrounded by a 'diffusion' flame of length of 150 mm. As the liquid traverses the flame, it is stripped off the organic and solvent components. The metal-oxide clusters generated in the hot zone ultimately coalesce into oxide nano-sized particles onto a copper substrate, which is kept below the nebulizer tip, and whose distance with the burner tip is adjustable. The nano particles thus produced are scrapped and then collected for the analysis. It is important to control the following variables : (*a*) 'flame parameters', (*b*) 'droplet residence time', and (*c*) 'capture distance'. This is necessary in order to obtain uniform sized nano particles. This process has also been used to produce bulk quantities of nano-crystalline titania, zirconia and alumina [16].

5. Chemical Route

This particular subject is quite extensive, and there is a tremendous amount of literature on the precursor materials for alumina, and hence only some of them will be mentioned here. Ramanathan et al [17] have prepared α-Al_2O_3 powder at 1400°C by using urea and $AlCl_3$ as precursor materials. Borsella et al [18] synthesized α-Al_2O_3 fine powder from the gas-phase precursors at 1200°-1400°C. Ding et al [19] have synthesized α-Al_2O_3 at 1250°C via the reaction $2AlCl_3 + 3CaO \rightarrow Al_2O_3 + 3CaCl_2$. Yu et al [20] synthesized ultrafine particles of α-Al_2O_3 at 1200°C. In a comprehensive study, Sun et al [21] have prepared α-Al_2O_3 fine powder by using ammonium aluminium carbonate hydroxide (AACH) of **5 nm** as precursor materials at only 1050°C without any seed, and then down to 850°C by using 5% of α-Al_2O_3 **(100 nm)** as seed crystals, and the final materials had particles with **30 nm-150 nm** in size. Fanelli and Burlew [22] have reported synthesis of a fine particle of transition alumina by the treatment of aluminium sec-butoxide in sec-butanol. However, due to the need for high temperatures and pressures as pre-requisites for the chemical reaction process, the direct formation of α-Al_2O_3 via a hydro-thermal method has not been commercially exploited.

Finally, Pati et al [23] have reported a novel chemical route for the preparation of nano-crystalline single-phase α-Al_2O_3 with an average particle size around **25 nm.** In this case, sucrose and PVA are used as the fuels and the medium to keep the metal ions , *i.e.* Al^{3+} ions, in homogeneous solution. This leaves sufficient flexibility for the system to distribute the metal ions uniformly throughout the liquid medium before setting to a 'solid mass'. In order to ensure this, the amount of sucrose in the precursor solution has been optimized (at 4 mole). The route involves 'dehydration' of the solution of the metal ions with sucrose and PVA, which is followed by 'decomposition' of the polymer matrix. After drying of the carbonaceous mass, it is ground to a fine powder forming the precursor, which when heat-treated at maximum 1150°C results in nano-sized alumina powder. The particle size is found by these workers to decrease with increasing sucrose content, which is an interesting observation in the alumina precursors [23].

The above sums up some of the details of the so-called 'other methods or techniques' of preparation of nano-crystalline alumina powder from different precursor materials. The following section will give details of preparation of nano particles by high energy attrition milling route.

3.3. NANO MATERIAL PREPARATION

The alumina powder from Alcoa (USA) was used as a starting material. This material had specific surface area of 10.5 m^2/g, which was measured by Surface Analyser of Micromeritics (USA). The average initial particle size was found to be 0.65 μm, which was measured by Malvern Autosizer 2C of Malvern Instruments Ltd. (UK).

3.3.1. Attrition Milling

For the preparation of nano particles of alumina, a commercial high power Attrition Mill [Model - PRIS of Feinmath Technik GmbH (Germany)] was employed. **Figure 2.1** shows the Attrition Mill used in this work. It consists of a water-cooled 500 mL steel vessel and a motor driven steel agitator. In contrast to a conventional ball mill, the balls in an Attrition Mill are agitated by a series of 'stirring arms', which are mounted on a motor shaft. The grinding balls having diameter of 3 mm made of zirconia were used to reduce the effect of undesirable impurities. This Attrition mill is very efficient in reducing the particle size to a nano range. The details of such type of Attrition mills for laboratory experiments and the other type of batch Attritors used for some commercial purpose, *i.e.* large batch processing, are already given in **section 2.2.1.**

The milling was done with 3 mm diameter zirconia balls in five different liquid media. The five liquids were : ethylene glycol, *n*-hexane, butanol, xylene and water with surface active agent. In a preliminary experiment, a grinding time of 10 hours was found to be suitable for reducing the particle size to nano range in a particular liquid medium. By fixing this grinding time of 10 hours, the ratio of the weight of the zirconia balls to alumina to get nano particles was also studied in these five liquid media. A suitable ratio of 10 : 1 was found out, which was then fixed to study the effect of grinding time in hours in all the five liquid media of grinding.

The final particle size in each of the above experiment was measured by Malvern Autosizer 2C. After the optimum conditions for grinding were estimated, the transmission electron microscopy was carried out on the nano sized alumina particles in case of grinding in water with surfactant for 10 hours in a Transmission Electron Microscope (Model - 400CX of JEOL, Japan).

3.3.2. Nano Particles of Alumina

The initial condition of the as-received alumina particles is presented in terms of the distribution of particle size in **Figure 3.1,** which has a mean size of **650 nm** and the distribution is also found to be quite narrow. **Figure 3.1** clearly shows that the alumina particles with a relatively narrow range of initial particle size is present.

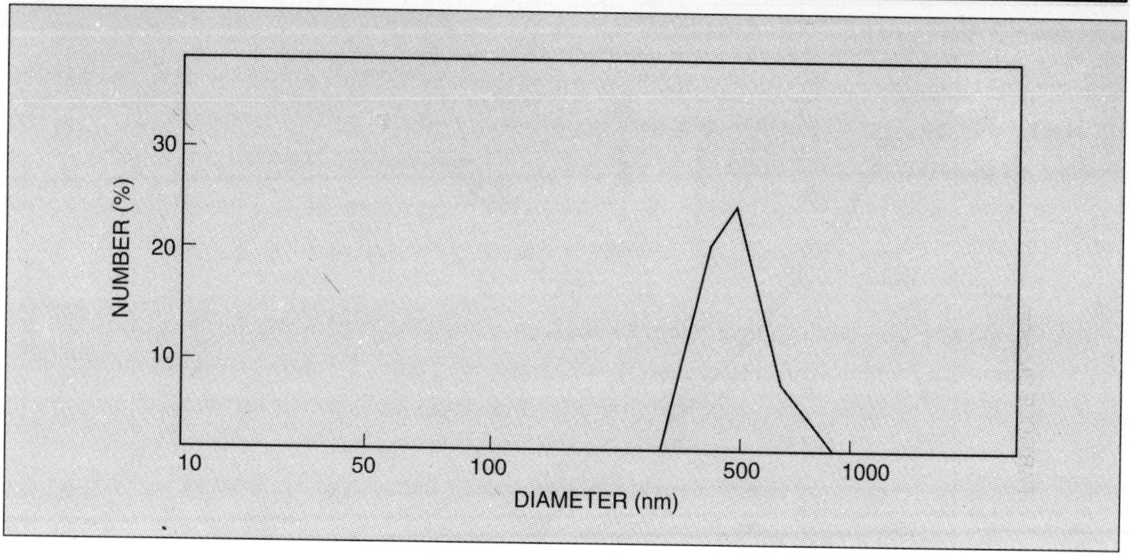

Figure 3.1: Particle size distribution of as-received alumina powder.

Figure 3.2 shows the effect of the particle size on the weight ratio of grinding balls to alumina in order to obtain nano sized particles in five different liquid media and for a fixed grinding time of 10 hours. It is seen that the particle size decreases quite sharply as the weight ratio increases upto about 5 : 1, and then the rate of decrease of particle size is relatively slower, which ultimately goes to a sort of saturation value at a ratio of 10 : 1 for all the five liquid media of grinding. This ratio is considered to be optimum for alumina with a mean particle size of **50 nm.**

The curves in **Figure 3.2** represent the particle size curves in five different liquid media of grinding. The decrease in particle size is the most pronounced for 'water with surface active agent', which is very distinct compared to the other four liquids, where the effect of reduction of particle size is quite similar. In terms of this optimum ratio, the next best liquid is ethylene glycol, where the lowest particle size is over **120 nm,** and that for the least efficient grinding liquid like xylene is over **150 nm.** Surely, this is in contrast to 'water with surfactants', where the grinding efficiency is the highest for an optimum weight ratio of 10 : 1 in order to obtain the lowest nano sized alumina particles of mean size of **50 nm.** Therefore, this figure clearly gives us the information regarding an optimum ratio for an optimum liquid of grinding at the same time to achieve the goal of obtaining nano sized alumina particles.

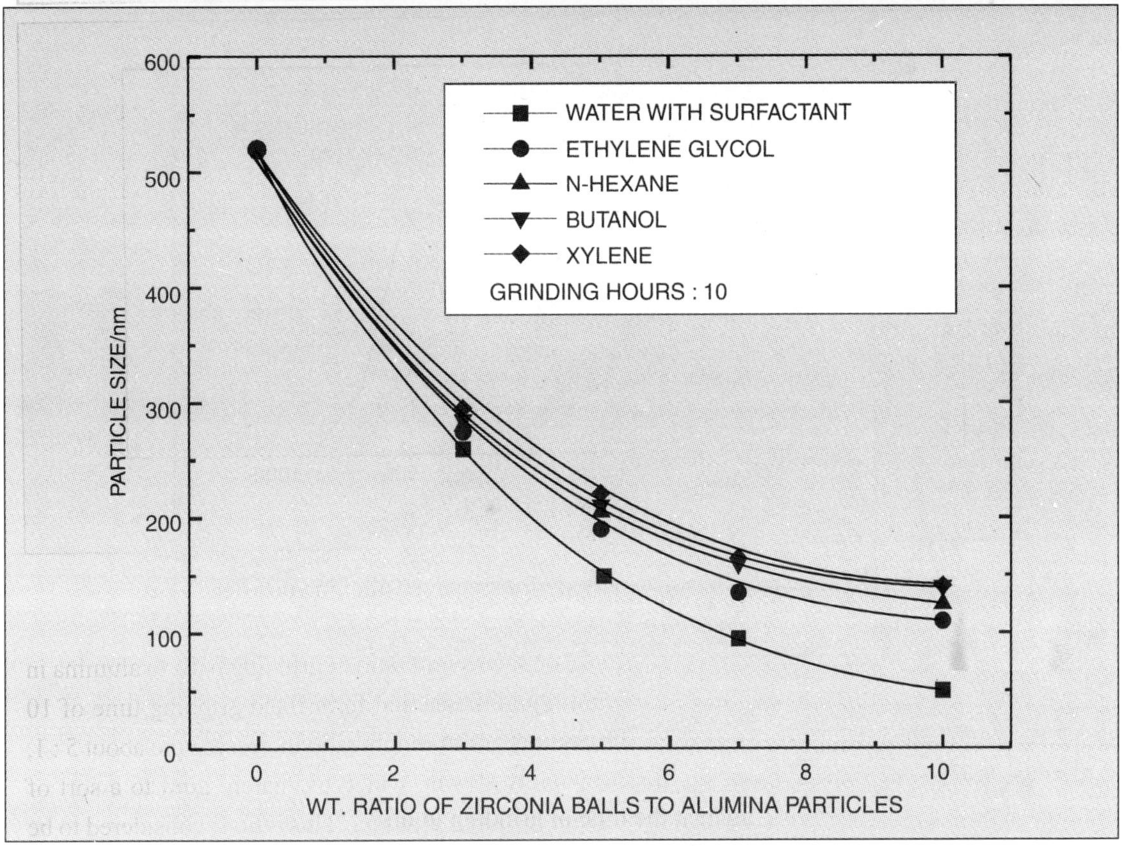

Figure 3.2: Particle sizes of alumina after attrition milling in five different fluid media against the wt. ratio

Figure 3.3 shows the particle size variation with the grinding time in hours for an optimum weight ratio of grinding balls to alumina of 10 : 1 in five different liquid media. It is seen that the particle size decreases quite sharply as expected as the grinding time increases upto about 7 hours, and then the rate of decrease of particle size is relatively slower, which ultimately goes to a sort of saturation value at about 10 hours for all the five liquid media of grinding. This grinding time is considered to be optimum for alumina with an average particle size of **50 nm.**

As before, the curves in **Figure 3.3** represent the particle size curves in five different liquid media of grinding. The decrease in particle size is the most pronounced for 'water with surface active agent', which is quite distinct (not exactly like **Figure 3.2**) compared to the other four liquids, where the effect of reduction of particle size is quite similar. In terms of the optimum grinding time, the next best liquid is ethylene glycol, where the lowest particle size is about **80 nm,** and that for the least efficient grinding liquid like xylene is just below **100 nm.** This is in contrast to water with surfactants, where the grinding efficiency is the highest for an optimum grinding time of 10 hours in order to obtain the lowest nano sized alumina particles of mean size of **50 nm.** Therefore, this figure again gives us the information regarding an optimum grinding time for an optimum liquid of grinding at the same time to achieve nano sized alumina particles.

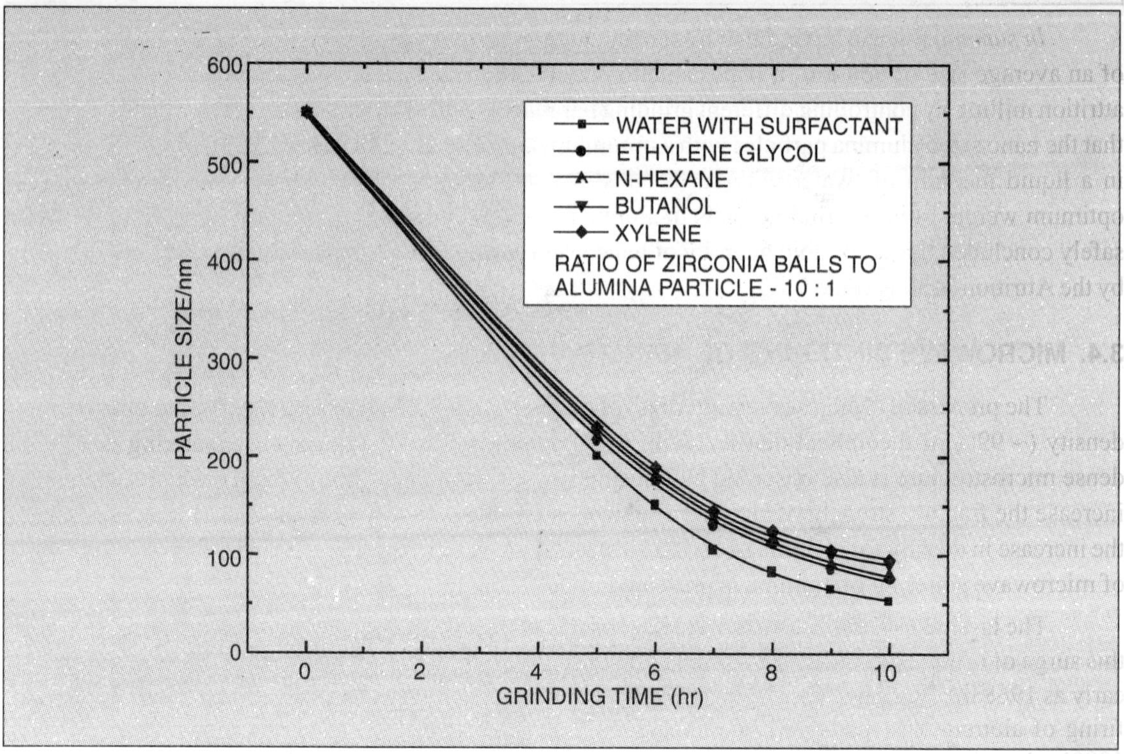

Figure 3.3: Particles sizes of alumina after attrition milling in five different fluid media against the time of grinding in hrs.

Figure 3.4 shows the distribution of nano sized alumina particles for attrition milling in 'water with surfactants'. This figure clearly shows the effect of variation of the liquid media on grinding in order to obtain nano sized alumina particles, as depicted above in **Figures 3.2–3.3.**

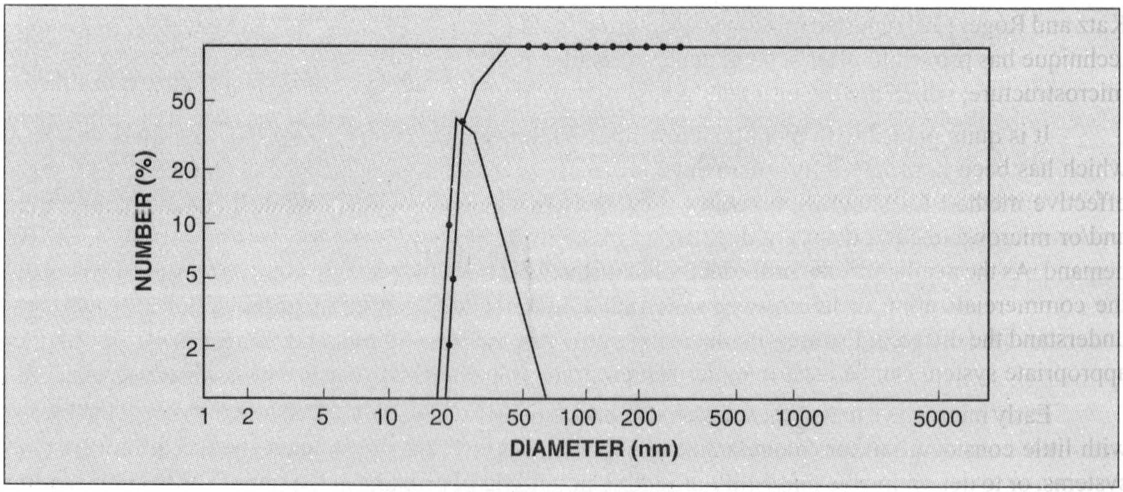

Figure 3.4: Particles size distribution of alumina after attrition milling in water with surfactants.

In summary, it can be said that by starting with the initial particles of alumina from Alcoa (USA) of an average size of **650 nm,** it is possible to obtain nano sized particles of a mean size of **50 nm** by attrition milling by controlling different grinding parameters and then optimizing them. This work shows that the nano sized alumina particles with a minimum particle size of **50 nm** can be achieved by grinding in a liquid medium of 'water with surfactants' for an optimum grinding time of 10 hours with an optimum weight ratio of grinding balls to alumina as 10 : 1. After this optimization process, it can be safely concluded that a sizeable quantity of nano sized alumina particles can be prepared economically by the Attrition Milling route.

3.4. MICROWAVE SINTERING OF NANO PARTICLES

The process of "Microwave Sintering" of nano-sized alumina powders is helpful to obtain a high density (~ 99% of theoretical density) with MgO doping. Under the microwave sintering process, a dense microstructure is also observed by transmission electron microscopy (TEM), which is known to increase the fracture strength of sintered alumina. Such a dense microstructure can be observed due to the increase in density for alumina materials sintered at a higher temperature. Some details of the process of microwave sintering of alumina is presented in this section.

The last half of 1980's showed a great deal of interest in the area of microwave sintering. Despite this surge of recent activity, the first pioneering experiments on microwave sintering were performed as early as 1968 by Tinga and Voss [24]. The real activity began to accelerate by Sutton [25] for drying and firing of alumina castables, and by Schubring [26] for sintering of ceramics. In the late 1970's and 1980's, the microwave heating and sintering of uranium oxide was reported by Hass [27], for ferrites by Krage [28] and for 'Alumina-Silicon Carbide Composites' by Meek et al [29]. But a very few literature existed on alumina sintering by the microwave route [3].

Alumina ceramics are normally densified by a conventional process of sintering, which requires higher temperatures (~ 1600°C) and longer duration. Thus, the manufacturing cost naturally becomes very high, since the energy cost is increasing by the day. In order to make the production of alumina economically viable, *i.e.* lesser cost, the microwave sintering process should be very useful. In 1989, Katz and Roger [30] reported microwave sintering of alumina. Compared to conventional methods, this technique has proved to offer several advantages like attainment of the desired phase and fine-grained microstructure, which are the ultimate 'goals' of sintering any ceramic material.

It is quite plausible to mention a few lines on the considerable amount of developmental work, which has been in progress for microwave furnaces, now that it is clear that microwave heating is an effective method for sintering ceramics. The companies with manufacturing experience in furnaces and/or microwave equipment are developing research and industrial microwave furnaces to meet the demand. As the availability and affordability of microwave furnaces improve, this in turn would stimulate the commercialization of microwave sintering, especially for advanced ceramics. It is important to understand the different features available in microwave furnaces and to make comparisons, so that the appropriate system can be used in materials research, product development, and production.

Early microwave materials research was performed using systems built 'in house' and/or designed with little consideration for 'manufacturing expenses'. It is difficult to compare the features of these systems, or to determine the feasibility of scaling up microwave processes from the literature. Recently, microwave system designs have improved, becoming more user friendly and cost effective. In any comparative work, different 'approaches' to microwave furnace design should be discussed with the description of the important 'operating parameters', and then a direct comparison should be made by sintering known ceramic 'standards' in different microwave furnaces, and also commercially available microwave furnaces should be used.

Shulman and Walker [31] investigated different types of microwave systems which included

1. Several sizes of multimode 2.45 GHz,
2. Hybrid microwave + electric,
3. Hybrid microwave with 2 frequencies, and
4. Millimeter-wave.

It was found that microwave firing parameters were readily transferred between different microwave systems, if similar refractory packages were used. The equivalent properties and microstructures could be achieved, by using different microwave systems. However, temperature measurements may not be comparable. Energy savings could vary between the systems and depend on the (*a*) Insulation, (*b*) Load size, and (*c*) Type of ceramic. There is now an excellent selection of microwave furnaces available for research and production. It is anticipated that for material scientists, the research focus will shift from fabricating furnaces to developing useful processes. Some companies are currently developing microwave processes for many types of advanced ceramics and exploring the feasibility and challenges of scaling up [31].

Shulman et al [32] also studied the 'scaling-up' of microwave processes for production, which requires the systematic study of the following :

1. Uniformity of microwave sintered materials, and
2. Reproducibility of material properties for given processing parameters.

In this study, a microwave power versus time recipe for 'sintering zirconia ceramics' to full density was successfully transferred from a 1.1 kW modified 'kitchen microwave' to a large 3 kW chamber (35″ dia × 54″ length). The effects of increasing the load and refractory box volume, as well as the box position were investigated. The uniformity and reproducibility of the density, hardness, and microstructure were explored [32].

Finally, Shulman's group [33] investigated alumina and zirconia materials, which were sintered to > 99% theoretical density in less than 1 hour total cycle time. The study investigates the effects of input power and cycle time on heating rates, microstructure, density and mechanical properties, *i.e.* hardness and MOR. The properties are compared to conventional sintering conditions, which resulted in similar densities. In addition, the effects of various sintering conditions, *e.g.* (*a*) Location of sample within the microwave chamber, *i.e.* hot spots, (*b*) Number of samples within a given insulation box, (*c*) Number, shape and position of various SiC susceptors, and (*d*) Aging effects of the insulation as well as susceptors, were also investigated. Additionally, some materials from a rapid prototyping process, *i.e.* gelation casting, were investigated and interesting results obtained [33].

3.4.1. Microwave Sintering Route

The merits of microwave sintering of ceramics are

(*a*) Very Rapid Heating, and

(*b*) High Densification achieved at relatively low temperature.

In this process, microwave energy interacts with the material at the molecular level. The ability of the molecules to follow the rapid reversal of the 'electric field' results in the conversion of the 'electro-magnetic' energy into heat within the irradiated material. The dissipation of power by a ceramic material in an electric field is given by the following equation

$$P = (2\pi f \sigma)(E^2/2) \tan \delta$$

where, P is the power dissipated, f is the frequency of microwave generator, σ is the dc conductivity, E is the electric field strength, and tan δ is the 'loss tangent' of the dielectric material.

It is readily apparent from the above equation that the microwave energy absorbed within a given material depends on its 'effective dielectric loss' and the distribution of electric field within the material due to its quadratic dependence. Here, the depth of the microwave penetration into the material is quantified by defining the 'skin depth', which is the distance at which the electric field falls to 37%. The skin depth is calculated by the following formula :

$$\text{Skin Depth} = \frac{1}{(\pi f \mu \sigma)^{1/2}}$$

where, μ is the dielectric permittivity.

The microwave heating is fundamentally different than conventional heating for sintering of ceramics, where heat energy flows from outside the material towards the interior of the matrix. This may cause some type of inhomogeneity of heat distribution (*i.e.* thermal gradients inside and outside the material) resulting in differential density arising out of the inadequate distribution of the required phase for strength development. However, in the microwave heating, the heat energy is generated 'internally' within the core of the material instead of coming from the external heating source.

As a result, an 'inverse' temperature profile is developed from centre or core, *i.e.* to the outside surface of the material, in microwave heating in comparison with the conventional sintering. Due to this 'reverse' flow of heat, the microwave heating makes it possible to sinter small and large shapes very rapidly and uniformly, and also to reduce the 'thermal stresses' due to 'thermal gradients' that cause cracks during the fabrication process.

The microwave heating is sensitive to the amount and the type of impurities present in the material. If the material contains some 'impurities', which reasonably absorb in the microwave region of the electro-magnetic spectrum, then the heating will be impaired. However, high purity alumina, which normally requires higher temperatures in conventional sintering, is highly transparent to 2.45 GHz microwave frequency region at room temperature that is usually used in the sintering process through 'microwave route'.

At the above frequency domain, the 'skin depth' is of the order of one meter, which is much more than the usual shapes being sintered in special ceramics. In contrast, for 'lossy' material with dispersion behaviour, such as silicon carbide, the 'skin depth' is of the order of microns, and in most cases, the microwaves are absorbed before they can fully penetrate the 'platelets' used in this type of study.

The details of the material preparation with alumina from Alcoa (USA) as a starting material is already given in **section 3.2.1.** with the details of Attrition Milling. The dopants used were magnesium oxide from E-Merck (Germany) of grade - Gr., as a sintering aid during the milling. The details of the 'grinding parameters' on the final particle size of **50 nm** alumina particles are also given in **section 3.2.1.** After measuring the particle size in an Autosizer 2C, the particles were also checked under a transmission electron microscope, as described in the **section 2.6.3.** The electron micrograph is shown in **Figure 3.5** to show the distribution of nano-particles.

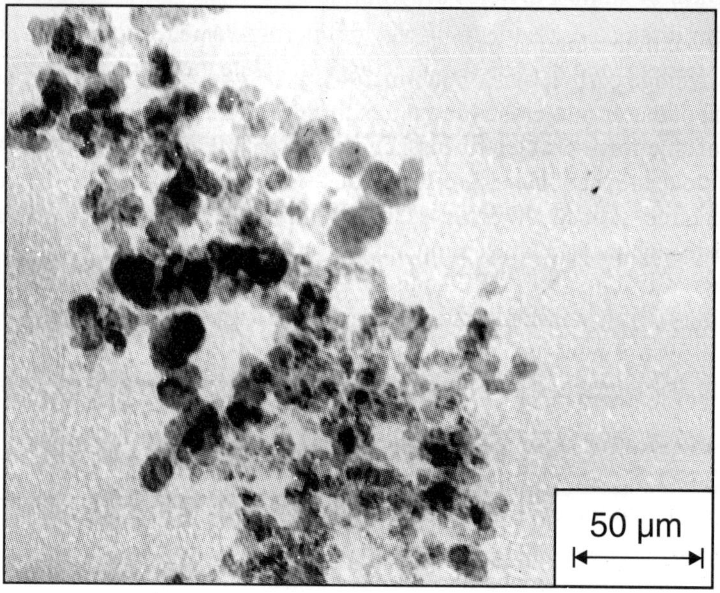

Figure 3.5: TEM photo of nano particles of alumina powders.

3.4.2. Sample Preparation from Nano Particles

The milled nano-crystalline alumina powder was mixed in acetone, with 1% Oelic acid in solution as a 'pressing aid', in order to form a slurry. After drying the slurry at 80°C, the mixture was passed through a sieve of mesh size 30 to form a 'free flowing' granules, which could be easily compacted under isostatic pressure. These granules are then packed into a rubber mould usually needed in isostatic pressing and compacted at an isostatic pressure of 210 MPa. The tubes were formed of length of 45 mm with inner diameter of 10 mm and outside diameter of 20 mm. These compacts were processed in such a way in order to have a high green density, *i.e.* 64% in this case, which is fundamentally important in order to achieve a high sintered density.

3.4.3. Sintering Procedures of Nano Particles

The sintering of this compacted alumina tube is carried out in a microwave oven at 1400°C and 1500°C from 30 to 120 minutes. The microwave energy is supplied by using an optimised power of 1350 ± 10% Watts magnetron operating at a single frequency of 2.45 GHz. In order to increase the microwave energy absorption by the alumina compacts and also to ensure an uniform temperature, a furnace cavity is designed from ceramic wool boards (formed under vacuum) after giving it a special treatment to withstand high temperature. The green alumina tubes are loaded in alumina tray and packed with α-silicon carbide powder, and kept in the furnace cavity in the microwave oven.

After sintering, the tubes are characterized by different measurements like density by simple Archemedes principle, microstructure by scanning electron microscope (SEM) and fracture strength by Universal Testing Machine in order to see the effect of microwave sintering on different physical properties of nano-sized alumina particles.

3.4.4. Sintering Data of Nano Particles of Alumina

The nano-crystalline alumina particles have been prepared by optimising different grinding parameters having average particle size of **50 nm,** as shown in the TEM micrograph in **Figure 3.5.** The microwave sintering of these nano-crystalline alumina particles show interesting densification behaviour as a function time of sintering between 30 and 120 minutes at 1400°C and 1500°C, as shown in **Figure 3.6.** It is seen that the densification at 1500°C is much faster than at 1400°C, and thereby leading to a maximum density of 3.93 g/cc for 120 min [*i.e.* 99% of theoretical density (TD)] for the former case. The maximum density achieved at lower temperature of 1400°C for 120 min is only 3.57 g/cc (*i.e.* 90% of TD), indicating a reduction of 9.2%. While the rate of densification is quite steady till 60 minutes for 1500°C, that for 1400°C it is constant upto 75 minutes, showing both the effects of temperature and time of sintering of **50 nm** nano-crystalline alumina particles. After these times, the sintering rate decreases with time showing a sort of saturation near 120 minutes, *i.e.* the highest time used in this work [3].

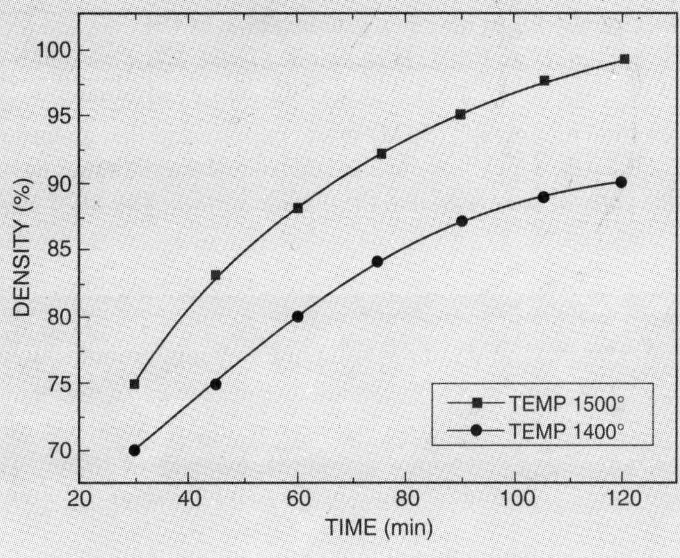

Figure 3.6 : Densification behaviour of nano particles of alumina at 1400°C and 1500°C.

The diametral fracture strength of the samples sintered at 1500°C is 390 MPa, whereas that for 1400°C is only 210 MPa, indicating a reduction of 46.2%. This is where the 'quality improvement aspect' of nano-crystalline particles is clearly observed with respect to microwave sintering, *i.e.* a density reduction of 9.2% corresponds to a reduction of fracture strength of 46.2%. This is ascribed to the formation of dense microstructure with increasing density from 90 to 99% by increasing the sintering temperature from 1400 to 1500°C [3].

3.5. CHARACTERIZATION

3.5.1. Electron Microscopy

For the characterization of nano particles of alumina and consequently on the sintered material, both scanning electron microscopy (SEM) and transmission electron microscopy (TEM) are essential. The details of the working principle of these techniques are given in **section 2.6.**

3.5.2. Sample Preparation for TEM and SEM Study

For SEM study, a small piece of the sintered alumina sample is mounted in resin for grinding and polishing. This mounted sample is initially ground by α-silicon carbide powder of grit size 200 and 400. The sample is then polished by 0.5 μm alumina powder in a vibratory polishing machine, and finally in fine alumina paste. The optically smooth polished surfaces are etched with a typical mixture of nitric and hydrofluoric acids to reveal the intricate details of the microstructure of the sintered material. Extreme care is taken at every step to get full details of the microstructure.

For TEM study, a cylindrical sample of 3 mm diameter and 1 mm height, which is a size suitable for this study, is cut directly from the bulk sintered material. The sample is prepared by mechanically thinning to ≈ 75 μm, followed by dimpling and subsequent low-energy (*i.e.* 5 to 6 kV) and low angle (*i.e.* 15°) Argon ion beam milling. The films are then examined in a transmission electron microscope at different regions, which is operated at an accelerating voltage of 100 keV.

The transmission electron microscopy (TEM) picture of nano sized alumina particles with an optimum particle size of **50 nm,** which has been obtained by 'Attrition Milling' in a fluid medium of grind of 'water with surfactants' for an optimum grinding time of 10 hours and for an optimum weight ratio of grinding balls to alumina as 10 : 1, is shown in **Figure 3.7.** This figure shows some narrow distribution of particle size, as also observed in case of the starting material.

The scanning electron micrograph (SEM) shows uniform and fine grained microstructure, with an average grain size of **3.2 μm,** which was obtained due to microwave sintering owing to the uniform heating from the 'inside core' of the material to the outside surface. The SEM photograph is shown in **Figure 3.8 [3].**

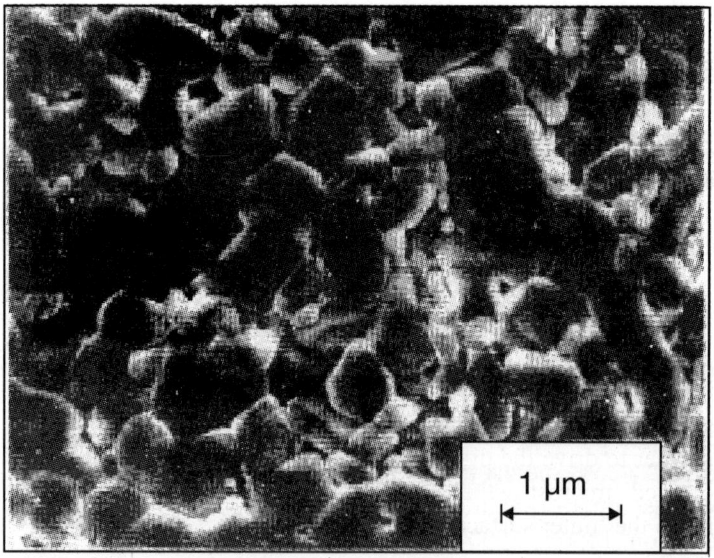

Figure 3.7: TEM photo of sintered alumina after microwave sintering at 1500°C for 120 min.

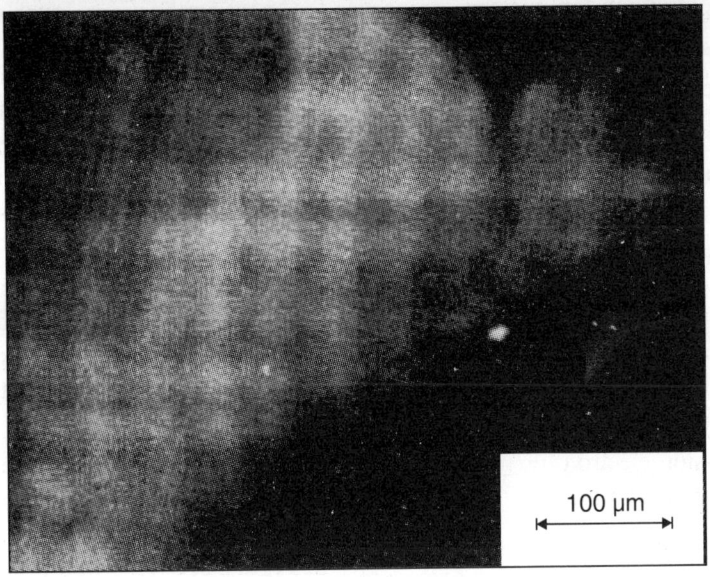

Figure 3.8: SEM photo of sintered alumina after microwave sintering at 1500°C for 120 min.

In this particular work, the grain-boundary region has been thoroughly studied through transmission electron microscope for alumina sintered at 1500°C for 120 min, as shown in **Figure - 3.8.** No liquid phase has been found at the grain-boundaries. Thus it could be concluded that the sintering mechanism of **50 nm** nano-crystalline particles of alumina doped with MgO is a solid-state sintering process. The mechanism is likely to be as follows

$$Al_2O_3 \Leftrightarrow 2Al^{3+} + 3O^{2-}$$
$$MgO \Leftrightarrow 2Mg^{2+} + 2O^{2-} + V_0$$

In case of sintering of alumina, the oxygen ions are the slow moving species. The creation of vacancies by the addition of MgO occurs due to the replacement of Al atom by Mg atom. These vacancies through the diffusion process help to get better sintering by uniform heating by microwave system [3]. Since the flow of heat is from the within the material due to the interaction of nano particles containing alumina molecules with the high frequency in the microwave domain, the microstructure is developed in a much better fashion, thereby giving a better mechanical property.

In summary, it can be stated that nano-sized alumina powders can be produced by attrition milling by carefully controlling various grinding parameters and eventually optimizing them. The sintering of nano-sized particles of alumina can be achieved near to its theoretical density by a small addition of magnesium oxide in order to drive the diffusion rate through the microwave route, which was clearly evident by a 'fine-grained' microstructure. The sintering of alumina by microwave heating process from the 'inside core' to the 'outer surface' through the addition of MgO is a solid-state process.

3.6. Wear Materials and Nano Composites

Normally, the 'wear' can be subdivided into two distinct categories :

A. **Erosive Wear** in which the particles in a gaseous or liquid medium strike a surface, and

B. **Frictional Wear** where two materials are in sliding contact.

There are some important material properties for 'wear applications', such as

(*a*) Hardness,

(*b*) Fracture Toughness,

(*c*) Thermal Conductivity,

(*d*) Chemical Inertness, and

(*e*) Corrosion Resistance.

There are many ceramic materials that possess these properties, which make them suitable for a range of applications such as

(*a*) Orthopaedic Implants,

(*b*) Thread Guides,

(*c*) Seal Rings,

(*d*) Valves,

(*e*) Metal Extrusion Dies, and

(*f*) Chutes in Material Handling Systems.

Alumina ceramics have an excellent wear-resistance and it is also cost effective, *i.e.* higher wear-cost index, and hence they are used in many applications as wear parts. Although there are a variety of other ceramic materials such as :

(*a*) Silicon Carbide (SiC),

(*b*) Partially Stabilised Zirconia (PSZ),

(*c*) Zirconia Toughened Alumina (ZTA),

(*d*) Tungsten Carbide (WC), and

(*e*) Silicon Alumina Oxy-Nitrides (SIALONS).

These interesting materials also have superior wear resistance, but still alumina has been the "material of choice" for many applications primarily because of its 'Low Cost' and 'Ease of Fabrication'.

For example in a comparative sense, an Alumina Seal costing Rs. 400 would cost approximately Rs. 1200 in Silicon Carbide and Rs. 1600 in Tungsten Carbide. However, standard alumina ceramics do not meet the more stringent requirements of 'high wear resistance' applications. As a consequence, the materials with increased wear performance, such as SiC, Sialons and ZTA, are now being routinely used as wear parts, despite their higher costs for many hi-tech applications.

Therefore, clearly, there is a 'need' that existed in the search of materials in order to bridge the gap between Low Performance-Low Cost and Hi-Tech-High Cost. There is a question of material strategy here, wherein one should look for newer materials that would perform properly at a reasonable cost. This has led to a new thrust in order to improve the wear resistance of alumina based ceramics. The toughness and wear resistance have been increased through the use of some materials such as

(*a*) SiC Fibres,

(*b*) Whiskers, and

(*c*) Platelets or Particles.

But there are certain 'related' problems like

(*a*) Health Issues related to Whiskers,

(*b*) Anisotropy of Properties (*i.e.* axial dependence),

(*c*) Difficulties with Processing and Densification, and finally

(*c*) Creation of some Defects.

The consequential effect is that the 'increases in toughness' are not always commensurate with the 'increases in fracture strength'. That brings us to the newer development of nano-composite ceramics, which might be in the limelight in the search of 'real' materials.

3.6.1. Nano Composite Ceramic Materials

A new family of materials called 'Nano-Composites' have enhanced mechanical properties. For the application as structural ceramic materials, a nano-composite represents a ceramic matrix composite with the grain size of the matrix in the range of several microns and containing a second phase of nano-sized **(< 100 nm)** grains.

During the past, many systems with matrix-inclusion have been investigated such as

(*a*) Alumina, (*b*) Magnesia, (*c*) Silicon Nitride, (*d*) Mullite, and (*e*) Cordierite as matrix materials.

The second phase materials used as the 'inclusion-phase' are

(*a*) Silicon Carbide, (*b*) Silicon Nitride, (*c*) Titanium Nitride and (*d*) Metallic Particles.

But, the system which has attracted more attention is **'Alumina-Silicon Carbide'.**

An initial study shows that the mechanical strength for this new class of materials increases from 360-390 MPa to 1450-1500 MPa on addition of 5% SiC to an Al_2O_3 matrix. In case of fracture toughness it is observed that it increases from 3.6 to 4.7 $MPa.m^{1/2}$. The creep resistance, thermal shock resistance and hardness also seem to show some improvements [34].

However, there remains some difficulties in getting uniform distribution of nano-sized SiC in the alumina matrix. Some improvements are obviously possible by processing the material through the Attrition Milling route in order to achieve high density with a dense microstructure, as done for the nano particles of both silicon carbide and alumina, as depicted in the previous sections.

3.6.2. Nano Composite Alumina Ceramics

Both Al_2O_3 and SiC powders were as-received commercially as already explained in the previous sections. The processing route was through Attrition Milling, as also described for both silicon carbide and alumina. By the use of proper dispersing agents, the distribution of SiC was ensured within the alumina matrix, which was a very important aspect for the preparation of the nano composite materials for high performance applications. Some care has to be taken for the processing of nano composite materials in order to ensure a 'high degree of homogeneity'.

Although hot pressing technique has been used so far for the fabrication, *i.e.* sintering, of nano composites based on aluminous materials, only pressureless sintering method has been used [34], like it has been done in case of microwave sintering of high density alumina (see **section 3.4.3.**) [3], and in the processing of high density silicon carbides (see **chapter 2)** [4, 5], with nearly the theoretical density achieved in both these cases.

With the help of 'sintering additives' that allowed pressureless densification of nano-composites at industrially acceptable temperatures, a preliminary study on the sintering behaviour of Al_2O_3 and some Al_2O_3-SiC systems was undertaken. The Al_2O_3 material was fully dense on sintering in a normal furnace at 1540°C, whereas the Al_2O_3-4.5 vol% SiC nano-composite material required sintering at 1680°C in order to obtain almost full density, *i.e.* 97.5%.

However, the addition of 1.1 wt% Y_2O_3 (99.9% purity) as a sintering additive had a significant effect with > 98.5% density achieved at temperatures as low as 1530°C. The TEM photo (not shown here), which has already been shown in **Figures 3.7** and **3.8,** indicate a fine-grained microstructure [34]. This type of inter-connected microstructure is very helpful. Although the observed densification and the microstructural aspects are quite encouraging, but a better microstructure can be still obtained for a better 'wear' property by a combining various types of 'sintering additives' in order to reduce the sintering temperature further as well as to get better abrasion-resistance. This was recently done quite extensively by an intercontinental group at Ireland [35].

O'Sullivan et al [35] found out that Al_2O_3-SiC nanocomposites, produced by industrially acceptable processes including pressureless sintering, had higher 'wear' resistance than alumina and, in turn, the nano-composites doped with a combination of different 'sintering aids' showed higher 'wear resistance' than undoped nano-composites. At lower SiC region, the 'wear resistance' increased with SiC content up to 3 vol% with only little improvements observed above this limit. In erosive wear tests, these workers found out that the doped materials had 'wear resistance' 30 times higher than that of alumina. It has been proposed that nano-sized SiC particles located at Al_2O_3-Al_2O_3 grain boundaries contribute significantly to increased wear resistance through pinning and strengthening of the grain boundaries, which is an interesting observation by this group [35]. Some more interesting investigations on the line of improving mechanical properties of alumina nano-composites are given below.

Although pressureless sintering is now popular for making alumina nano-composites with SiC, this work needs some mention, since the material was prepared differently than the conventional technique. In this work, the hot-pressing temperature, matrix grain size, fracture mode and the distribution of SiC were studied to reveal the effect of SiC inclusions on the microstructure of Al_2O_3. The Al_2O_3-SiC composite powder was prepared by the 'precipitation' method [36].

The nano-scale SiC particles were coated with Al_2O_3. After hot pressing, most of the nano-scale SiC particles were randomly located within the Al_2O_3 grains. The inclusions of SiC raised the hot-pressing temperature, decreased Al_2O_3 grain size, and inhibited the abnormal grain growth of Al_2O_3. The inter-granular fracture for Al_2O_3 was transformed to the intra-granular fracture for Al_2O_3-SiC nano-composites due to the addition of SiC. The observed microstructure changes increased the mechanical properties of Al_2O_3-composites by 40% [36].

Recently, the wear and other mechanical properties of nano-composites have been increased by engineering the reinforcement of SiC platelets, so this work needs some mention where also a new technique was used for material preparation. In this work, BaO-Al_2O_3-2SiO_2 (BAS) glass-ceramic powders were prepared by sol-gel technique. The SiC platelet reinforced BAS glass-ceramic matrix composites with high density and uniform microstructure were fabricated by hot-pressing. The effect of additional crystalline seeds on Hexagonal to Monoclinic phase transformation of BAS was studied [37].

The effects of SiC platelet content on the microstructure and mechanical properties of the composites were also investigated. The results showed that the flexural strength and fracture toughness of the BAS glass-ceramic matrix composites can be effectively improved by the addition of SiC platelets. The main toughening mechanism was crack deflection, the platelets pull-out and the bridging. The increased value of flexural strength is contributed to the load transition from the matrix to SiC platelets [37].

This particular work is also special in that alumina was already toughened by stabilized zirconia. In this work, Al_2O_3 ceramic matrix composites toughened with 3Y-TZP (yttria-containing tetragonal zirconia polycrystals) and oriented hexagonal shape α-SiC platelets were fabricated by using slip casting and cold isostatic pressing. The effect of platelet aspect ratio and volume fraction on fracture toughness, and mode and microstructure of the final composite were examined. The fracture toughness of the reinforced-composites was evaluated using indentation in four-point bending test. The scanning electron microscopy and transmission electron microscopy were used to determine the microstructural details and dominant toughening mechanisms, which occurred in these materials [38].

The toughness measurement tests and detailed observations of microstructures and fracture surface profiles have led to the conclusion that multiple-toughening behaviour via transformation toughening, microcracking, crack deflection, load transfer and platelet debonding and pull-out, as well as thermal residual stresses have a significant contribution in improving the fracture toughness. A fracture toughness value as high as 11.2 MPa . $m^{1/2}$ was achieved for the materials sintered at 1600°C for 3 h with a platelet addition of 30 vol% [38].

3.7. NANO PARTICLES OF ZIRCONIA

Like alumina, as explained in the previous sections, zirconia is also one of the most important ceramic materials with diverse field of applications. Hence, the interest of preparing nano-crystalline zirconia has also grown during the past many years. The researchers have developed various processes to synthesize nano-sized zirconia powder without any doping agent. The growing interest on nano-sized zirconia is due to its excellent mechanical properties like flexural strength and fracture toughness, making it also a candidate for its use as ceramic engines and in many other ceramic components in industrially severe conditions as wear applications. Among various ways of preparing nano-sized zirconia powder for such diverse applications, the sol-gel route is the most chosen one, and an extensive amount of work has been done in the field. In this section, a brief review of the work done mainly on sol-gel process will be highlighted. The details of the sol-gel process are given in the last **chapter - 8.**

3.7.1. Applications of Zirconia

The popularity of zirconia has grown due to its high mechanical strength, fracture toughness and hardness. This is the reason why zirconia is a well-known material for 'structural' applications [39]. In order to make a good quality 'tetragonal' polycrystalline zirconia ceramics, the initial powders should have the following properties :

(*a*) Homogeneity in composition,

(*b*) Small particle size,

(*c*) Narrow particle size distribution, and

(*d*) Low level of agglomeration for better sintering [40].

Here, only the sol-emulsion-gel [41] and the well-known sol-gel methods [42] will be described for the synthesis of nano-crystalline zirconia particles. But, there are other processes for the same purpose as follows :

1. Inert Gas Condensation [43],
2. Non-Isothermal Synthesis of Yttria-Stabilized Zirconia Nano-Powder [44],
3. Spray Flame Technique at High Production Rate [45],

4. Carbon Nano-Tube Template Method [46],
5. Mechano-Chemical Synthesis Route [47], and finally
6. Heating of Alcohol Aqueous Salt Solutions [48].

Whatever be the processing route, the nano-sized zirconia powder materials cover a wide range of applications and definitely maintains its position as one of the most important ceramic materials. In the earlier days, the application of zirconia and zirconia-bearing materials had limitation in that they were used as 'foundry sand' and 'abrasion resistent' products [49]. These materials were also used as

(*a*) Fuel Cells,
(*b*) Resistive Heating Elements, and
(*c*) Jewelry Items.

This is mainly due to their properties such as high oxygen ion conduction and high refractive index [50-52]. Now, it has also found some new applications such as :

(*a*) Buffer Layer for Superconductor Growth [53],
(*b*) Damage Resistant Optical Coatings [54],
(*c*) Gate Dielectrics (as an important electrical material) [55-56], and finally
(*d*) Oxygen Sensors (the most important applications) [57].

It is known that zirconia also has good wear-resistance and thermal shock resistance properties. These properties coupled with high flexural strength and fracture toughness, i.e. critical stress intensity factor, make it an ideal candidate for its use as :

(*a*) Automobile Components for Ceramic Engines,
(*b*) Wire-Drawing Dies, and
(*c*) Many Types of Cutting Tools.

Furthermore, its low thermal conductivity together with its reasonably higher thermal expansion coefficient make it highly suitable for its application as 'thermal barrier coating' material in many thermal applications [50, 58].

Finally, an interesting work on zirconia nano powders needs mention here. The zirconia powders containing 12 mol% ceria have been prepared by the co-precipitation technique in order to obtain nano-sized powders with suitable sinterability and reduced grain size in the sintered ceramic by means of this 'solution technique' in its simplest route, *i.e.* without using any milling or other special procedure. In order to accomplish this aim, some processing variables have been systematically studied. For comparison purposes, specimens of the same composition have been prepared by the powder mixing technique. With the optimization of some processing variables, it was possible to obtain apparent densities of 98% of the theoretical value, 100% of tetragonal phase, and **500 nm** of average grain size, which is little higher in the nano range [59].

3.7.2. Synthesis of Nano Particles of Zirconia

As said earlier, there are various techniques for the preparation of nano-crystalline zirconia particles. However, the most common and versatile techniques are

A. Sol-Emulsion-Gel Technique, and
B. Sol-Gel Method.

Although the latter topic is discussed in some details in the **chapter - 8,** some salient features will be given here with reference to zirconia along with that of the first technique in the following subsections.

3.7.2.1. Sol-Emulsion-Gel Technique

The use of alkoxides in the preparation of ceramic powders has not always been successful due to its high cost. In any processing technique, the economy is quite an important factor. So, after a search of so-called inexpensive chemicals, it was found that the 'zirconyl nitrate' is relatively cheaper precursor and it can be easily used for the preparation of zirconia nano powder. The chemical reaction in ammonia solution is expressed as

$$ZrO(NO_3)_2 + H_2O + 2NH_3 = ZrO_2 + 2NH_4NO_3$$

In this process, the nitrogeneous compounds are preferred over chloride compounds due to the fact that they lead to a non-polluting and easy burn-out. This process is basically a combination of both sol-gel and emulsion methods [60].

Here, the Zr ions are entrapped in the water droplets and stabilized in an organic solvent by the addition of ammonia gas. It is through the process of 'heterogeneous distillation', water can be removed, thus stabilizing the zirconia gel droplets. The remaining unreacted water is also removed by heterogeneous distillation [61-62].

First of all, the sol is prepared by using zirconyl nitrate and deionised water. Then, the gel formed is heated to around 90°C with continuous stirring until a stable 'suspension' of fine colloidal particles is obtained. In another beaker, a surfactant like 'Tween - 80' (chemical name - poly-oxyethelene sorbitan mono-oleate) is dissolved in xylene. The sol is then added to this 'xylene solution', which leads to the formation of small 'emulsion droplets' in the reaction mixture, and stabilized due to the presence of the surfactant.

At this stage, suspension is agitated, when the ammonia gas is bubbled through this suspension, and then the 'emulsion droplets' transform into the 'gel'. In the next step, the ultra-sonic agitation of the solution is again needed to prevent any 'agglomeration' during the gelation process. Finally, the reaction mixture is refluxed and the washing is done by xylene several times before separating the powder by a centrifuging system. The whole process of the 'sol-emulsion-gel' is depicted in **Figure 3.9,** where all the details of the progression of the process is shown. Finally it should be mentioned that the selection of a 'surfactant' for a particular emulsion depends on the hydrophobic-lyophilic balance (HLB) number of the particular surfactant, and the composition of the suspension [63].

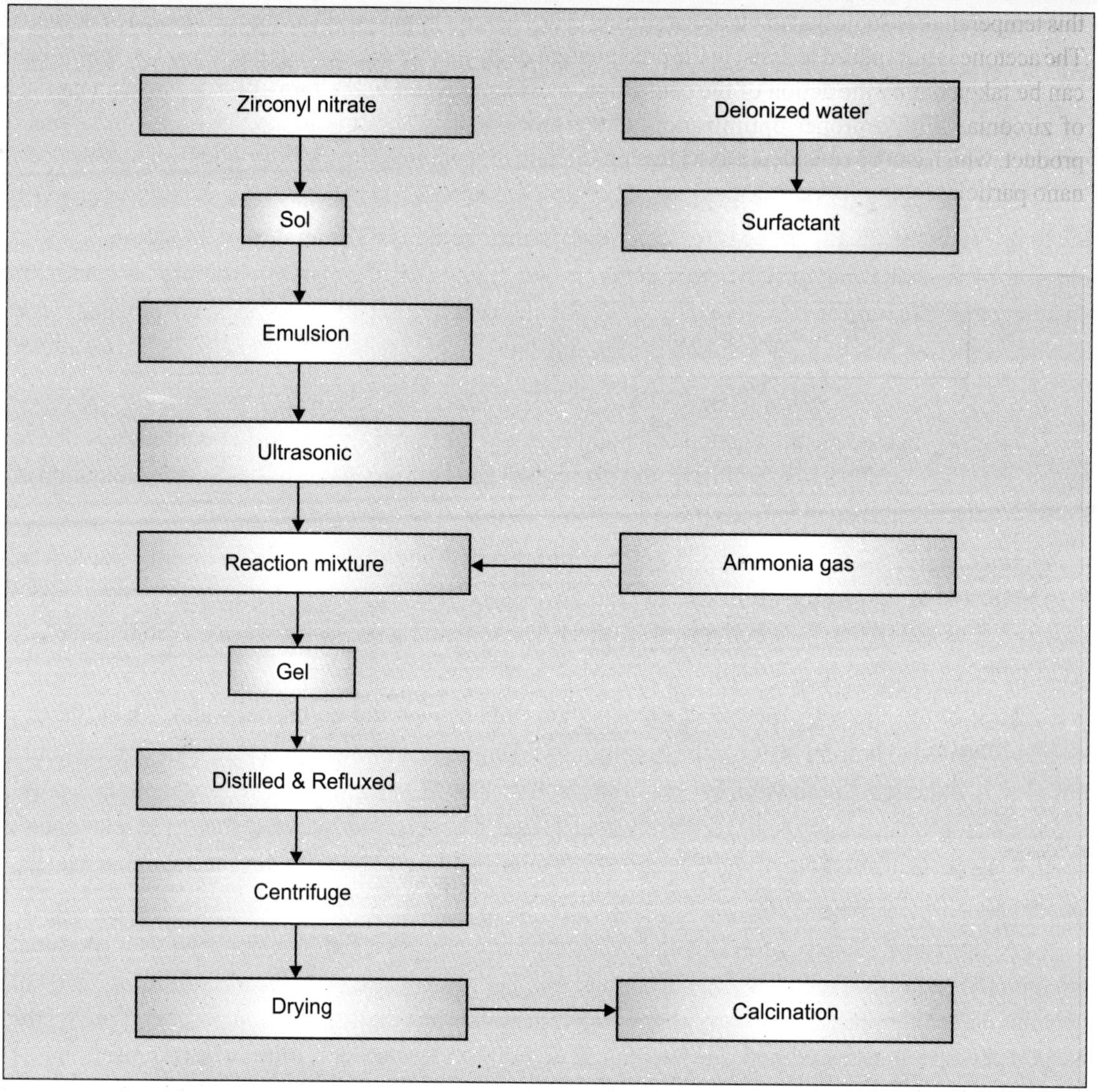

Figure 3.9: The flow diagram for the sol-emulsion-gel process. [68]

3.7.2.2. The Sol-Gel Technique

This technique of synthesis is based on non-hydraulic sol-gel reaction mechanism between zirconium iso-propoxide and zirconium chloride at 340°C. By this process, it is possible to synthesize a reasonably large quantity of uniform-sized nano particles without any size selection process. In this case, 'zirconium iso-propoxide propanol complex', $Zr[OCH(CH_3)]_4.(CH_3)_2\,COOH$, and zirconium chloride are added to extremely pure and degassed tri-octyl phosphine oxide, *i.e.* TOPO, at 60°C in argon atmosphere. Then, the temperature of the system is increased to 340°C and the whole reaction mixture at this temperature is subjected to vigorous stirring. The reaction mixture is then cooled, dried, and degassed. The acetone is now added to assist in the precipitation of the nano-sized zirconia particles. The 'precipitate' can be taken out by the action of the centrifuge, in order to result in the formation of the nano particles of zirconia. With a proper optimization of the process, it is possible to synthesize a few grams of product, which can be considered as a large production in 'colloidal chemical' synthesis of non-dispersed nano particles.

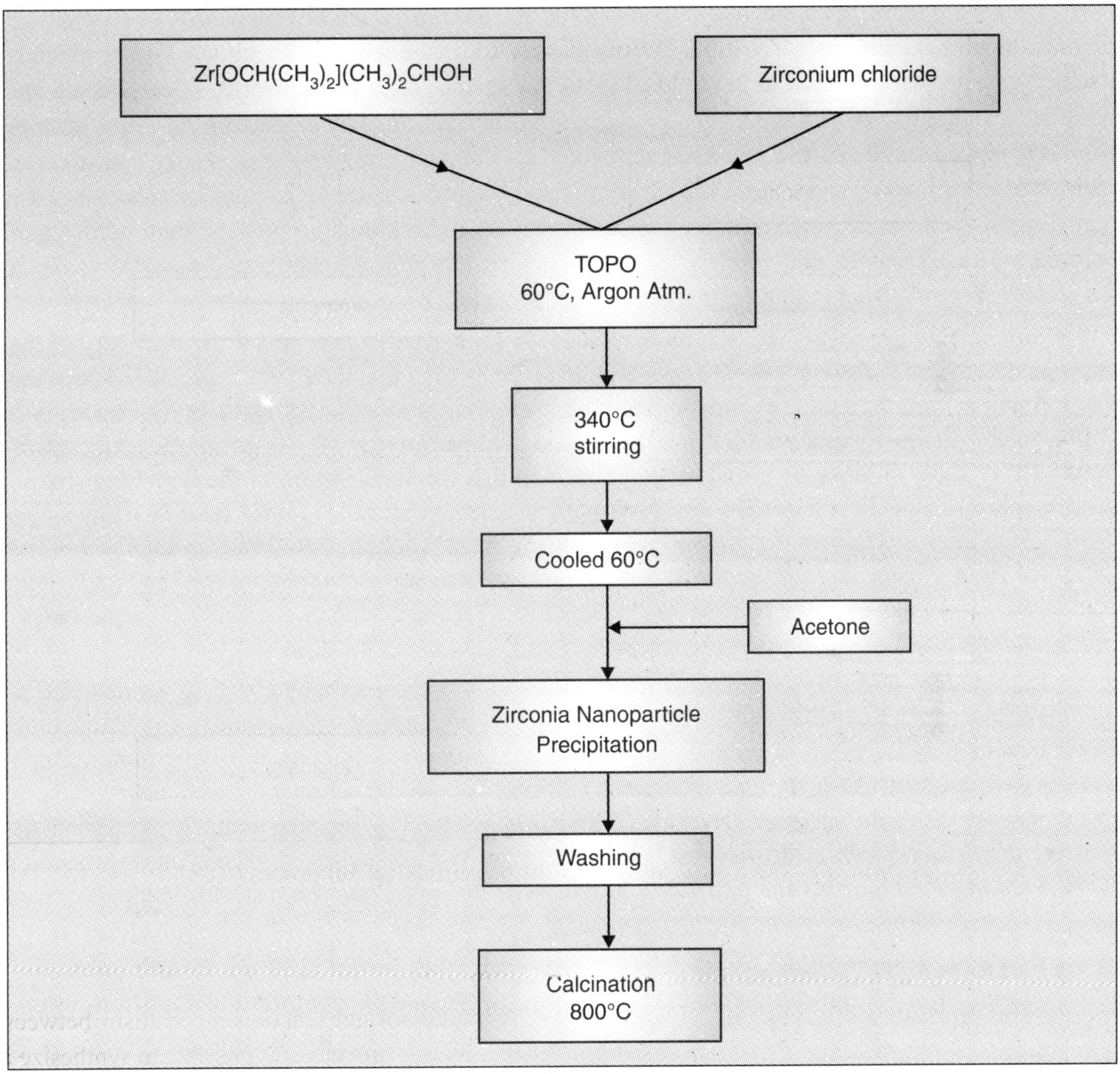

Figure 3.10: The flow diagram for the sol-gel process. [68]

The entire process of sol-gel is shown in **Figure 3.10,** which clearly shows various steps involved. Some points still need to be mentioned here. In this sol-gel process, the use of water forms hard agglomerates, and hence they should be avoided as much as possible, if not completely eliminated. The final product on drying is obtained in the powder form. After a controlled heat-treatment process, the yellow colour of powder changes to white.

3.7.3. Phase Transformation in Nano Particles of Zirconia

Compared to the materials like silicon carbide and alumina, there is a peculiarity or specialty in zirconia. In zirconia, there is a structural transformation from monoclinic to tetragonal phase, which is a subject of intensive research in the field of materials science all over the world, since zirconia has many interesting and important application as engineering materials. In order to speed up the transformation process to make a stabilized zirconia in the tetragonal crystalline form, many dopants like calcia, ceria, yttria, etc. are commonly used. Since the advent of nano-materials based on zirconia or rather since the beginning of the so-called discovery of various processes and methods to prepare nano particles of zirconia, we have been naturally thinking on how best we can exploit the nano properties of zirconia on this stabilization of 'structural phase transformation' process.

Simple classical ideas like 'thermodynamics' is applied to enquire about the difficulty in the retention of tetragonal phase at lower temperature [64]. It has been found by Subbarao et al [65] that the transformation from tetragonal to monoclinic zirconia is 'martensitic' in nature, which is basically a 'diffusionless' thermal transformation involving 'cooperative movement' of the atoms taking place within a very short distance. It has also been observed that the stabilization of the high temperature tetragonal form cannot be achieved by any quenching mechanism below the transformation temperature [65-67]. However, it is known that the surface energy of the monoclinic form is higher than that of the tetragonal form. Hence, the 'difference in surface energy' can play an 'important role' in that it must become more pronounced as the particle size becomes smaller and smaller, *i.e.* towards the nano-range. This is an important 'clue' for scientific research.

Like in many scientific phenomena involving 'criticality', there must be a 'critical' particle size of zirconia below which this 'difference in the surface energy' precludes the tetragonal (T) $\rightarrow$ monoclinic (M) transformation. But, there is another problem in that if the 'transformation' involving finite volume changes is suppressed, like in the case of zirconia particles dispersed in another ceramic matrix, there is a kind of strain inside the structure. Thus, this 'strain energy effect' associated with the 'expansion' of transforming zirconia particles also become important. Lange [64] has suggested that the 'micro-cracking' of the matrix around the zirconia particles and the 'twinning' involving the particles themselves can cause a 'neutralization of a part of this 'strain energy'.

Agarwal's group at BHU has done some important work in this field for the overall energy balance during 'T $\rightarrow$ M' transformation [68]. They have considered the thermodynamics of the system as $\Delta G_{Tot} = -\Delta G_{chem} + \Delta U_S + \Delta U_{Strain}$ with ΔU_S being the total change in surface energy and ΔU_{Strain} is the total strain energy. These workers also considered various parameters of importance, and came up with an equation for the critical particle size for zirconia (D_C) as follows :

$$D_C = \frac{6(\gamma_M - g_t\gamma_t + g_T\gamma_T + g_c\gamma_c)T_0}{\Delta H_{chem}(T_0 - T) - T_0[(1 - F_c)/F_c]\Delta U_{Strain}} \quad ...(3.1)$$

where, γ_M, γ_T = specific surface energies for monoclinic and tetragonal phases, t and c represent the twinning and microcracking for the geometric factor (g) and surface energy (γ), T_0 = equilibrium temperature, *i.e.* $T_0 - T$ = degree of undercooling, and F is a factor by which ΔU_{Strain} is reduced.

The above equation explains the complex interaction of various factors like temperature, strain energy, microcracking and twinning effects, and finally of the particle size. It also shows that the zirconia particles of suitable size range, *i.e.* possibly in the nano range, and dispersed in a properly selected matrix can retain tetragonal form on cooling below its bulk phase transformation temperature. Then, it can be made to transform under the influence of an applied 'tensile stress', which would eventually reduce the strain energy of the matrix, *i.e.* ΔU_{Strain}. Thus, the particles of a given size, which help in the retention of the tetragonal form even on cooling, can be made to transform to monoclinic form by applying stresses in excess of the critical value.

It has been shown by Lange [64] that the above action leads to the 'absorption' of energy in the form of 'elastic strain', which can be obviously related to the strain of transformation. Hence, the 'absorption' can be brought about just at the point of crack extension., which eventually contribute towards the enhancement of fracture toughness of the zirconia materials, as said earlier, since the whole process depends on some 'criticality, the critical size depends on the 'constraints' that can be produced by providing the following conditions :

1. The dopant oxide is present in the concentration range, which is less than that needed for the complete 'stabilization' of the cubic phase.
2. The cubic phase is heat-treated in order to develop a two-phase ceramics.

3.7.4. Characteristics of Nano Particles of Zirconia

The particles synthesized through the sol-gel route normally have an uniform size distribution with very small nano particles, *i.e.* between **5 nm** to about **40 nm.** The crystalline nature of these particles has been investigated in details by XRD [68]. It is noteworthy that the nano particles with the tetragonal structure closely resemble the cubic phase, which also have a relatively low density on the 'surface trap state' [69]. As emphasized earlier in the **chapter 2** for SiC, a higher green density is always congenial for better sintered density of a ceramic material. The variation of the green density is shown in **Figure 3.11** as a function of applied pressure. It increases steadily up to 700 MPa linearly, then it increases rather slowly up to about 850 MPa, and afterwards the green density almost becomes constant at about 52% of the theoretical density with further increase of the applied pressure [68].

It should be noted that for nano particles of alumina with an average size of **50 nm,** a green density of 64% was achieved with an applied pressure of 210 MPa (see the **section 3.4.2**) [3], and that for nano particles of SiC with an average size of **37 nm** was 62% with 250 MPa (see the **section 2.3.1**) [4-5]. Therefore, interesting data are obtained when we work with nano particles of different materials under different conditions of preparation, which give us a lot of insights into the 'world of materials' in order to be able to prepare various types of newer kind of materials for a diverse field of applications.

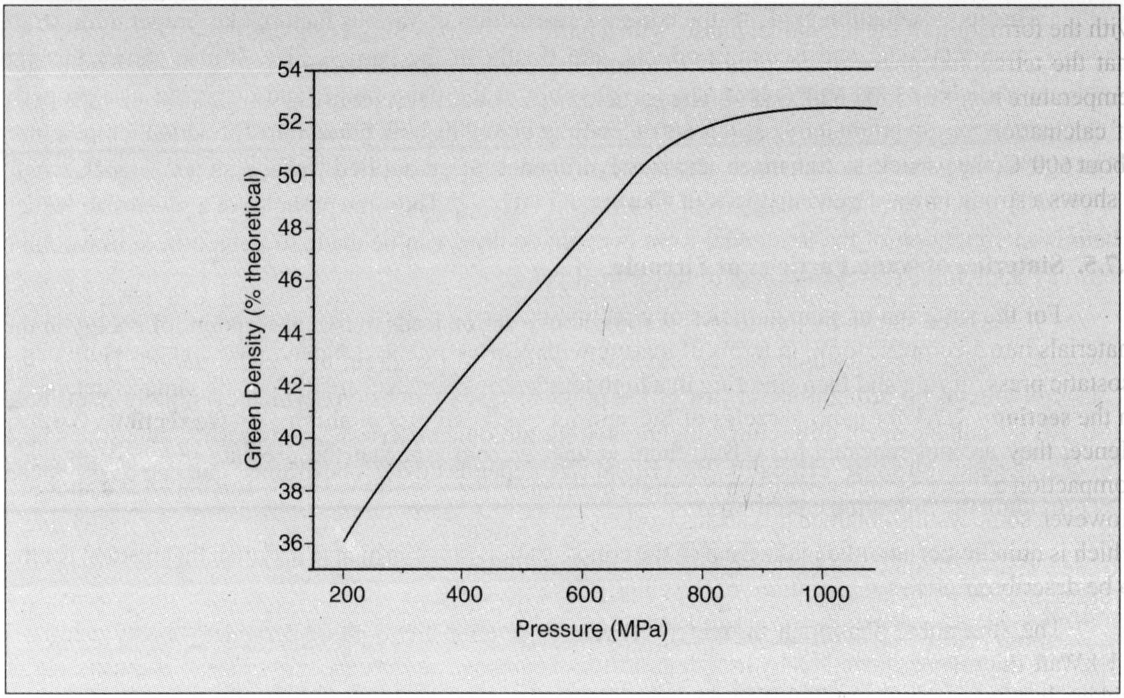

Figure 3.11: Green density vs. applied pressure for gel derived precursors [68].

It was found that the nano powder, which is prepared by the sol-emulsion-gel method, was amorphous in nature. When this nano powder is heat-treated at above 500°C, they become fully crystallized

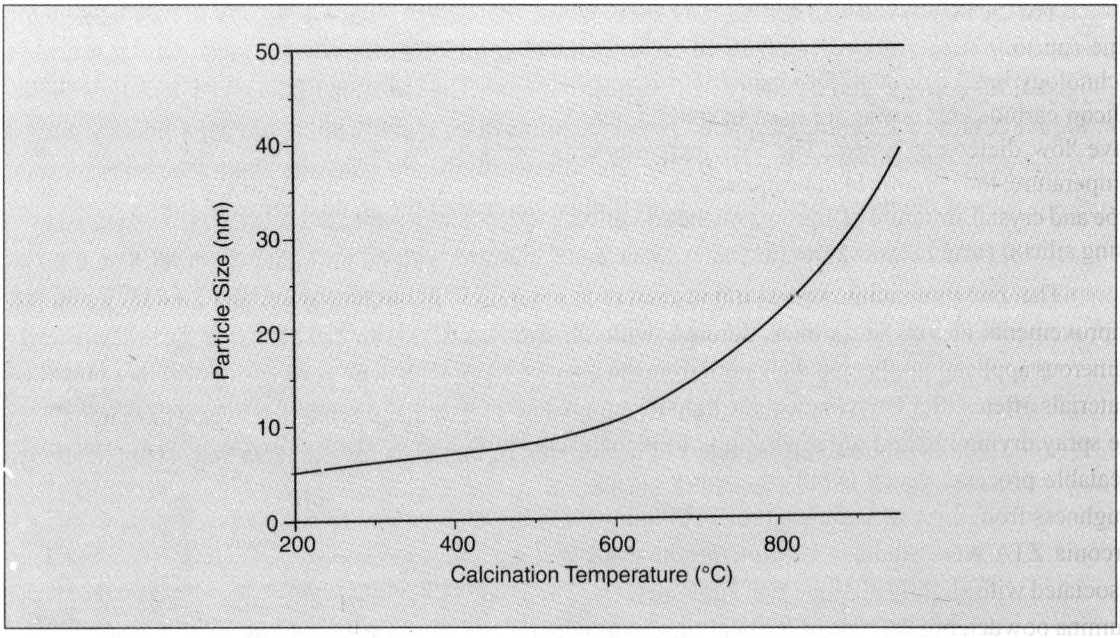

Figure 3.12: Variation of particle size with calcination temperature for zirconia gel [68].

with the formation of the tetragonal phase, with a particle size range of **4 nm - 10 nm** [68]. It is known that the tetragonal phase of zirconia is thermodynamically metastable, and it is formed within the temperature range of 550 - 650°C [41]. The particle size of nano particles of zirconia is a strong function of calcination temperature above about 600°C, which is depicted in **Figure 3.12.** It is seen that upto about 600°C, the particle size changes very slowly from **5 nm** to about **10 nm,** but after this temperature, it shows a strong upward trend up to about **40 nm** with increasing temperature up to about 1000°C [68].

3.7.5. Sintering of Nano Particles of Zirconia

For the sintering of nano particles of zirconia, the procedures for the preparation of the powder materials into a compact form, in terms of mixing with green binders, compaction at high pressure in a isostatic press, drying and then sintering in a high temperature furnace, are almost the same as detailed in the **section - 2.7.1** for nano particles of SiC and for nano particles of alumina in the **section - 3.4.3.** Hence, they are not repeated here. But, there is one exception in that the pressure required for the compaction of nano particles of zirconia is quite high, as already mentioned in the **section - 3.7.4** above. However, some results obtained by various workers in this field on microwave sintering of nano zirconia, which is quite important like in the case of sintering of nano particles of alumina as shown above, need to be described here.

The zirconia and alumina samples were sintered to over 99.5% of theoretical density using a 1.1 kWatt microwave generator is less than 45 minutes, or 1.5 hours 'cold to cold'. The density and vickers hardness were comparable to dense conventionally sintered materials, which were fired for a longer time of 12 to 16 hours. The container volume and susceptor position were found to have an impact on the heating rate and uniformity. The relationship between the applied power, the heating rate, temperature and properties of the material were worked out, along with the scale-up for microwave sintering of zirconia for larger amount of materials preparation [70].

For the purpose of cost saving, the time of sintering is also important. The savings in processing time (up to 90%) and energy (20-80%) were expected in microwave sintering of ceramics, as this technology breaks through into industrial firing processes, *i.e.* a hybrid microwave system. Typically, silicon carbide susceptors are used to initiate heating from room temperature, where many ceramics have 'low dielectric losses'. The loss increases with temperature, and at some "kick in" transition temperature, the ceramic load heats preferentially over the susceptors. In this work, the effect of dopant type and crystal structure of zirconia on the so-called 'kick in' temperature of transition was explored by using silicon carbide susceptors [71].

The zirconia toughened alumina (ZTA) is a 'composite system', which offers significant improvements in toughness over alumina, with superior hardness compared to zirconia. There are numerous applications that would benefit from the use of ZTA. Unfortunately, the commercially available materials often suffer from inadequate transformation toughening and processing flaws associated with the spray drying method of preparation of the materials. In this work, the feasibility of using a simple 'scalable process' which avoids the spray drying step was explored. The ZTA bars with optimized toughness from this process were broken in four-point-bending test. The fracture surfaces for 40 vol.% zirconia ZTA were studied. The fracture analysis indicated that the strength was limited by cracks associated with slightly alumina rich agglomerates. The Weibull modulus was affected by the source of alumina powder, but the type of 'critical flaw' was not affected, which is an important observation [72].

As explained in the **section 3.7.3** above, the stress-induced transformation of zirconia from tetragonal to monoclinic phase offers an important contribution to toughening of ceramic composites.

The stabilization of the tetragonal phase is affected by the zirconia content and its grain size, and by stabilizing agents such as yttria. The zirconia phases in zirconia toughened alumina (ZTA) with several volume levels of zirconia (5 to 40 vol%) and yttira additions (0 to 3 mol%) were analyzed by quantitative X-Ray diffraction. The combined effect of zirconia and ytrria on hardness, toughness, and retention of the tetragonal phase was investigated. The amount of yttria required to stabilize the 'optimum mixture' of tetragonal and monoclinic zirconia was determined for 5 vol% zirconia [73-74].

REFERENCES

1. M. S. Datta, A. K. Bandyopadhyay and B. Chowdhury, 'Nanocrystalline silicon carbide particles by attrition milling route' , Indian J. Physics-A, 78 (2004) 257.
2. M. S. Datta, A. K. Bandyopadhyay and B. Chowdhury, 'Preparation of nano α-SiC crystalline particles by attrition grinding', Interceram, 53 (2004) 242.
3. M. S. Datta, A. K. Bandyopadhyay and B. Chowdhury, 'Microwave sintering of nano particles of alumina', Trans. Ind. Inst. Metals, 55 (2002) 531.
4. M. S. Datta, A. K. Bandyopadhyay and B. Chowdhury, 'Sintering of nano crystalline α-silicon carbide doping with aluminium nitride', Bull. Mater. Science, 25 (2002) 121.
5. M. S. Datta, A. K. Bandyopadhyay & B. Chowdhury, 'Sintering of nano crystalline α-silicon carbide by doping with boron carbide', Bull. Mater. Science, 25 (2002) 181.
6. G. D. Zhan, J. Kuntz, J. Wan, J. Garay and A. K. Mukherjee, 'A novel processing route to develop a dense nanocrystalline alumina matrix (<100 nm) nanocomposite material', J. Am.Ceram. Soc., 86 (2003) 200.
7. J. Zhao, L. C. Stearns, M. P. Harmer, H. M. Chan, G. A. Miller and R. F. Cook J. Am. Ceram. Soc., 76 (1993) 503.
8. H. J. Hwang, T. Sekino, K. Ota and K. Nihara, J. Mater. Sci., 31 (1996) 4617.
9. H. Gleiter, Prog. Mater. Sci., 33 (1989) 223.
10. K. Lai Hing, R. G. Wheeler, W. L. Wilson and M. A. Duncan, J. Chem. Phys., 87 (1987) 3401.
11. C. J. Sandroff, J. P. Herbision, R. Ramesh, M. J. Andrejco, M. S. Hedge, D. M. Hwang, C.C. Chang and E. M. Vogel, Science, 245 (1989) 391.
12. K. Baba, N. Shohata and M. Yonezawa, Appl. Phys. Lett., 54 (1989) 2309.
13. D. W. Hoffman, R. Roy and S. Komarneni, Ceram. Bull., 62 (1983) 375.
14. P. M. Kumar, P. Borse. V. K. Rohatgi and S. V. Bhoraskar, 'Synthesis and structural characterization of nano-crystalline aluminium oxide', Mater. Chem. and Phys., 36 (1994) 354.
15. S. Ramesh, E. Sominska, B. Cina, R. Chaim and A. Gedan, 'Nanocrystalline γ-alumina synthesized by sonohydrolysis of alkoxide precursor in the presence of organic acids : structure and morphological properties', J. Am. Ceram. Soc., 83 (2000) 89.
16. K. Varatharajan, S. Dash, A. Arunkumar, R. Nithya, A. K. Tyagi and B. Raj, 'Syntheis of nanocrystalline α-Al_2O_3 by ultrasonic pyrolysis', Mater. Res. Bull., 38 (2003) 577.
17. S. Ramanathan, S. K. Roy, R. Bhat, D. D. Upadhaya and A. R. Biswas, J. Alloys Compd., 243 (1996) 39.

18. E. Borsella, S. Botti, R. Giorgi, S. Martelli, S. Turtu and G. Zappa, Appl. Phys. Lett., 63 (1993) 1345.
19. J. Ding, T. Tsuzuki and P. G. McCormick, J. Am. Ceram. Soc., 79 (1996) 2956.
20. Z. Yu, Q. Zhao, Q. Zhang and W. C. Xuebao, Wuji Cailiao Xuebao, 9 (1994) 475.
21. X. Sun, J. Li, F. Zhang, X. Qin, Z. Xiu and H. Ru, 'Synthesis of nanocrystalline α-Al_2O_3 powders from nanometric ammonium aluminium carborate hydroxide's, J. A. Cera. Soc., 86 (2003) 1321.
22. A. J. Fanelli and J. V. Burlew, J. Am. Ceram. Soc., 69 (1986) C174.
23. R. K. Pati, J. C. Ray and P. Pramanik, 'A novel chemical route for the synthesis of nanocrystalline α-Al_2O_3 powder', Mater. Lett., 44 (2000) 299.
24. W. R. Tinga and W. A. G. Voss, 'Microwave Power Engineering', Academic Press, New York (1968).
25. W. H. Sutton, 'Microwave Processing of Materials', Vol. 124, Materials Research Society, Pittsburg, (1988) p. 287.
26 N. W. Schubring, Am. Ceram. Soc., Electronics Division Fall Meeting at Grossinger, New York, September (1983).
27. P. A. Hass, Am. Ceram. Soc. Bull., 58 (1979) 873.
28. M. K. Krage, Am. Ceram. Soc. Bull., 60 (1981) 1232.
29. T. T. Meek, R. D. Blake and J. J. Petrovic, Ceram. Egg. Sci. Proc., 8 (1987) 861.
30. J. D. Katz and R. D. Blake, Am. Ceram. Soc. Bull., 68 (1989) 1221.
31. H. S. Shulman and W. J. Walker, Jr., 'A comparison of microwave furnaces for sintering ceramics', Presented at the 3rd World Congr. on Microwave & Radio Frequency Applications, Sydney, Sept.(2002).
32. H. S. Shulman, M. L. Fall, W. J. Walker Jr., T. A. Treado, S. J. Evans, M. Marks and M. L. Tracy, 'Sintering uniformity and reproducibility with 2.45 GHz microwaves in an industrial sized chamber', Presented at the 3rd World Congr. on Microwave & Radio Frequency Applications, Sydney, Sept.(2002).
33. H. S. Shulman, H. Giesche and O. Ugurlu, 'Mapping applicator uniformity : microwave sintering of alumina and zirconia', Presented at the 3rd World Congr. on Microwave & Radio Frequency Applications, Sydney, Sept. (2002).
34. M. S. Dutta, unpublished work.
35. D. O'Sullivan, 'New alumina-based ceramic materials for abrasive and wear applications', Matl. Ireland Res. Centre, Univ. of Limerick, Ireland.
36. H. Z. Wang, L. Gao and J. K. Guo, 'The effect of nanoscale SiC particles on the microstructure of Al_2O_3 ceramics, Ceram. Intl., 26 (2000) 391.
37. J. C. Gu, F. Ye, Y. Zhou and T. C. Lei, 'Microstructure and mechanical properties of SiC platelet reinforced $BaOAl_2O_{32}SiO_2$ (BAS) composites', Ceram. Intl., 26 (2000) 855.
38. C. Kaya, F. Kaya, P. A. Trusty, A. R. Boccaccini and M. Marsoglu, 'On the toughening mechanisms of SiC platelet-reinforced Al_2O_3/Y-TZP nano-ceramic matrix composites', Ceram. Intl., 25 (1999) 359.
39. D. J. Green, R. H. J. Hannick and H. V. Swain, 'Transformation toughening of ceramics', CRC Press, USA (1080) pp. 5.

40. W. Li and L. Gao, 'Nano $ZrO_2(Y_2O_3)$ particles processing by heating of ethanol-aqueous salt solutions', Ceram. Intl., 27 (2001) 543.

41. S. D. Ramamurthi, Z. Xu and D. A. Payne, 'Nanometer-sized ZrO_2 particles prepared by a sol-emulsion-gel method', J. Am. Ceram. Soc., 73 (1990) 2760.

42. J. Joo, T. Yu, Y. W. Kim, H. M. Park, F. Xu, J. Z. Zhang and T. Hyeon, 'Multigram scale synthesis and characterization of monodisperse tetragonal zirconia nanocrystals', J. Am. Chem. Soc., 125 (2003) 6553.

43. G. Skandan, 'Processing of nanostructured zirconia ceramics', Nanostr. Mater., 5 (1995) 111

44. O. Vasilkiv and Y. Sakka, 'Non-isothermal synthesis of yttria stabilized zirconia nano powder through oxalate processing : I, Characterization of Y-Zr oxalate synthesis and decomposition', J. Am. Ceram. Soc., 83 (2000) 2196.

45. H. K. Kamler, L. Madler S. E. Pratsinis, 'Flame synthesis of nano particles', Chem. Engg. Technol., 24 (2001) 583.

46. C. N. R. Rao, B. C. Satishkumar and G. Govindaraj, 'Zirconia nanotubes', Chem. Commun., (1997) 1581.

47. P. G. McCormik, T. Suzuki, J. S. Robinson and J. Ding, 'Nanopowders synthesized by mechanochemical processing', 13 (2001) 1579.

48. W. Li, L. Gao and J. K. Guo, 'Synthesis of yttria-stabilized zirconia nano particles by heating of alchol aqueous salt solution', Nanostr. Mater., 10 (1998) 1043.

49. M. J. Mayo, J. R. Seindensticker, D. C. Hague and A. H. Carim, 'Surface chemistry effects on the processing and superplastic properties of nano-crystalline oxide ceramics', Nanostr. Mater., 11 (1999) 271.

50. T. Chraska, A. H. King and C. C. Brendt, 'On the size dependent phase transformation in nanoparticulate zirconia', Mater. Sci. & Engg., A286 (1995) 169.

51. M. Gell, 'Application opportunities for nanostructured materials and coatings', Mater. Sci. & Engg., 204 (1995) 246.

52. H. Gleiter, 'Materials with ultrafine microstructure : Retrospectives and Perspectives', Nanostr. Mater., 1 (1992) 1.

53. J. M. Phillips, 'Substrate selection for high temperature superconducting thin films', J. Appl. Phys., 79 (1997) 1829.

54. N. Mansour, K. Mansour, E. W. V. Stryland and M. J. Soileau, 'Diffusion of colour centres generated by 2-proton absorption at 532 nm in cubic ZrO_2', J. Appl. Phys., 67 (1990) 1475.

55. G. D. Wilk, R. M. Wallace and J. M. Anthony, 'High-K gate dielectrics : current status and materials properties consideration', J. Appl. Phys., 89 (2001) 5243.

56. V. V. Afanasev, M. Houssa, A. Stesmans and M.M. Heyns, 'Electron energy barrier between (100) Si and ultra-thin stacks of silica, alumina and zirconia insulators', Appl. Phys. Lett., 78 (2001) 3073.

57. C. Leon, M. L. Lucia and J. Santamaria, 'Correlated ion hopping in single crystal yttria stabilized zirconia', Phys. Rev. B, 55 (1997) 882.

58. R. Neilsen and T. W. Chang, 'Zircomium and zirconium compounds', 'Uhlman's Enclyopedia of Industrial Chemistry', Wiley-VCH Verlag, Weinheim (Germany), A28 (1996) pp. 543.

59. E. N. S. Muccillo and D. M. Ávila, 'Synthesis and characterization of submicron zirconia-12 mol% ceria ceramics', Ceram. Intl., 25 (1999) 345.

60. P. Reyen, H. Bastius and M. Fielder, 'The use of emulsion in the preparation of ceramic powders', 'Ceramic Powders', Ed. P. Vincenzini, Elsevier, Amsterdam (The Netherlands), (1983) pp. 499.

61. H. K. Bowen, W. E. Rhine and T. Kanai, 'Preparation of $2ZrO_2 - Yb_2O$ by emulsion methods', 'Ceramic Transactions' Vol. 1A : Ceramic Powder Science II, Eds. G. L. Messing, E. R. Fuller Jr. and H. Hausner, Am. Ceram. Soc. (USA), (1988) pp. 119.

62. D. D. Perrin and W. L. F. Armrego, 'Purification of Laboratory Chemicals', Pergamon Press, USA, (1988) pp. 11.

63. K. Shinoda and S. Frigberg, 'Emulsion and Solubilization', Wiley, USA, (1986) pp. 56.

64. F. F. Lange, 'Transformation toughening - Part 1 : Size effects associated with the thermodynamics of constrained transformation', J. Mater. Sci., 17 (1982) 225.

65. E. C. Subbarao, H. S. Maiti and K. K. Srivastava, 'Martensitic transformation in zirconia, Phys. Stat. Solidi, A21 (1974) 9.

66. G. K. Bansal and A. H. Heuer, 'Martensitic phase transformation in zirconia : I, Metallographic evidence', Acta. Metall., 20 (1972) 1281.

67. G. K. Bansal and A. H. Heuer, 'Martensitic phase transformation in zirconia : II, Crystallographic aspects', Acta. Metall., 22 (1974) 409.

68. P. Barrick, R. Lodha, R. Pyare and G. N. Agarwal, 'Zirconia - Part II : A review of nanoparticle systhesis by sol-gel', Indoceram, 41 (2004) 7.

69. H. Gleiter, 'Nanostructured materials : Basic concepts and microstructure', Acta. Mater., 48 (2000) 1.

70. M. L. Fall, H. S. Shulman, L. Wolfe, EPL and W. J. Walker Jr., 'Sintering wear parts with microwave heating', Presented at the 104th Annual Meeting and Exposition of the Am. Ceram. Soc., St. Louis (USA), April, (2002).

71. M. Moeller, L. H. Therm, H. S. Shulman and H. Giesche, 'A novel approach to understanding microwave heating of zirconia', Presented at the 104th Annual Meeting and Exposition of the Am. Ceram. Soc., St. Louis (USA), April, (2002).

72. H. S. Shulman, J. Varner, W. Votava and A. Clarke, 'Fractographic study of zirconia toughened alumina', New Zealand Ceram. Trans., 122 (2001) 145.

73. M.Samandi, H. S. Shulman, J. Young and A. Clarke, 'Millimeter-wave sintering of 20 vol% zirconia toughened alumina', 2nd World Congr. on Microwave and Radio Frequency Processing, Orlando (USA), April, (2000).

74. H. S. Shulman, M. J. Ryan, N. I. Baxter and I. W. M. Brown, 'Optimization of yttria additions in zirconia toughened alumina', J. Aust. Ceramic Soc., 34 (1998) 127.

Chapter 4

Mechanical Properties

PREAMBLE

In order to talk about the mechanical properties of sintered materials based on nano particles, it is important to describe briefly about the fracture mechanics of such materials. In this chapter, a brief account is given on how the famous Griffith theory is revisited. By limiting the integration within the zone where fracture occurs in brittle materials, based on a sinusoidal approximation as used in the Griffith formulation of theoretical strength of such solids, a modification of the equation for theoretical strength can be achieved. This modified equation for theoretical strength has been developed using the linear Hook's law near to the maximum of applied stress by taking a small variation in the spatial elongation. This equation gives rise to a multiplication factor, which has to be used to predict the theoretical strength when micro-flaws are present in the brittle materials. This theoretical result is discussed in terms of the available data on fused silica.

4.1. THEORETICAL ASPECTS

There has been a tremendous amount of work done in the field of fracture mechanics in brittle ceramic materials and glasses, particularly on their theoretical strength for a long time [1 - 7]. This is mainly based on Griffith equation based on the formation of an elliptical crack and its main plank is the utilization of material parameters or constants, which are measurable, in designing suitable materials for various important applications [2]. An insight should be obtained on the nature of theoretical strength based on a sinusoidal approximation in the stress vs. spatial elongation curve.

The theoretical strength (σ_{Th}) of a 'body' is the stress required to separate it into "two parts", with the separation taking place simultaneously across the cross-section. To estimate σ_{Th}, let us consider 'pulling' on a cylindrical bar of unit cross section. The "force of cohesion" between the two planes of atoms varies with their separation, after the inter-atomic spacing (*a*). As shown in **Figure - 4.1,** a part of the curve is approximated by the sinusoidal relation as [3]

$$\sigma = \sigma_{Th} \sin\left(\frac{2\pi X}{\lambda}\right) \qquad \ldots(4.1)$$

This equation represents the so-called governing equation of stress (σ) against the spatial elongation (X). The work per unit area to separate the two planes of atoms is then calculated by the integral of the curve between X = 0 and X = $\lambda/2$ as

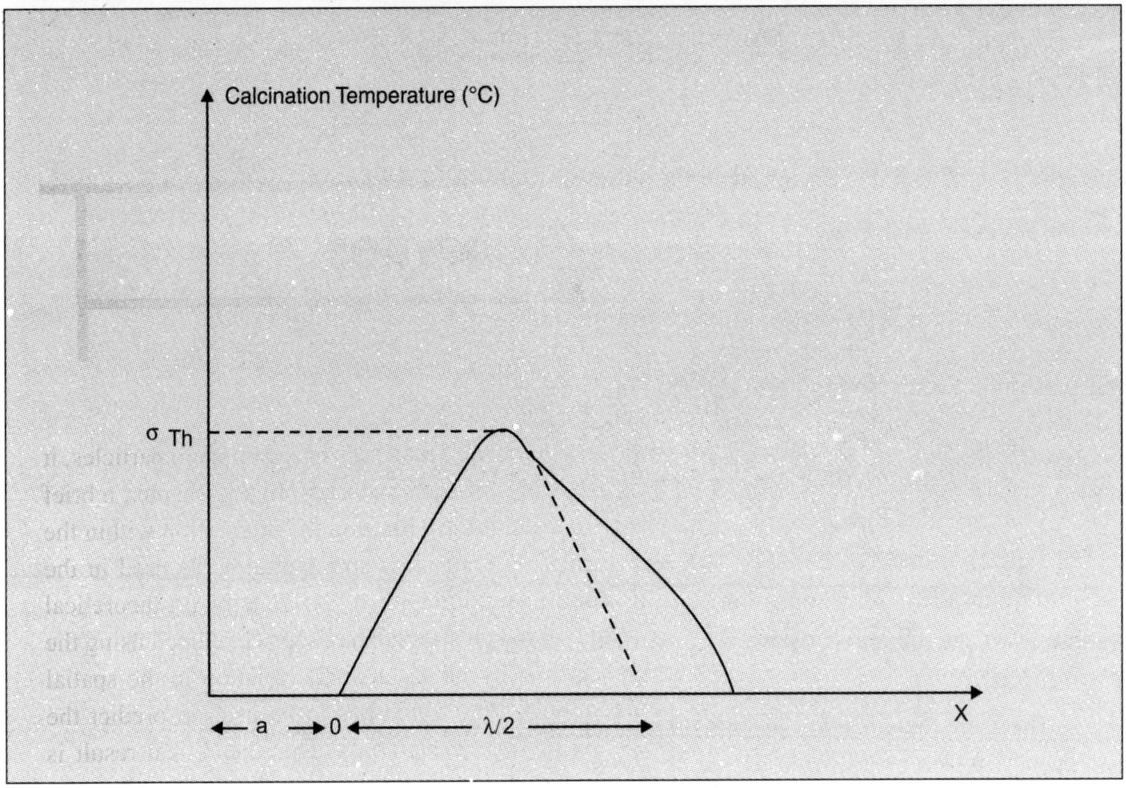

Figure 4.1: Applied stress vs. spatial elongation curve.

$$\int_0^{\lambda/2} \sigma_{Th} \sin\left(\frac{2\pi X}{\lambda}\right) dX = \sigma_{Th}\left(\frac{\lambda}{\pi}\right) \qquad ...(4.2)$$

This work or energy is then equated with the surface energy (2γ) of the two newly created surfaces, giving :

$$\sigma_{Th} = \left(\frac{2\pi\gamma}{\lambda}\right) \qquad ...(4.3)$$

For the initial part of the curve near the equilibrium spacing (a), as in **Figure 4.1,** the Hook's law has been expressed as

$$\sigma = E\left(\frac{X}{a}\right) \qquad ...(4.4)$$

where, E = Young's modulus. For this small part of X of this curve, from equation (4.1), the following relation has been deduced

$$\left[\frac{d\sigma}{dX}\right]_{X=0} = \frac{2\pi\sigma_{Th}}{\lambda}\cos\left(\frac{2\pi\gamma}{\lambda}\right) \approx \frac{2\pi\sigma_{Th}}{\lambda} \quad ...(4.5)$$

This is done at X = 0. By equating this with $\frac{d\sigma}{dX}$ obtained from equation (4.4), the following relation has been obtained as

$$\frac{2\pi}{\lambda}\cdot\sigma_{Th} = \frac{E}{a} \quad ...(4.6)$$

and substituting equation (4.6) into equation (4.3), we get

$$\sigma_{Th} = \left(\frac{E\gamma}{a}\right)^{1/2} \quad ...(4.7)$$

Typical values of $E = 3 \times 10^{11}$ dynes/cm^2, $\gamma = 10^3$ ergs/cm^2 and $a = 3 \times 10^{-8}$ cm, and $\sigma_{Th} = 10^{11}$ dynes/cm^2 as per equation (4.7). If $\lambda \approx a$, then σ_{Th} = E/5 to E/10 as per equation (4.6). For window glass, the strength is 10^4 psi, σ_{Th} = E/1000, and for alumina ceramics, the strength is 5×10^4 psi, σ_{Th} = E/1000.

Therefore, between the theoretical predictions and the actual experimental values, there is a discrepancy, which needs to be solved. It should be mentioned that involving material's constants (E and γ) and half of the elliptical crack length (c), Griffith's criterion of the maximum strength at which the material fails on cracking is based on the above equation. Hence, this equation certainly merits careful attention. Moreover, in line with Griffith's concept of micro-flaw formation, the reduction of theoretical strength also merits further attention.

Therefore, changing the 'limit of integration, and a very small 'fluctuation' of δ around the λ value, we arrive at the revised value of the theoretical strength after which the material will crack. This is found out as [8]

$$\sigma_{Th} = \sqrt{\frac{\frac{E\gamma}{a}}{\frac{\pi\delta}{\lambda} + \frac{2\pi^2\delta^2}{\lambda^2}}} \quad ...(4.8)$$

Therefore, it is clearly seen that the above equation will put an incremental effect on the theoretical estimate of maximum strength with respect to a simple Griffith criterion $(E\gamma/a)^{1/2}$ involving the material parameters, with "a" replaced by half of the elliptical crack length. It is known for a long time that Griffith criterion of predicting and eventually designing the right materials, only through measurable material properties like surface energy (γ) and elastic modulus (E), has been very popular, since the equilibrium inter-atomic distance (a) is more or less known for all the crystalline brittle ceramic materials, and even approximately for glasses.

4.1.1. Data Analysis of Theoretical Strength

It should be pointed out that if we put δ/λ to be much less than 1/4, then it would be possible for us to predict the correct theoretical strength of brittle materials. Therefore, this equation (4.8) can be used to precisely do this prediction by adjusting the value of δ/λ . For example, for three different values of $\delta/\lambda = 0.1, 0.01, 0.001$, we have to multiply Griffith value (under the square root sign) with 1.40, 5.47 and 17.10 respectively. In the literature, very often, there is a factor of $\sqrt{2}$ in the Griffith's value. In the first case, our assumption of taking the value $\delta/\lambda = 0.1$ gives rise to a multiplication factor of 1.40 (close to $1.414 \approx \sqrt{2}$).

The above analysis will help us analyse a variety of materials with different values of the ratio of δ/λ (non-dimensional value) to fit the experimental value with that of the theoretical estimate. Since both the values of δ and λ are not measurable, it is always better to take a ratio to estimate the strength as per equation (4.8).

Let us take the example of a common glass, where the value of σ_{Th} is 14 GPa as per equation (4.7), but as the experimental values are always lower, Griffith [2] put forward a new equation of $\sigma = (2E\gamma/\pi L)^{1/2}$, where L = length of the micro flaws, which were considered to reduce the strength, as in many other brittle materials. As per this revised equation, Griffith [2] postulated that even micron (10^{-6} m) sized flaw can reduce the observed strength of the glass by a factor of 100. In such situations, the ratio of δ/λ has to be still lower so that the multiplication factor is higher. Actually, this ratio clearly indicates the presence of micro-flaws.

Finally, an example of fused silica, whose $\gamma = 1.75$ J/m^2 and E = 72 GPa and taking $a = 1.6 \times 10^{-10}$ m, we find a theoretical value of strength as per equation (4.7) as 28.1 GPa, whereas the experimental value is 24.1 GPa. The close similarity of these values clearly indicates that it does not take the 'micro-flaws' into account. The theoretical value should be much higher. By multiplying the Griffith value with 1.40 (*i.e.* $\delta/\lambda = 0.1$), we estimate the strength value as per equation (4.8) as 39.34 GPa. This discrepancy (or even more discrepancy) will actually justify the presence of the 'micro-flaws' in fused silica, which is a known fact [7]. However, an analysis can be based on the estimated value of strength as 28.1 GPa.

Table 4.1 : Theoretical strength at different values of flaw size of fused silica [σ_{Th} = 28.1 GPa as per equation (4.7)]

a = L (flaw size in 10^{-10} m)	*σ_{Th} estimated Griffith Rev. Equ. $(E\gamma/\pi L)^{1/2}$*	*Level of Crack*	*Optimum Ratio of δ/λ to get 28.1 GPa [eqn. (4.8)]*	*Multiplication Factor needed to get σ_{Th}*
1.6	15.84	Quantum	705.0000×10^{-4}	1.77
16	5.01	Nano	95.8000×10^{-4}	5.60
160	1.58	Nano	10.0600×10^{-4}	17.74
1600	0.50	> Nano	1.0110×10^{-4}	56.11
16000	0.16	Granular	0.1011×10^{-4}	177.44

As per the revised equation of Griffith involving micro-flaws, if we take the size (L) of the flaws at the quantum level, *i.e.* the value of "*a*" in equation (4.7), the theoretical strength goes down by almost half to 15.84 GPa. As the size of the flaw increases to a level normally considered in the micron level, the value goes down by a factor of 100, as shown in **Table 4.1,** as also mentioned above. This necessitates the inclusion of the ratio δ/λ in the calculation of theoretical strength, which should also be in consonance with the data on fused silica on the probable flaw size, as shown in **Table 4.1.**

The last two columns clearly shows the need of a multiplication factor, when the concept of micro flaws needs to be introduced, which is calculated from our equation (4.8) taking smaller values of δ/λ ratio. It is seen that as the level of micro flaws goes to an "usual" granular level, the value of δ/λ ratio becomes still smaller, and the need for a higher multiplication factor arises, as shown in the last column of the above table.

It is pertinent to mention here that although the data for fused silica are fitted here, the information shown in the above table can be obtained on a variety of other ceramic brittle materials in order to be able to explain the discrepancy between theoretical and experimental values of strength for effective design.

In summary, it can be said that the modification of the basic equation on theoretical strength can be achieved, within the context of a sinusoidal approximation in the applied stress vs. spatial elongation curve, by assuming a small spatial variation and by changing the limit of integration in the energy formulation for crack formation. This modification yields a ratio of this variation giving rise to a multiplication factor, which can correctly predict the theoretical strength of brittle ceramic materials. The available data on fused silica has been fitted with this new model and found to be effective in explaining a lower observed strength due to the presence of micro-flaws. Many such data on other brittle ceramic materials can be fitted in future to give it a comprehensive shape.

Finally, it is pertinent to mention here that the above modified Griffith equation can be also utilized to determine the theoretical strength, based on certain assumptions of crack length, of sintered compacts of nano-sized silicon carbide and alumina particles. However, the analysis of these materials are done in the usual line of interpretation of fracture strength with the hope that the above formalism might be used by other researchers in the field of nano-materials to derive more information on this emerging field with some degree of refinement, if needed.

4.2. STRENGTH OF NANO CRYSTALLINE SIC

Preamble

The mechanical properties like flexural strength and the fracture toughness have been studied on a sintered silicon carbide, which is prepared by the pressureless sintering route from the nano crystalline silicon carbide particles of an Acheson type α-SiC that is processed by high energy attrition grinding route. The average flexural strength is found to be 390 MPa and the average fracture toughness is found to be 4.3 MPa . $m^{1/2}$. This gives some insights as well as interest in such type of study.

4.2.1. The Basic Concepts

It is seen in the **chapter 2** that the sintering behaviour of nano particles of SiC is very interesting indeed [9-11]. The non-oxide ceramics are attractive candidates for structural materials because of their high temperature strength, which makes them highly suitable for the applications in ceramic engines and gas turbines [12]. The silicon carbide seems to be particularly well suited because of its high temperature strength and intrinsic resistance to oxidation. However, silicon carbide ceramic has a strong covalent bonding character and a small amount of sintering aids such as AlN or B_4C are usually necessary to make dense materials [9, 10]. Usually, the dense silicon carbide materials have lower strength compared to silicon nitride materials. The latter experience a degradation of strength at 1200°C–1300°C, while silicon carbide materials do not show any decrease of strength up to 1500°C. Moreover, the reliability of silicon carbide ceramic in high performance applications is still in doubt. The silicon carbide, like other ceramics, shows a 'wide scattered distribution' in flexural strength. This 'distribution' of strength is a matter of concern and hence an important topic to be studied.

The strength distribution of silicon carbide varies while obeying Weibull statistics [13], as it is typical for many other ceramic materials. Most of earlier work is based on a 4-point bending test. Depending on the manufacturer and the type of silicon carbide, the average room temperature strength varies between 350 and 550 MPa. For such cases, the Weibull modulus which is a measure of the 'dispersion' in strength varies between 6 and 15 depending on the strength of the ceramic materials [11].

Contrary to the observed behaviour in metals, the room temperature strength of the sintered α-silicon carbide increases, as the exaggerated grain growth is intentionally introduced by a suitable heat-treatment [14]. However, the comparisons between the grain sizes should be made between two different materials having a uniform distribution of 'small grain sizes' versus a uniform distribution of 'large grain sizes'. In this study, with the isolated rise to a 'mixed grain-size' matrix, it should be pointed out that the mechanical strength of the sintered silicon carbide remains unaltered at temperatures up to 1500°C, which has a definite advantage in the high temperature applications such as gas turbines and ceramic engines (see the **section 2.1**). A further increase of strength has been observed in this material, when heated in the argon atmosphere [15, 16].

Schwetz and Lipp [17, 18] also studied the effect of the dopants on the flexural strength. They observed that the materials doped with aluminium had a higher strength than that doped with boron, and with the improvement in the processing parameters, it would be as strong as the 'hot-pressed' SiC materials. They also found that the production of the materials with ultrafine grains occurred in a very narrow range of sintering temperature (2050-2075°C). The exaggerated grain growth at temperatures above 2075°C was the cause of the reduction in strength of boron doped materials. The materials sintered with aluminium nitride, which acted as a grain growth inhibitor, showed a higher strength [17, 18].

Moreover, as said earlier, the reliability of silicon carbide materials in high performance applications is still in doubt. The silicon carbide, like other ceramics, shows a wide distribution in strength. One common way of characterizing the variability of strength of brittle materials is by using the Weibull modulus [13]. A high Weibull modulus material (*i.e.* $m > 20$) will give a narrow distribution of fracture strength, whereas a low Weibull modulus material (*i.e.* $m < 5$) has a wide distribution in fracture strength, and hence a low reliability [19-22]. However, a number of workers indicated that the structural reliability of ceramics is primarily dependent on the 'flaw size distribution' during the processing [23-28]. The studies on the strength and Weibull data of silicon carbide materials, which are prepared from the nano-crystalline particles, are somewhat limited in the literature [11].

The ability of a material to 'resist' cracking and once cracks start appearing, its ability to 'resist' further crack propagation leading to catastrophic failure, can be characterized by the 'manner of crack initiation and propagation', which is a function of fracture toughness The fracture toughness remains more or less constant with temperature, for the 'hot-pressed' and sintered silicon carbide, varying between 3 and 5 MPa.$m^{1/2}$. However their experimental data were limited to relatively 'large grain' size region [29]. Kodama and Miyoshi [30] studied the fracture behaviour of the 'fine-grained' silicon carbide materials, which were prepared by 'hot-pressing' with AlN additives. They observed a dependence of the fracture toughness on grain size. The maximum fracture toughness of 5.1 MPa.$m^{1/2}$ was observed for a grain size of **700 nm,** and for grain sizes that are 'finer' or 'coarser' than **700 nm,** the fracture toughness decreased. This was larger than that of the ordinary silicon carbide materials with the grains having micrometer grain-sizes. They also found that the maximum fracture toughness was observed mainly in the inter-granular mode. Dutta [31] observed that the mechanical failure of silicon carbide materials occurred mainly due to the defects introduced during the fabrication of the materials These defects are normally 'surface flaws', 'inclusions', 'voids', having a very large diameter of **100 mm.**

The above literature survey then gives us the 'basic concepts' by adequately covering almost all aspects of fracture mechanical behaviour of silicon carbides that are doped with various additives, particularly boron carbide and aluminium nitride, which are relevant for the present study. However, a glimpse of the Weibull statistics need to be given here, since it is the most fundamental aspect for the understanding of the 'dispersion' behaviour of fracture strength in materials. It should be clearly pointed out that the theoretical discussion on silicon carbides, which are given here and in the previous section on fracture strength and theoretical strength, are equally relevant for many other materials including nano-materials.

4.2.2. Weibull Theory

It is pertinent to write about this theory due to its tremendous importance in interpreting the fracture mechanical data of materials. The determination of the 'reliability' of the material components is based on Weibull's 'weakest link theory', which is very simple and is based on an assumption that "a part is like a chain of many links". If any link (*i.e.* small element of the part) fails, then the whole chain or the part has failed. Similarly, if any small volume in a material part is sufficiently stressed to cause a crack, the part will generally fail. Thus, the key is to determine if any of the elements in a part is likely to fail. Since the strength properties of the materials are variable due to the 'random distribution' of flaws, there is a variation of strengths of different elements of a part. Thus, the strength of various elements can be considered to have a 'statistical distribution' with the values below and above some characteristic strength.

If an element is subjected to some stress, there is a certain probability that the local strength of the material will be exceeded. As the number of elements in a chain is increased, the probability that "a weak link will occur and cause a failure" also increases. Similarly, the probability of a material component is a function of volume of the material, which are subjected to various stress levels. By combining the 'probability of failure' of all the elements, the probability of failure of the total part can be determined. That's a simple concept !

One common way of characterizing the variability of strength of brittle materials is by using the theory of Weibull [13], who established that the function, which would describe the 'cumulative probability of failure' of all the elements of a part of N elements, can be expressed as

$$P = 1 - \left[\sum_{i=1}^{N} \left(\frac{\sigma - \sigma_u}{\sigma_0} \right)^m \frac{V_1}{V_0} \right] \qquad ...(4.9)$$

where, P is the 'fracture probability' at stress σ, σ_0 is a normalising parameter, which is defined as the characteristic stress at which a volume of the material (V_0) would fail in an uniaxial tension, σ_u the threshold stress *i.e.* the minimum stress for which the fracture can occur, and *m* the Weibull modulus. For a fixed volume and shape of the samples, the equation (4.9) can be written as

$$P = 1 - \exp\{-[(\sigma - \sigma_u)/\sigma_0]^m\} \qquad ...(4.10)$$

For a conservative estimate, the threshold stress is taken to be zero. Thus the equation (4.10) can be written as

$$P = 1 - \exp\{-(\sigma/\sigma_0)^m\} \qquad ...(4.11)$$

A high Weibull modulus material ($m > 20$) will give a narrow distribution of the fracture stresses, whereas a low Weibull modulus material ($m < 5$) has a wide distribution of the fracture strength, and hence a low reliability, as pointed out earlier [11].

Here, the studies on the strength properties of nano-crystalline α-silicon carbide ceramics with the addition of boron carbide and aluminium nitride are described. The strength of the materials is characterised in terms of the baseline strength, the strength distribution and the microstructure.

4.2.3. Stress Intensity Factor

First of all, a little background on the theoretical aspect of fracture toughness should be given before the measurement technique is discussed. It is well established that the brittle fracture in ceramic materials, viz. silicon carbide, occurs through the propagation of 'pre-existing' flaws. The most common 'fracture-inducing' flaws are micro-cracks. The stress intensity at the crack tip, which is denoted by K_1, is related to the 'applied stress' by the equation as

$$K_1 = \sigma Y \sqrt{a} \qquad ...(4.12)$$

where, Y is a dimensionless parameter, which depends on the geometry of loading and the crack configuration, and '*a*' is the half of the crack length. Here, K_1 is called the 'stress intensity factor'. Actually, it is the driving force at the crack tip. At the time of the fracture, it is reasonably assumed that the normal stress reaches the 'critical value', which is just sufficient for the 'rupture' of the atomic bonds. Consequently, the 'stress intensity factor' (K_1) will also attain its 'critical value'.

It should be pointed out here that, like many other physical phenomena, there is also a "criticality" in the 'stress intensity factor', which surely depends on the 'velocity of crack propagation'. If we make a plot of crack velocity vs. K_1, then we would normally notice 'three zones'. In the first zone, the crack velocity increases moderately, then it increase slowly or remains almost constant (*i.e.* sub-critical growth), and finally at a critical value of K_1, the crack velocity moves upwards very fast → leading to fracture or failure of a brittle material [21].

Thus, the "critical stress intensity factor" (K_{1C}) or the 'fracture toughness' is a measure of the stress required to initiate a rapid crack propagation. Alternatively, it can be said that K_{1C} is a measure of the 'resistance' of a brittle material to crack propagation. K_{1C} can be easily considered as a 'material parameter' provided certain conditions are fulfilled. The fracture toughness has been expressed in terms of the 'surface energy' and the Young's modulus, which are measurable material parameters, by Irwin [5] as

$$K_{1C} = (2\,\gamma\,E)^{1/2} \qquad \text{...(4.13)}$$

This equation forms the basis of evaluation of the fracture mechanical parameters of interest. The above equation for 'critical stress intensity factor' was derived by assuming an infinite size of the samples. For the sample of finite size, this equation becomes

$$K_1 = \sigma\,\sqrt{\pi a}\; f(a/w) \qquad \text{...(4.14)}$$

where, w = width of the sample, a = crack length, σ = stress. The function $f(a/w)$ has to be known before K_1 can be determined. Obviously, $f(a/w)$ approaches $\rightarrow$ 1 for small values of (a/w). For the samples with larger (a/w) values, as mentioned earlier, the general equation is written as :

$$K_1 = Y\,\sigma \qquad \text{...(4.15)}$$

where, Y is a polynomial in (a/w). The factor $\sqrt{\pi a}$ is incorporated in Y. The general equation was derived by assuming a crack inside the material.

4.3. PREPARATION FOR STRENGTH MEASUREMENTS

4.3.1. Nano Powder Preparation and Characteristics

The details of the procedure to prepare nano powders of Acheson type α- and β-silicon carbides with both boron carbide and aluminium nitride as dopants with 1 wt% carbon are given in **chapter 2.** The method of preparation of nano particles of SiC, purification, mixing with green binders and pressing in a isostatic press at 250 MPa and then subsequent sintering at 2050 - 2100°C for 15 min under vacuum atmosphere (3 mbar) are all given in **chapter 2** in details. The same materials are characterized for their mechanical properties in order to see the effect of introducing nano particles of SiC on these properties of the final sintered material. Here, only the following details of the strength measurements will be given, since these details are important for analysis of the mechanical data concerning nano-materials.

4.3.2. Strength Measurement

4.3.2.1. Flexural Strength

The measurement of flexural strength requires the following steps :

1. Preparation of the Samples,
2. Introduction of 4-Point Bending Samples, and
3. Measurement of the Fracture Load.

The measurement of the flexural strength at room and high temperatures was made by four point bending method using high temperature Bending Strength Tester (Model - 422-S, Netzsch, Germany). It consists of the following main parts :

(*a*) The Frame,

(*b*) Electrically Heated Split Furnace (up to 1400°C),

(*c*) The Loading Device,

(*d*) Measurement System with Inductive Displacement Transducer,

(*e*) The Recording and the Control Console.

The test bars of size of 45 × 4.5 × 3.5 mm^3 were machined with a diamond wheel. Each surface of every sample was ground by silicon carbide powder (– 400 mesh). The bars were finally polished on diamond lapping disc with 1 μm diamond paste to a roughness of 0.01 mm or less. The tensile edges were chamfered (45°) with a similar abrasive.

The 4-point bending tests were carried out on a 40 mm outer span and 20 mm inner span width at a transverse speed of 0.05 mm/min at room temperature and also at high temperatures. On attainment of the temperature, the load was applied at 1.25 N/sec. The measurements were carried out from room temperature to 1200°C in air. The deflection was recorded by the X-Y recorder until the fracture occurred. From the fracture load and the sample dimension, the strength was calculated utilizing the following equation :

$$\sigma = (3P_f \cdot d)/(B\, w^2) \qquad ...(4.16)$$

where, P_f = fracture load, d = bending arm = $(L_1 - L_2)/2$, L_1 = major span , L_2 = minor span, w = specimen height, B = specimen width.

4.3.2.2. Fracture Toughness

A large number of methods of measurement of fracture toughness and such other parameters have been standardized for metals and alloys in recent years. These methods are normally suitable for ambient laboratory conditions. Different methods utilize different configurations. In the present work for sintered nano-crystalline particles of silicon carbide, the measurement of fracture toughness was made by 4-point 'Single Edge Notch Bend' (SENB) specimen technique.

The load required for causing the fracture of the sample was measured by the above high temperature 'Bending Strength Tester', having a length 1600 mm x width 750 mm × height 2100 mm. The bars of size (45 mm × 4.5 mm × 3.5 mm) were notched with a diamond blade of 0.2 mm thickness (100 μm diameter), after grinding in a 25 micron diamond wheel.

The notch dimensions were measured with the help of a traveling microscope. As a result of using 0.2 μm thick blade for notching, the notch width varied from 0.31 to 0.39. The value of $a/w \approx$ 0.25 - 0.40 mm, where a = notch length and w = notch width. The notched samples were put into the furnace. On attainment of the desired temperature, the load was applied on the 'Bending Strength Tester' at the speed of 1.25 N/sec, and the fracture load was noted. The fracture toughness (K_{IC}) was calculated by using the following equation as

$$K_{IC} = (3P_f d)/(Bw^2)[1.99 - 2.47(a/w) + 12.97(a/w)^2 - 23.17(a/w)^3 + 24.8(a/w)^4] \qquad ...(4.17)$$

At least 30 samples were tested at each temperature for a better accuracy of the fracture toughness data in order to give a better interpretation of the fracture mechanical behaviour of the sintered silicon carbides.

4.4. MECHANICAL PROPERTIES

Many details about the mechanical properties are given in the previous sub-sections along with some details of different experimental techniques, including some descriptions of the Weibull's statistical theory, which are all relevant for the present study in the interpretation of the data of fracture strength,

Weibull modulus and fracture toughness, *i.e.* critical stress intensity factor. Here, first of all, a comparison will be made between the mechanical data of both α- and β-silicon carbides in terms of their flexural strength, and then the mechanical data of α-silicon carbide will be presented in order to be able to correlate them with their microstructural behaviour.

4.4.1. Comparison of Mechanical Data of α- and β-SiC

The flexural strength of both α- and β-silicon carbides, which are doped with boron carbide + carbon and aluminium nitride + carbon, has been studied. For boron carbide doping, the flexural strength of α- and β-silicon carbides are found to be 390 and 330 MPa respectively. Such a behaviour of the flexural strength of two types of silicon carbide could be interpreted by the sintered density and the polytype formation, as described in various sections of the **chapter 2.** The grains of these silicon carbides doped with boron carbide are mostly elongated in shape. The sintered density of α-silicon carbide is 3.18 g/cc, which corresponds to 99% of theoretical density, whereas the density of β-silicon carbide is 3.10 g/cc that corresponds to 96.5% of the theoretical density. Therefore, the higher density of α-silicon carbide compared to that of β-silicon carbide could be one reason for a higher flexural strength of the former material. But there could be other reasons related to material properties.

As β-silicon carbide contains 3C-polytype, which is a thermodynamically unstable phase, the growth of the crystals in a particular direction has been very fast compared to that of α-silicon carbide according to the energy considerations. The very fast growth of the elongated crystals during the sintering of β-silicon carbide is expected to 'entrap the pores', and thus it might have prevented the densification to be completed. Therefore, as mentioned in the **section 2.7.3,** two types of porosities have been identified : (*a*) some within the grains, and (*b*) others between the grain-boundaries. The length of the crystal of α-silicon carbide is 5.7 μm, whereas that of β-silicon carbide is 12 μm. This should explain why the flexural strength of α-silicon carbide is higher than that of β-silicon carbide, which are both sintered with the same dopant of the same concentration.

As shown later from the microstructure, the fracture mode in α-silicon carbide is found to be inter-granular, which is also similar in β-silicon carbide. This type of fracture presumably originates from the surface, and propagates through the crystals through inter-granular porosities. The flexural strength of both these materials remained constant from room temperature up to 1400°C. This invariance of strength against temperature can be attributed to the fact that there is neither any liquid phase nor any glassy phase in the grain boundaries, as shown later for α-silicon carbide.

Therefore, the lower value of strength for β-silicon carbide can be attributed to the larger grain size (*d*), which is related to the flexural strength by Petch equation [24] as

$$\sigma_y = \sigma_x + kd^{-1/2}$$

where, σ_x is the base strength for the single crystal and k is a constant. Thus, the flexural strength of β-silicon carbide is found to be lower due to the presence of such larger grains, *i.e.* 12 μm.

4.4.2. Flexural Strength of α-SiC

There are many factors which affect the strength of ceramic materials such as porosity, grain size, grains shape and surface conditions etc. **Figure 4.2** shows a SEM micrograph of a polished and etched surface of the dense α-silicon carbide. The grains had mostly tabular shapes. The average grain size was found to be **5700 nm.** The **Figure 4.3** shows the flexural strength as a function of temperature. No variation of strength data with temperatures was found up to 1400°C. The average flexural strength was found to be 390 MPa. The **Figures 4.4 and 4.5** show the flexural strength as Weibull modulus (m) at room temperature and at 1400°C. The number of specimens was 30 in each case. The Weibull modulus was 11.3 at room temperature and 13.3 at 1400°C. The Weibull modulus is found to increase at higher temperature presumably due to healing of surface cracks, which were induced during the cutting of the samples by high speed diamond blades, *i.e.* the curing of the surface damages at higher temperature.

The Weibull modulus for β-silicon carbide (not shown here) was found to be 10.9 and 12.9 at room temperature and 1400°C respectively. The low value of '*m*' for β-silicon carbide could be due to the presence of pores of different sizes and also due to a relatively wider distribution of flexural strength at room temperature. The increase of Weibull modulus by about 20% for both these materials from room temperature to 1400°C might be due to the 'healing of surface cracks' and possibly due to the 'release of residual stresses'.

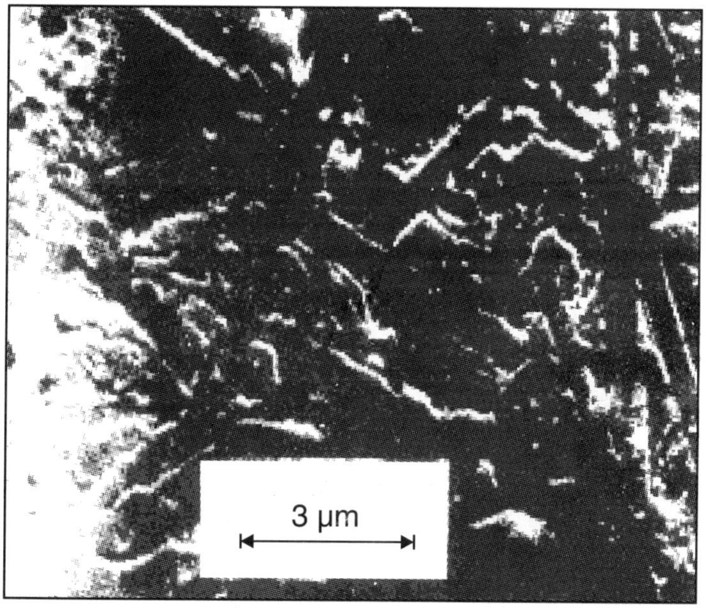

Figure 4.2: SEM photo of etched and polished surface of a dense α-silicon carbide (dopant : 0.5 wt% boron carbide + 1 wt% carbon)

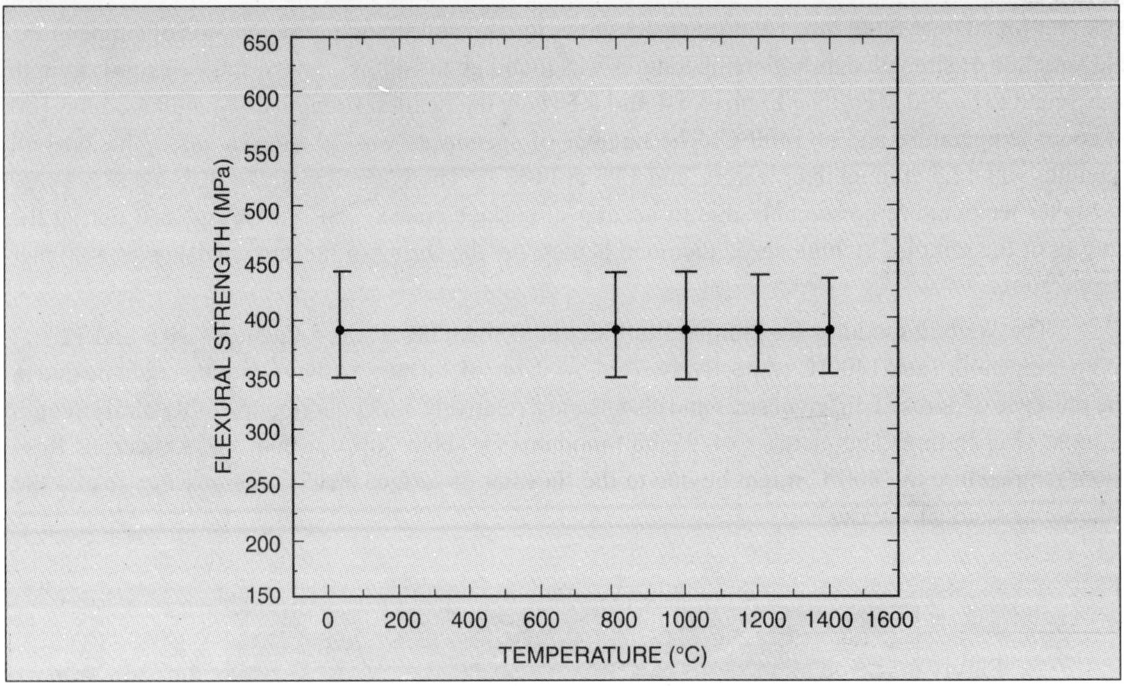

Figure 4.3: Flexural strength as a function of temperature of α-silicon carbide (dopant : 0.5 wt% boron carbide + 1 wt% carbon)

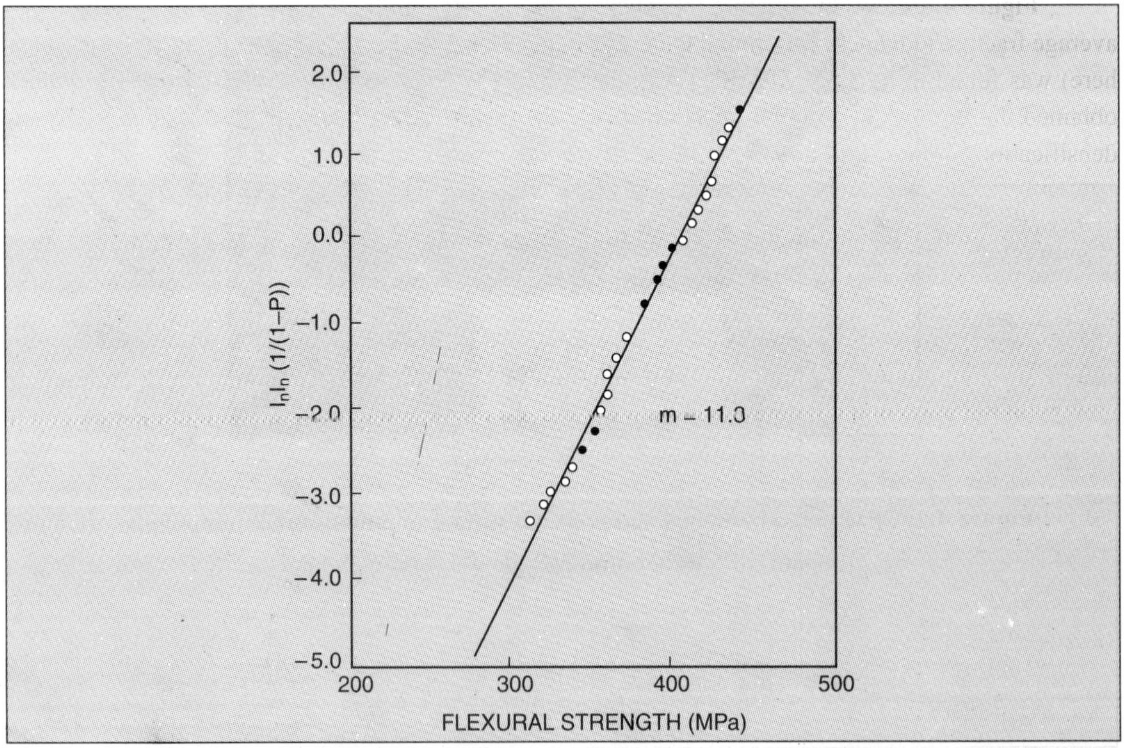

Figure 4.4: Weibull modulus of flexural strength at room temperature for α-silicon carbide (dopant : 0.5 wt% boron carbide + 1 wt% carbon)

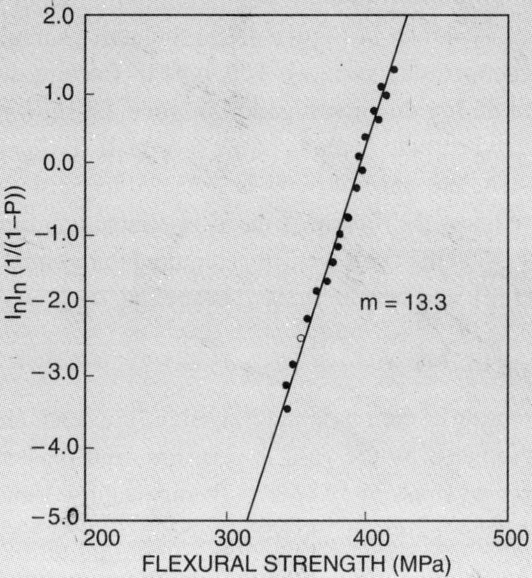

Figure 4.5: Weibull modulus of flexural strength at 1400°C for α-silicon carbide (dopant : 0.5 wt% boron carbide + 1 wt% carbon)

Figure 4.6 shows the fracture toughness of α-silicon carbide as a function of temperature. The average fracture toughness was found to be 3.95 $MPa.m^{1/2}$. The value for β-silicon carbide (not shown here) was found to be 3.55 $MPa.m^{1/2}$. A higher value of fracture toughness for α-silicon carbide is obtained due to the fine grained microstructures obtained after sintering and due to higher level of densification compared to that of β-silicon carbide.

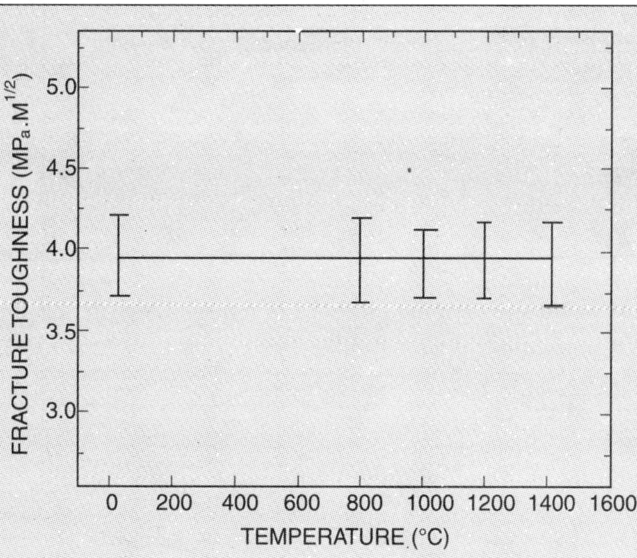

Figure 4.6 : Fracture toughness as a function of temperature of α-silicon carbide (dopant : 0.5 wt% boron carbide + 1 wt% carbon)

4.4.3. Microstructure

There are some pores as shown in **Figure 4.7** in the above samples, and the fracture mainly originate from the crack generated due to pores. This type of fracture, which is caused by an intra-granular pore, was observed during this investigation. **Figure 4.8** shows a SEM micrograph of the fracture surface of the α-silicon carbide in terms of the propagation of cracks and the fracture caused by the inter-granular pores. The presence of pores is expected to cause the formation of tiny cracks. The propagation of cracks requires the rupture of the inter-atomic bonds at the tip of the crack. Thus, the 'stress concentration' (σ_{max}) at the tip of a crack is required to be equal to the 'theoretical stress' to pull the atomic bonds apart [32], and when a stress σ is applied, σ_{max} is given by

$$\sigma_{max} = 2\sigma(c/\rho)^{1/2} \qquad ...(4.12)$$

where, ρ is the radius of curvature of the crack and c is half of the crack length. In this case, the radius of curvature of the crack produced by the void is very low, and thus σ_{max} might be equal to the theoretical stress (σ_{Th}) at applied stress of 390 MPa (average value). Hence, the average value of the flexural strength, as observed above, is also 390 MPa.

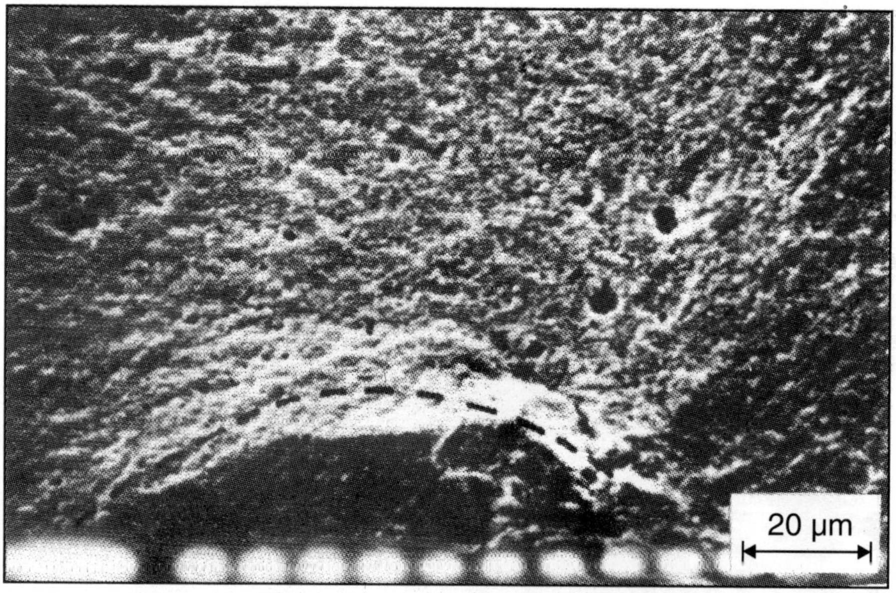

Figure 4.7: SEM photo showing the presence of a large void (the origin of the crack).

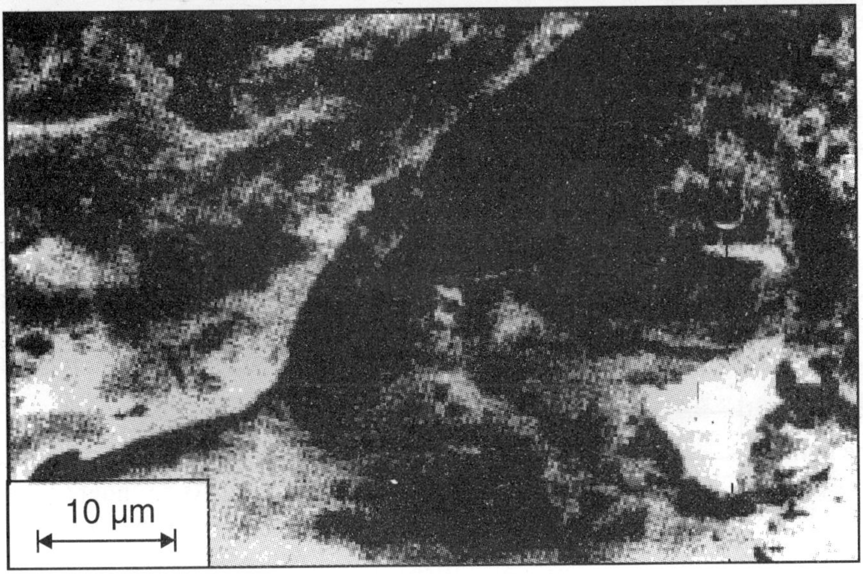

Figure 4.8: SEM photo showing crack propagation through the sintered sample.

The SEM micrograph of the fracture surface of α-silicon carbide is shown in **Figure 4.9,** where the fracture mode is found to be inter-granular. The grain boundary is very clearly revealed. The sharp grain boundaries could indicate a very thin grain boundary between the grains. **Figure 4.10** shows the grain boundary where two grains are meeting at an arbitrary angle, as observed by transmission electron microscopy. No glassy phase has been observed along the grain boundaries.

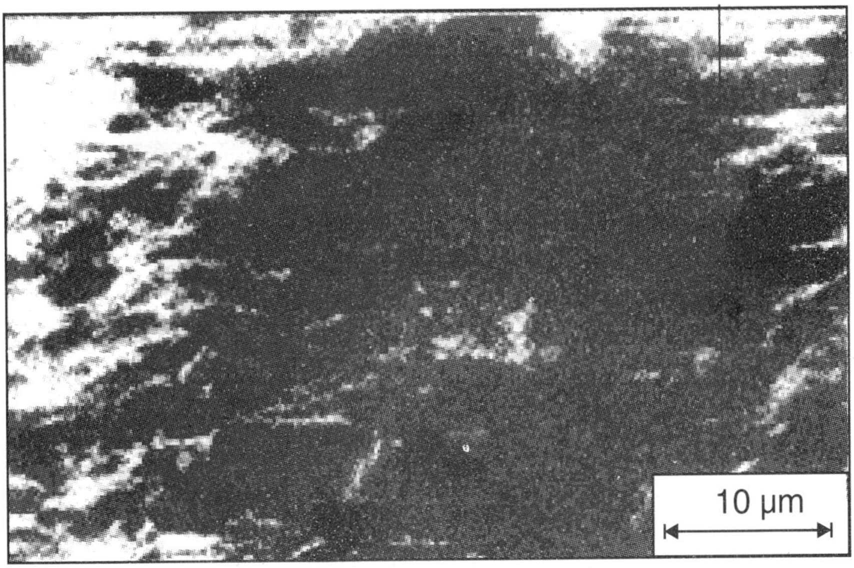

Figure 4.9: SEM photo of the fracture surface of α-silicon carbide.

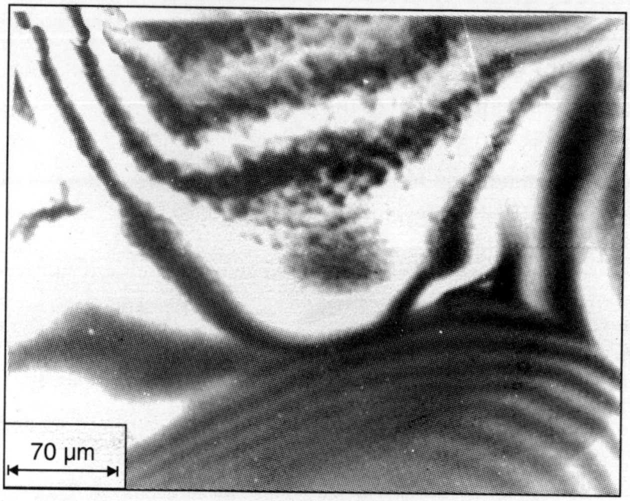

Figure 4.10: TEM photo showing the grain boundary of the sintered α-SiC.

In order to examine the effect of boron carbide, TEM photo is taken of a α-silicon carbide sample containing 3 wt% boron carbide and 1 wt% carbon, which was sintered at 2050°C for 15 minutes under vacuum (3 mbar). It is seen from **Figure 4.11** that sintered silicon carbide contained boron in its structure, excess boron carbide remained as a separate phase not at the grain boundary. Thus, for 0.5 wt% boron carbide doping level, all boron carbide and carbon must have been diffused uniformly into the silicon carbide grains. The absence of glassy or any other phase at the grain boundary is revealed by the electron diffraction pattern shown in **Figure 4.12,** whereby the presence of 'double dots' clearly indicate the inclusion of boron carbide inside the main SiC structure. This is also proven by the constant flexural strength up to 1400°C. It may be concluded that the sintering of silicon carbide with the addition of boron carbide and carbon is a solid state sintering process, which is clearly evident from the microstructural study and which is also supported by the flexural strength data.

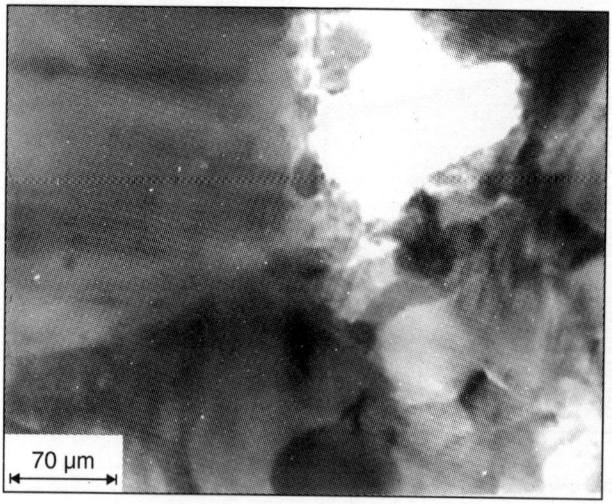

Figure 4.11: TEM photo of sintered α-SiC doped with 3 wt% B_4C and 1 wt% carbon.

For the mechanical properties of α-silicon carbide doped with 2 wt% aluminium nitride and 1 wt% carbon, the results are quite similar to those obtained for boron carbide doping in terms of microstructure involved in the interpretation of the data. However, the flexural strength is higher in case of AlN doping, since the value is 550 MPa, which is about 41% higher than that of boron carbide doping. The average fracture toughness is found to be 5.35 MPa.m1/2, which is higher by about 35% than that of boron carbide doping. The Weibull modulus (m) is also found to be little higher at 12.5 (10.6% higher) and 15.1 (13.5% higher) than those of boron carbide doping at room temperature and 1400°C respectively.

Figure 4.12: Electron diffraction pattern showing double spots for solid solution.

From the observed SEM and TEM micrographs as well as the electron diffraction pattern showing double dots (like in the case of boron carbide doping), it has been clear that AlN has entered into solid solution inside the structure of SiC. The grain boundaries showed no glassy phase or any other phase, indicating that the sintering is a solid-state diffusion process, which is also evident by the invariance of flexural strength with temperature up to 1400°C.

In summary, it can be said that α-silicon carbide doped with 0.5 wt% of B_4C + 1 wt% of carbon or 2 wt% AlN + 1 wt% carbon had a flexural strength of 390 MPa and 550 Mpa respectively by 4-point bending test. No degradation of strength occurred up to a temperature as high as 1400°C in both the cases of doping. The intra-granular fracture was found to occur. No glassy phase or any other phase was found at the grain boundary in both cases of doping The Weibull modulus of the sintered α-silicon carbide was 11.1 at room temperature and 13.3 at 1400°C for boron carbide doping, and those for AlN doping are 12.5 and 15.1 respectively, indicating a narrow distribution of flexural strength. This strength behaviour of sintered SiC with nano-crystalline particles show that it can be fruitfully used as 'ceramic engine components' and for many other engineering applications.

REFERENCES

1. E. Orowan, Inst. of Engg. & Shipbuilders in Scotland, 42 (1945-6) 165.
2. A. A. Griffith, Phil. Trans. of Royal Society, Series A, 221 (1920) 163.
3. W. D. Kingery, H. K. Bowen and D. R. Uhlmann, Introduction to Ceramics, John Wiley & Sons, New York (1976) pp. 783.
4. S. M. Wiederhorn, J. Am. Ceram. Soc., 50 (1967) 407.
5. G. R. Irwin, Encyclopedia of Physics, Vol. IV, Springer-Verlag, Berlin (1958) pp. 551.
6. W. B. Hillig and R. J. Charles, High Strength Materials, Ed. V. F. Zackay, John Wiley, New York (1965) pp. 682.
7. A. K. Varshneya, Strength of Glass in Chemistry of Glasses, Ed. A. Paul, Chapman & Hall, London (1982) pp. 140.
8. A. K. Bandyopadhyay and P. C. Ray, to be published in J. Mater. Sci.., (2005).
9. M S Datta, A. K. Bandyopadhyay, B. Chaudhuri, 'Sintering of nano crystalline α-silicon carbide doping with aluminium nitride', Bull. Mater. Sci., 25 (2002) 121.
10. M. S. Datta, A. K. Bandyopadhyay and B. Chaudhuri, 'Sintering of nano crystalline a-silicon carbide by doping with boron carbide', Bull. Mater. Sci., 25 (2002) 181.
11. M. S. Datta, A. K. Bandyopadhyay and B. Chaudhuri, 'Strength and microstructure of sintered α-silicon carbide preparaed from nano-crystalline particles doped with boron carbide' Trans. Ind. Ceram. Soc., 61 (2002) 75.
12. M. S. Datta, A. K. Bandyopadhyay and S. Maitra, 'Nano crystalline sintered α-silicon carbide for special applications', Indoceram, 39 (2002) 19-22.
13. W. Weibull, 'A statistical distribution function of wide applicability', J. Appl. Mech., 18 (1951)193.

14. J. D. Hong, R. F. Davis and D. W. Newbury, 'Self-diffusion of silicon-30 in alpha-SiC single crystals', J. Mater. Sci., 16 (1981) 2485.

15. M. Srinivas, S. G. Seshasdri and G. W. Weber, 'Microstructural effects on the mechanical properties of hexoloy Sa silicon carbide', Presented at the 82nd Annual Meeting of Am Ceram. Soc., Chicago, (1980).

16. M. Srinivas and R. H. Smoak, 'Elevated temperature strength and fracture toughness determination of sintered alpha silicon carbide', Presented at the Intl. Conf. on Fracture Mechanics in Engineering Applications, Bangalore (India).

17. K. A. Schwetz and A. Lipp, 'The effect of aluminium and boron sintering additives on the properties of dense sintered alpha silicon carbide', Special Ceramics (Vol. 10), Ed. P. Popper, Academic Press, New York, (1979) pp. 150.

18. K. A. Schwetz and A. Lipp, 'Dense sintered shaped articles of polycrystalline alpha silicon carbide and process for their manufacture', U. S. Patent, 4,230,497, (1980).

19. E. H. Kraft and S. W. Freiman, 'Mechanical response of high performance silicon carbide', 2nd Intl. Conf. on Mechanical Behaviour of Materials, Boston, (1976).

20. J. J. Mecholsky and S, W, Freiman, 'Determination of fracture mechanics parameters through fractographic analysis of ceramics', 'Fracture Mechanics Appllied to Brittle Materials', Ed. S. W. freiman, Am. Soc. for Testing of Materials, Philadelphia (USA), (1979) pp. 135.

21. R. W. Rice, 'Strength/grain size effects in ceramics', Proc. Brit. Ceram. Soc., 20 (1972) 205.

22. S. M. Wiederhorn, 'Crack propagation in polycrystalline ceramics', 'Ultrafine Grain Ceramics', Eds. J. J. Burke, N. L. Reed and V. Weiss, Syracuse Univ. Press, New York (USA), (1970) pp. 317.

23. S. Bortz, 'Reliability of ceramics for heat engine applications', 'Ceramics in High Performance Applications III, Eds. E. M. Leno, R. N. Katz and J. J. Burke, Plenum Press, New York (USA), (1983) pp. 445.

24. A. G. Evans, 'Structural reliability : A processing dependent phenomena', J. Am. Ceram. Soc., 65 (1982) 127.

25. A. G. Evans, 'Consideration of inhomogeneity effect in sintering', J. Am. Ceram. Soc., 65 (1982) 497.

26. F. F. Lange, B. I. Davis and I. A. Aksay, 'Processing related fracture origins III, Differential sintering of ZrO2 agglomerates in Al_2O_3/ ZrO_2 composite', J. Am. Ceram. Soc., 66 (1985) 497.

27. S. Dutta, 'Densification and properties of α-silicon carbide', J. Am. Ceram. Soc., 68 (1985) C269.

28. S. Dutta, 'Microstructure and property characterization of sintered Si_3N_4, SiC, SIALON', J. Am Ceram. Soc., 65 (1982) C2.

29. A. Ghosh, M. G. Jenkins, K. W. White, A. S. Kobayashi and R. C. Bradt, 'Elecvated temperature fracture resistance of a sintered a-silicon carbide', J. Am. Ceram. Soc., 72 (1989) 242.

30. H. Kodama and T. Miyoshi, 'Study of fracture behaviour of very fine-grained silicon carbide ceramics', J. Am. Ceram. Soc., 73 (1980) 3081.

31. S. Dutta, 'Strength distribution in commercial silicon carbide materials', J. Am. Ceram. Soc., 81 (1988) C474.

32. V. Raghavan, 'A First Course in Materials science', Prentice Hall India (P) Ltd., New Delhi, (1974) pp. 30.

Chapter 5

Magnetic Properties

PREAMBLE

Whenever we talk about magnets, we immediately think of 'metal' magnets. However, ceramic magnets called 'ferrites', based on the spinel structure (see later), have also been extremely popular for various technological applications. This encompasses a gamut of magnetic materials, which are of great interest to the materials scientists and engineers as well as to the electrical and electronics engineers dealing with various devices. While a complete book can be devoted to such a vast field, a very short description is given in this chapter on the basics of magnetism and then various types of materials, *i.e.* spinel based materials, are discussed. This is done in order to bring the ferrites, *i.e.* magnetite in this case, into focus in the form of nano-sized materials and their peculiar properties in the nano range within a diamagnetic glassy matrix.

5.1. INTRODUCTION

Let us recall Michael Faraday's important discovery, and immediately we can say that the magnetic fields arise from a change in electric field. This change might be produced due to the linear movement of electronic charge *e* (current) or due to the spin of *e*, or the change of the electric field strength E, *i.e.* an electromagnetic wave. The magnetic flux or induction (B) arises from the presence of the two terms as

A. Magnetizing Field Strength H [Ampere/meter], and

B. Magnetic Dipoles M [A/m].

The following relation holds

$$B = \mu_0.H + \mu_0.M = \mu . H$$

where, $\mu_0 = 4\pi.10^{-7}$ [Volt/(Coulomb.meter)] = Magnetic Permeability in Vacuum, μ [V/(C.m)] = Effective Magnetic Permeability of Material, H [A/m] = Magnetic Field Strength, B [$V.s/m^2$] = Magnetic Induction.

If any material is subjected to an external magnetic field (H), it exhibits the "aligning" of elementary 'dipoles', *i.e.* due to the spinning of a negatively charged electron around a positively charged nucleus and obviously around itself. This "effect" is called magnetization (M), and it is expressed according to the following formula

$$M = n . P = n . a . H$$

Thus, we can write it as

$$M = [\text{Number of Magnetic Dipoles } (n)] \times [\text{Magnetic Moment of one Dipole (P)}]$$

where, P = Magnetic Moment of an Elementary Dipole [$A.m^2$], n = number of the elementary dipoles per unit volume [m^{-3}]. Hence, the following relation holds as

$$\alpha = P/H$$

With our knowledge in quantum mechanics, as described in the **chapter - 1,** the smallest elementary magnetic dipole P_e arises from the movement of an electron of charge $e = 1.6 \times 10^{-19}$C and of mass $m = 6.63 \times 10^{-31}$ kg).

$$P_e = e \,.\, h/(4\pi.m) = 9.3 \times 10^{-34}\ [A.m^2]$$

where, h = Planck's Constant = 6.63×10^{-34} J.s, $P_e = \mu_B$ = Bohr Magneton.

It is also known that the "permanent magnetic dipoles" arise when unpaired electrons exist in an atom, as in ions. The magnetic susceptibility is defined as a dimensionless quantity as the ratio of magnetization to magnetic field strength

$$\chi = M/H$$

Some materials exhibit "spontaneous magnetization" (M), *i.e.,* M > 0 even without an external magnetic field. Depending on the type of spin alignment, these are classified as

A. Ferromagnetic

B. Ferrimagnetic

C. Antiferromagnetic

Generally speaking, depending on the response of a material to the externally applied magnetic field, all materials can be classified into the above category of magnetic materials, apart from diamagnetics and paramagnetics. These are described very briefly in the following subsections.

5.1.1. Diamagnetics

For all materials, this effect arises due to the modification of electron movement in an external magnetic field (H). The modification is always such that the resultant magnetization counteracts the applied field, which is known as the Lorentz Law. This diamagnetic effect is very small but 'negative'. It is usually possible to be detected only if no permanent dipoles are present in the material, *i.e.* the material does not contain any unpaired electrons. The diamagnetic susceptibility is given as

$$\chi_D \approx -10^{-5} \text{ to } -10^{-6} < 0$$

5.1.2. Paramagnetics

A material is considered paramagnetic due to the alignment of the permanent dipoles arising out of m unpaired electrons in an external magnetic field H. The alignment of 'paramagnetic ions' is strongly opposed by the random thermal vibrations, *i.e.* quantized phonons, which give rise to the misalignment of these magnetic dipoles. This particular paramagnetic effect is very small but 'positive'. Here, it is usually possible to be detected only if no 'spontaneous' alignment of the magnetic dipoles, *i.e.* no ferromagnetic interaction of the unpaired electrons, takes place in the material. The paramagnetic susceptibility is given as :

$$\chi_P \approx +10^{-5} \text{ to } +10^{-6} > 0;$$

$$\chi_P = C/T\,,$$

which is the famous 'Curie Law'. Also, the magnetization can be expressed as :

$$M = m.n.\mu_s$$

5.1.3. Ferromagnetics, Ferrimagnetics and Antiferromagnetics

The effect of ferromagnetism arises due to the 'spontaneous' coupling of the permanent magnetic moments of the individual ions. A collective alignment results even in the absence of an external magnetic field within small regions of material called "domains", which are ~ 1-10 μm in size. The width of these domains play a vital role in the ferromagnetic property or rather in the 'spontaneous polarization' of the magnetic moments in the materials.

The ferromagnetism defines "positive spin interactions" when the spins are aligned in one single direction and also resulting in parallel alignment of the spins. The anti-ferro-magnetism defines "negative spin interactions" with the spins aligned in opposite directions and a complete cancellation of net magnetization results. The ferrimagnetism is actually antiferromagnetism with incomplete cancellation of magnetic dipoles due to spin alignment, or rather misalignment. (See **section 5.4.4**)

In magnetic materials, different types of spin alignment is possible, as shown by a diagram in the **section - 1.4.1,** which depends on a number of factors. This kind of spin alignment can happen if the following possibilities occur :

A. Two ions having different magnetic moments M_1 and M_2 interact negatively, but the cancellation is incomplete due to a difference in the moment, *i.e.,* $M_1 \neq M_2$

B. In the magnetic crystals, two sub-lattices interact, containing even similar ions, when $M_1 = M_2$, but the cancellation of the magnetic moments is incomplete due to the difference in the number of ions in the two sub-lattices, since the number of ions is also important.

The spontaneous spin alignment in ferrites is opposed by thermal vibrations in the lattice. At a critical temperature T_C, called the 'Curie Temperature', the ferromagnetic materials become paramagnetic. The magnetic susceptibility at $T > T_C$ is described by the 'Curie-Weiss Law' as :

$$\chi_P = C/(T - T_C)$$

Most ceramic magnets, *i.e.* ferrites, are ferrimagnetic with a crystal structure based on inverse or partially inverse spinels. So, the description of the 'spinels' need to be given here.

5.1.4. The Spinels

The spinel has the molecular formula AB_2O_4, and it has 8 molecules in its unit cell = $A_8B_{16}O_{32}$, A is a divalent ion (*e.g.* Mg^{2+}, Zn^{2+}, Fe^{2+}, etc.) and B denotes a trivalent ion (Al^{3+}, Fe^{3+}, etc.). In normal spinel structure, a unit cell contains 32 oxygen ions, 32 octahedral B sites, which are half filled by the B ions and 64 tetrahedral A sites that are 1/8 filled by the A ions. The typical example of some of the normal spinels are

$MgO.Al_2O_3 = MgAl_2O_4$, $ZnO.Fe_2O_3 = ZnFe_2O_4$, $FeO.Al_2O_3 = FeAl_2O_4$,

$CoO.Al_2O_3 = CoAl_2O_4$, $MnO.Al_2O_3 = MnAl_2O_4$, $NiO.Al_2O_3 = NiAl_2O_4$.

The inverse spinel has the general formula $B(AB)O_4$. The whole of A ions and half of B ions are located in the octahedral sites, and the other half of B ions are placed in the tetrahedral sites. Most of the "ferrites" have this inverse spinel structure. A typical example of some of the inverse spinels are

$MgO.Fe_2O_3 = FeMgFeO_4$, $NiO.Fe_2O_3 = FeNiFeO_4$, $CoO.Fe_2O_3 = FeCoFeO_4$,

$FeO.Fe_2O_3 = FeFeFeO_4 = Fe_3O_4 = Fe^{3+}(Fe^{2+}Fe^{3+})O_4$

(The Most Important Ferrite)

In the above spinels, there is only one type of divalent metal ion or only one kind of trivalent metal ion, but in reality, there are also mixed spinels showing interesting magnetic properties for various applications. The composition of such mixed spinels can vary. Obviously, there are various ways these multivalent ions can be distributed or re-distributed inside the spinel structure. The general formula is $(A_{1-x}B_y)(A_xB_{2-y})O_4$, *e.g.* Mg-Zn ferrite, Mn-Ni ferrite, etc. Some more details is necessary to be given for these types of mixed ferrites, as given below.

As said earlier, the ferrite of spinel structure, represented by AB_2O_4, have generated a lot of research interest due to their tremendous applications in nano science and technology. It is also known that the physical properties of spinel ferrites depend on the distribution of cations amongst the tetrahedral (A) and octahedral (B) sites, and the relative strengths of various kind of 'super-exchange interactions' via oxygens. It is noted that most of the works are confined on spinel ferrites MFe_2O_4 (M = Zn, Co, Mn, Ni etc.). In these spinels, B sites are occupied by Fe^{3+}. However, there are many spinels of different class and having no B site Fe^{3+}, but may be relevant in view of physics involved and the technological applications, *e.g.* the chromites MCr_2O_4 (M = Mn, Fe etc., ions) are one of such classes which exhibited many unusual magnetic properties like ferrimagnetism, half-metallic behaviour, colossal magneto resistance (CMR) effect etc.

One reason of predicting HFM in spinel is its typical 'spin structure' in A and B sites (sub-lattices), where the spins inside each of the A or B site are assumed ferromagnetically ordered, but two sublattices are antiparallel with respect to each other. For example, those electrons with 'spin up' are metallic, while those with 'spin down' are insulator. It has been proposed that $Mn[Cr_{2-x}V_x]S_4$ series might be a strong candidate for HFM (see later, Bhowmik [20] for important references). However, the experimental investigations indicate that bulk spinel $Mn[Cr_{2-x}V_x]S_4$ may not be stable in cubic spinel structure for $x > 0.6$, when the sample is prepared by a solid state method. By considering the 'site exchange of cations' in spinel nano particles with drastically different kind of behaviour in comparison with its bulk form and the stability of the structure, it is suggested that the production of nano particles may be an alternative approach for the stable structure of $Mn[Cr_{2-x}V_x]S_4$ series.

The magnetic and transport properties of MCr_2S_4 are quite stimulating. It is interesting to look for the similar kind of magnetic and transport properties in MCr_2O_4 (M = Mn, Fe etc.). Due to both (sulphide and oxide) spinels have been identical with cubic lattice structure, and both are ferrimagnet with T_C in the range (60°K-80°K) and non-collinear ferrimagnetic order starts below 20°K. The other genuine interests for investigating the $MnCr_2O_4$ spinel is to understand the role of strongly negative 'exchange interaction term', *i.e.* J_{BB} (Cr-O-Cr) interactions in controlling the magnetic properties in the chromite spinels.

In the spinel structure, the superexchange interactions will be important for the collinear configurations A-O-B with angle 125° and B-O-B with angle 90°. For other configurations, the distance between the oxygen ion and cation is too large to give rise to a strong A-O-B interaction. It can be said that direct (antiferromagnetic) interactions between B site cations are possible in compounds like $MnCr_2O_4$. In fact, many phenomena in recent years have been explained by the B site direct cation-cation interaction, as in $ZnCr_2O_4$, and similar to many other Cr-spinels, the non-collinear spin structure between B site Cr^{3+} moments has been confirmed by neutron diffraction experiment in $MnCr_2O_4$. Numerous attempts have been made to understand the non-collinear ferrimagnetic structure. However, the picture of spin configuration for chromites are not very clear even upto date (see the **section 5.4.4**).

The magnetic behaviour of the bulk and mechanically milled $MnCr_2O_4$ nano particles need to be studied in order to examine the possible change in B site spin configuration with the reduction in particle size and its correlation with the magnetic properties (see ref [20]). This sums up a short description of various types of spinels and there interesting magnetic properties which appear to be quite complex, but by making nano particles of different spinel structures and by studying their different magnetic properties as a function of both temperature and magnetic field, we might be able to reveal some important details, as shown in the subsequent sections.

5.1.5. Losses due to Eddy Currents in Magnetic Materials

Although in this chapter, we are no way connected with the 'losses' in ferrites, while dealing with nano materials. But, a brief description of 'losses' of ferritic materials is given here to highlight the point that this factor has to be taken into account while designing various electrical and electronic devices, and of course various compositions of magnetic materials.

The ferrites are good magnets due to their high magnetization M and good electric resistivity, *i.e.* high electrical insulation. If the electrical resistivity of a magnetic material is low, *e.g.* in magnetic metals, 'high frequency magnetic fields' would create 'high frequency electric fields', which induce 'loss currents', which are called 'eddy currents'. In some of the rare instances, the eddy currents can be useful, *e.g.* in induction heating in 'high frequency induction' furnaces for melting various metals including special alloys for hi-tech applications in aerospace and other industries. But, in most of the cases, the eddy currents cause a "loss of energy" and "undesirable heating".

The power lost in the magnetic material as heat = V . I = V^2/R, where R = resistance of the material, I = eddy current, and V = induced voltage ~ dB/dt. Let us compare the typical resistivity values for ferrites and metallic magnets : for cobalt ferrite, $CoO.Fe_2O_3$, the resistivity $\rho = 10^7$ ohm-cm, while for metallic iron (Fe), $\rho = 10^{-5}$ ohm-cm. Thus, for the same magnitude of induced voltage, the losses in the ceramic magnet (*i.e.* ferrite) will be 12 orders of magnitude smaller, as compared to the metallic magnet. This mean that the resistivities, and hence the 'power losses', differ by a factor as large as 10^{12} between metallic and ceramic magnets.

Since the power loss is significantly reduced in ferrites, this low loss factor is successfully used for making 'high frequency magnets', *i.e.* ferrites, in electronic equipment. It is useful to mention some of the applications of this important class of materials as follows

1. Audio-Visual Recording Heads (here, both the chemical resistivity and wear resistance of ceramics are also considered advantageous)
2. Deflection-Yoke Cores and Convergence Coil Cores (Television Circuitry)
3. Transformers and Magnetic Core Memories
4. Antennas based on Ferrites
5. Information Recording and Storage (*e.g.* Magnetic Tapes and Disks)

5.1.6. Structural Ordering of Ferrites

The preference of certain metal cations to octahedral B or tetrahedral A coordination in FCC oxygen lattice is dictated by the 'crystal field theory'. The particular preference of an ion for octahedral or tetrahedral coordination decreases in the following order

$$Cr^{3+} > Mn^{3+} > Ni^{2+} > Cu^{2+} > Al^{3+} > Co^{2+} > Mg^{2+} > Ti^{3+} > Fe^{2+} > Fe^{3+} > Mn^{2+} > Zn^{2+}.$$

Although our knowledge of magnetic properties and magnetic materials have grown up with time, the secrets of the extensive "class of ferrites" are sill largely hidden. Due to the above order, certain conclusions can be made as follows

A. Most of spinels which contain aluminum are normal, except for those of nickel and copper.

B. The proximity of the Fe^{2+} and Fe^{3+} ions indicates that they can easily interchange their positions between the A and B sites.

C. Nickel, cobalt, copper, and magnesium ferrites are "predominantly" inverse, *i.e.,* divalent ions prefer octahedral coordination, but in certain instances, these ions occupy normal positions as in mixed spinels.

D. Zinc spinel is a normal spinel.

E. Some sort of structural disorders have been detected in spinels, *e.g.,* a shift of oxygen and the metal cations in the B sites can create local permanent dipoles in ferrites, which could interact spontaneously in the anti-parallel orientation.

5.1.7. The Mechanism of Spontaneous Magnetization of Ferrites

Despite the separation of different metal cations by oxygen ions, a spontaneous magnetization and also a non-zero magnetic moment are shown by the ceramics magnets like ferrites. The reason for this behaviour should be explored. It can be considered that the electronic configuration of the outer shells of the elements is the most important in the structures of ferrites.

Each unpaired electron spin produces a magnetic moment, measured as 1 Bohr magneton ($1\mu_B$). In pure ferromagnetic materials such as metallic Fe, Ni, and Co, the spontaneous alignment of the spins takes place for the neighbouring ions, with a cooperative interaction of electrons that is necessary to result in the ferromagnetic effect.

Since the metal ions are separated by oxygen and consequently, there is a lack of direct exchange interactions, *i.e.* the overlapping of the wavefunctions of the atomic orbitals is absent. However, the 'spin interaction' still takes place in oxides of ferromagnetic metals, despite the above facts. It has been shown that cations in spinels can interact toward anti-parallel alignment of unpaired electron spins, despite the separating oxygen anions, which is known as the super-exchange interaction model.

This particular model has been proposed between the 3*d* orbitals of metal ions with unpaired electron spins and the 2*p* orbitals of oxygen ions. The distribution of ions, like Fe, Ni, Co, Mn, and Cu with unpaired outer electrons, between the tetrahedral (A) and octahedral (B) sites of the oxygen sub-lattice dictates the 'condition of criticality' of such super-exchange interactions. The potential energy diagram of such ions within a given atomic environment involving such interactions is very important for the equation of motion, which depends strongly on the type of site in which the cation sits.

For a particular ion to ion separation, the shape of the *p*-orbital of the oxygen ion suggests that this interaction is the strongest when the metal-oxygen bond angle ϕ is close to 180° :

A. For the interaction A-O-B, $\phi = 126°$ and 154°

B. For the interaction A-O-A, $\phi = 79°$

C. For the interaction B-O-B, $\phi = 90°$ and 125°

In the normal spinels containing iron, *e.g.* $FeAl_2O_4$ with all the Fe^{2+} ions occupying tetrahedral A sites and $ZnFe_2O_4$ with all the Fe^{3+} ions occupying octahedral B sites, both the Al^{3+} and Zn^{2+} have paired spins, since these ions are non-paramagnetic with $\mu_B = 0$. Hence, the A-B interaction cannot give any possible spin alignment. Moreover, the Fe-Fe interaction through A-A or B-B is too weak to give any alignment resulting in no net magnetic moment in normal spinels containing non-paramagnetic ions accompanied by iron.

But, in inverse spinels containing iron, an equal number of Fe^{3+} ions occupy the tetrahedral A sites and the octahedral B sites, and therefore the interaction (A-O-B) is obviously strong. As a result, there is an anti-parallel spin alignment with no 'net resultant magnetic moment' due to the iron ions. However, the inverse spinel can be doped with MO (where, M stands for a divalent metal ion). Here, M^{2+} ions have a preference for the octahedral sites, such that some of the octahedral B sites are occupied by M^{2+} ions. The strong (A-O-B) interaction of the M^{2+} ions, in the octahedral sites with the Fe^{3+} in the tetrahedral sites, results in a net uncompensated magnetic moment. This moment is proportional to the number of unpaired electrons in the M^{2+}, which means that all the Fe^{3+} alignment is still anti-parallel.

All polymorphs of iron oxide are based on cubic close packing (fcc) of oxygen, with cations filling octahedral (B) and/or tetrahedral (A) holes. Among the polymorphs are : Magnetite Iron Oxide $Fe_3O_4 = Fe_{24}O_{32}$, Gamma Iron Oxide $\gamma\text{-}Fe_2O_3 = Fe_{21.33}O_{32}$ and Wuestite Iron Oxide $FeO = Fe_{32}O_{32}$. The former oxide, *i.e.* magnetite, is a very important magnetic material.

It is an inverse spinel. It has 32 oxygens and 24 cations : 8 divalent ferrous Fe^{2+} ions in the available octahedral B sites, and 16 trivalent ferric Fe^{3+} ions distributed equally between the tetrahedral A and octahedral B sites. Thus, the net magnetic moment of magnetite results from the unbalanced spin alignment of Fe^{2+}. For different applications of ferrites, it is important to consider the magnetization and consequent hysteresis behaviour, *i.e.* B-H curve.

5.1.8. Magnetization of Ferrites and Hysteresis

Magnetization M is a function of the strength and direction of the magnetizing field H. From the point of view of the microstructure, increased magnetization results from growth of the domains within the grains of ferrite ceramics, which are oriented parallel to the applied magnetic field. The 'hysteresis loop' is the most important aspect of ferrite. The characteristics of the hysteresis loop are an effect of materials process engineering of ferrites. Various characteristics of the hysteresis loop are of particular importance, depending on the application for which the ferrite will be used. A typical diagram is shown in **Figure 5.1.**

The following parameters are important in the interpretation of the 'hysteresis loop' :

1. **Initial Permeability**

$$\mu_0 = \left(\frac{dB}{dH}\right)_{H=0}$$

This results from the 'reversible displacement' of the domain boundaries without rotating the domains. If H disappears, B disappears, and the domains return to their original random positions. It is important in the detection of weak magnetic fields, as in antennas, where the permanent (residual) induction is undesirable.

2. **Maximum Permeability**

$$\mu_{max} = \left(\frac{dB}{dH}\right)_{max}$$

This is the 'irreversible displacement' of the domain boundary without rotation of domains.

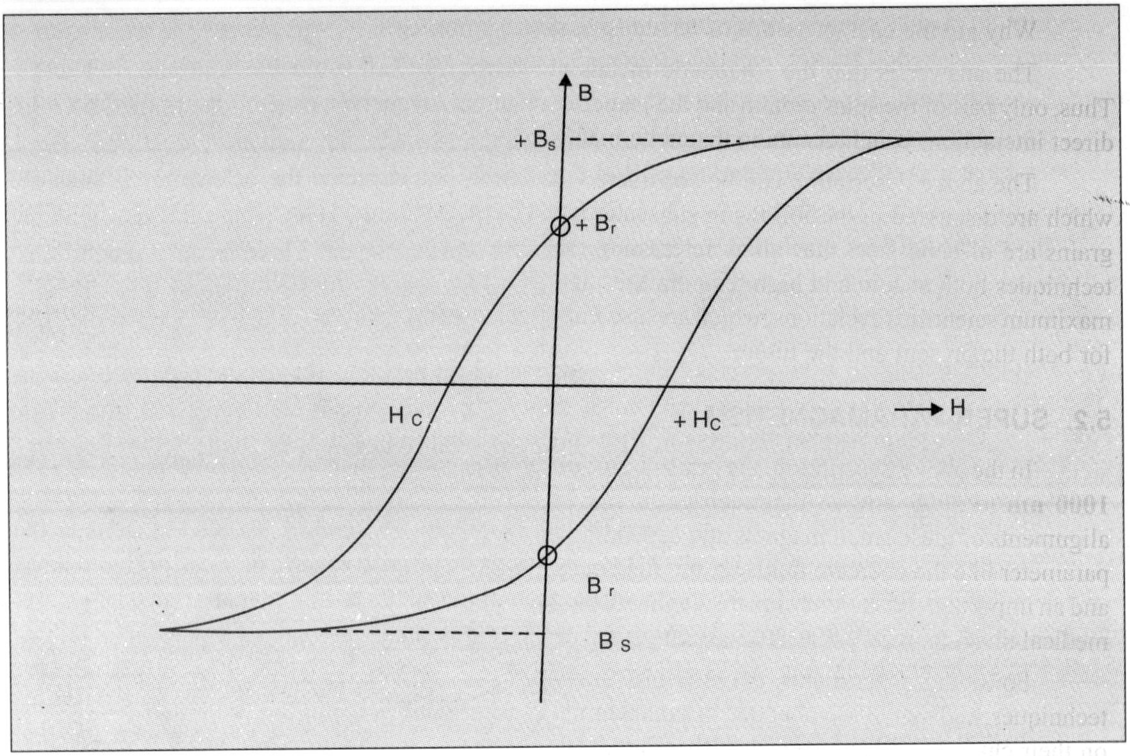

Figure 5.1 : Magnetic hysteresis curve of a typical magnet.

3. **Saturation Induction (B_S) and Saturation Permeability (μ_S)**

At this point of magnetization, all domains are rotated in the direction of the applied field. Both parameters are important for magnetic field generation and both parameters are 2-5 times lower for ferrites than for metallic magnets. Typically for ferrites, the saturation induction is 4,000 to 5,000 Gausses, and the permeability can vary widely from 2,000 to 18,000. The saturation induction for strong metallic magnets (cobalt-iron, silicon-iron, etc.) is about 20,000 Gausses, and the permeability 1,000 to 2,000. High-nickel magnets (iron-80%Ni) have relatively low saturation induction (7,500 Gausses), but very high permeability reaching 100,000.

4. **Remnant Induction (B_r)**

B_r is low for non-permanent soft ferrites, but is high for permanent magnet (hard) ferrites. Hence, the value of the remnant induction is used to decide on the type of magnet.

5. **The Area of Hysteresis Loop**

This area determines 'total energy' required for a full magnetization-demagnetization cycle. B.H represents a net loss in the system, usually in the form of heat. A "square" thin and tall loop is preferable in high-frequency applications, since the loss is low.

Additionally, $\Delta B = (B_S - B_r)$ determines the power losses in high frequency ferrites. The low coercive field for soft ferrites, *i.e.* $H_c < 80$ A/m, determines their application in dynamic fields. These ferrites could be used in high frequency inductors, transformers, switching cores for memory applications, or magnetic elements in microwave applications.

An analysis of ferrite's hysteresis loop brings us to this important question

Why are the characteristics of ferrite so different from those of magnetic metals?

The answer is that the magnetic metals are ferromagnetic, while ferrites are ferrimagnetic. Thus, only part of the spins remain and the 'super-exchange interaction' in ferrites is more difficult than direct interaction, which occurs in magnetic metals.

The above description can be considered sufficient to understand the behaviour of magnetite, which are discussed in the following subsections, wherein it will be evident that when the magnetite grains are of nano size, they show interesting magnetic properties. This is possible by a variety of techniques both at low and high temperatures and as a function of applied magnetic fields to give a maximum scientific revelations, which are also important to understand many technological applications for both the present and the future.

5.2. SUPER-PARAMAGNETISM

In the above description, the grain size of different ferrites is in the micron range, *i.e.* from about **1000 nm** to **5000 nm,** so that each grain can be considered as small magnet. Due to the various alignments of these small magnets and depending on various other parameters, an important hysteresis parameter like the coercive field (*i.e.* the field needed to demagnetize it) is decided which gives insight and an important indication for many applications. For many other delicate applications like sensors and medical imaging, a very fine grained magnetic particles are needed.

For doing experiments, the physicists prepare these fine-grained magnetic material by various techniques, and then it is embedded in a diamagnetic or inert carbon matrix in order to do further studies on their characterization. However, there is also an 'elegant way' of creating these small particles of magnetite at ease by controlling the time and temperature of heat-treatment of amorphous or glassy material containing iron oxides in order to precipitate the small crystallites in the nano range, *i.e.* ultra-fine particles of magnetite. This is the subject matter of this chapter, wherein the behaviour of the small grains of magnetite crystals in the nano range is created within a 'diamagnetic' basalt glass matrix.

Basalt forms a black homogeneous glass when cooled from the molten state. Subsequent re-heating above the annealing temperature results in the development of fine-grained glass ceramics [1]. In order to control the quality of glass-ceramics, it is important to study the nucleation and crystallization behaviour of this system, which has been done by 'Small Angle Neutron Scattering' (SANS), as described in the **section 5.7** [2-5]. Since the glass contains iron oxide and the glass-ceramic has magnetite as a major crystalline phase, the magnetic properties of this system are also interesting.

As the heat-treatment temperature or time, or both increase, the magnetic property of Fe ions participating into a magnetite phase shows 'different magnetic behaviour' depending on the particle size, since it depends on the heat-treatment schedule. This behaviour is also critically dependant on the 'temperature' at which the 'magnetic measurements' are taken [6-9].

These small nano particles of magnetite are "superparamagnetic" below the magnetic transition temperature of the bulk magnetite due to a 'fast' relaxation of the magnetization of the particles [10]. These alignments or re-alignments are constantly occurring inside the material depending on the temperature and the particle volume, but the time for this purpose known as the 'relaxation time' is very fast at room temperature at 300°K (*i.e.* short relaxation time). However, it becomes less faster, *i.e.* the relaxation time is longer when the material is cooled down towards liquid helium temperature at 4°K. It is of fundamental importance to detect these nano particles of magnetite showing super-paramagnetism with very fast relaxation to understand their interesting behaviour.

As said above, this also depends on the type and temperature of measurements. For example, at a given temperature below the so-called blocking temperature (T_B), the small nano particle might appear as 'ferrimagnetic' (slower relaxation), but at some other temperature higher than T_B, the same particle will show 'super-paramagnetism' (faster relaxation) [10]. The similar effect can be observed with the volume of the nano particle, *i.e.* the dimension of the nano particles of magnetite or any other magnetic phase. The smaller the particle, the faster is the relaxation so that the particles appear to be super-paramagnetic, and vice verse. In the nano range of particle sizes, it is quite obvious to be able to observe the super-paramagnetic behaviour, but a strong dependence of the magnetic behaviour on the 'criticality' of the particle size even within a narrow range of "nano sizes" is quite extraordinary, as described later in this chapter.

As shown in the **section 5.5,** the magnetic characteristics of the basalt glass were identified as due to pyroxene (Fe^{2+}, Mg) SiO_3 and to particles of magnetite distributed in small proportion inside the non-magnetic silicate matrix. In pyroxene, the iron ions having magnetic moment $\mu_{Fe^{2+}}$ seem to behave as 'free magnetic dipoles', which is well described by Langevin's theory of paramagnetism. According to Stoner [11], the value of $\mu_{Fe^{2+}}$ is approximately 540 emu/g of iron. According to the Langevin's theory, the 'specific magnetic moment' (M) of a given material is written as

$$M = N\mu^2 H / 3k_B T \tag{5.1}$$

where, N, H, k_B and T denote the number of Fe^{2+} ions, the applied magnetic field, the Boltzman's constant and the temperature respectively. In this case, as described in the **section - 5.5,** the value of N can be measured by the Lorentzian analysis of the 'intensity lines' of the Mössbauer spectra, so that the magnetic contribution of Fe^{2+} ions can be evaluated at every H and T value.

The magnetite has an 'inverse spinel' (cubic) crystal structure [12]. It is a typical ferrimagnetic material and the Curie temperature is 851°K with the saturation magnetization as 92 emu/g and 98 emu/g at 300°K and at 0°K respectively (*i.e.* bulk magnetite). However, if we deal with very small particles of magnetite, where each of them has a "single domain" structure and the thermal energy at the temperature of experiment can be sufficient to equilibrate the magnetization of 'assembly of particles' in a time short compared to that of this experiment, *i.e.* this experiment cannot 'arrest or rather detect' the particular (ultra-fine) particle. This behaviour has been termed "super-paramagnetic", as explained above in terms of the 'relaxation time'.

The 'group paramagnetic' behaviour of "single domain" small particles of magnetite, *i.e.* super-paramagnetic, shows a 'large magnetic moment'. The total 'magnetic moment' of such an assembly of non interacting magnetic particles above the so-called blocking temperature (T_B) contained in a specimen of volume (V_S) is given by :

$$M(H_{ext}, T) = \int_0^{V_s} N(V)\, M_S(T)\, V L(\alpha) dV \tag{5.2}$$

where, $M_S(T)$ is the saturation magnetization at temperature T, N(V) is the number of particle of volume V in the specimen, $\alpha = M_S(T)VH_{ext}/k_BT$ and $L(\alpha)$ is the well-known Langevin function.

It is important to mention here that certain 'asymtotic' behaviour of equation (5.2) is useful. When the single domain particle has an 'anisotropy' of cubic symmetry as in bulk magnetite, the magnetization can be described by the following expression for any direction of the applied magnetic field, as $H_{ext} \rightarrow 0$

$$M(H_{ext}, T) = \frac{M_S(T)^2 H_{ext}}{3k_B T} \int_0^{V_S} N(V)V^2\, dV = \frac{NM_S(T)^2 H_{ext} \overline{V}^2}{3k_B T} \quad ...(5.3)$$

and, as $H_{ext} \rightarrow \infty$

$$M(H_{ext}, T) = \int_0^{V_S} N(V)M_S(T)\, V dV - \frac{k_B T}{H_{ext}} \int_0^{V_S} N(V)\, dV$$

$$= NM_S(T)\, \overline{V} - \frac{Nk_B T}{H_{ext}} \quad ...(5.4)$$

where, $\overline{V}$ is the mean volume of the magnetite particles.

When M is independent of temperature, the equation (5.2) indicates that the magnetization (M) is a function of H_{ext}/T, *i e.* the magnetization curve at different temperatures should 'superpose' on a H_{ext}/T plot. This means that the magnetization of the super-paramagnetic particles adheres to the classical Langevin function as

$$M = M_S[\coth(\mu H_{ext}/k_B T) - (k_B T/\mu H_{ext})] \quad ...(5.5)$$

and it does not show any 'hysteresis loop' in the B vs. H plot.

For the magnetic particles in 'thermal equilibrium', *i.e.* in a super-paramagnetic state, the initial magnetic susceptibility is sensitive to the larger particles as per equation (5.3), whereas the 'approach to saturation' is governed by the smaller particles. When the magnetization curve at a given temperature shows a 'hysteresis loop', the remanence magnetization (M_R) should be a measure of the amount of magnetic material with particle volume greater than that in 'thermal equilibrium'. According to Jacob and Bean [13] in this case. M_R is qual to ½ M_S, where M_S is the value of the magnetization in the saturated state for an 'assembly' of independent particles, which are oriented at random.

For the 700 and 900 samples, the magnetization measurements were made between 4 – 300°K. The M vs. H_{ext}/T plot is shown in **Figure 5.2** only for the 700 sample. It is seen that the magnetization curves 'superpose' well both at 77°K and 300°K, and that they clearly adhere to Langevin function of equation (5.5) (see later in the **section 5.4** for detailed analysis). The shape of the curve is very similar to that obtained for small magnetite particles embedded in an inert carbonaceous matrix [14]. This confirms the 'super-paramagnetic' behaviour of the nano particles of magnetite even at 77°K. This also shows that the amount of 'ferrimagnetic' particles that are also of nano dimension and hence its contribution is quite small.

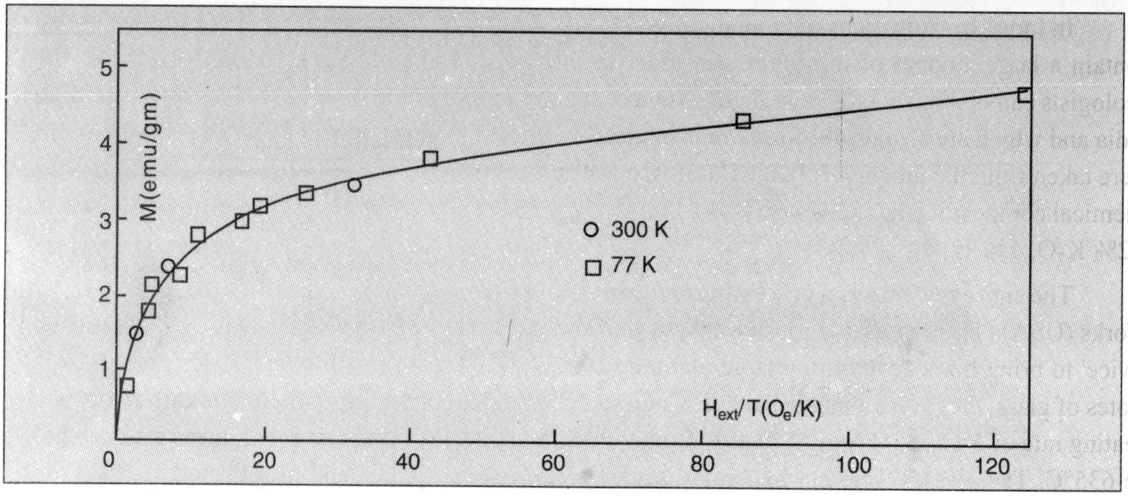

Figure 5.2 : Superposition of the magnetization curves as a function of H_{ext} /T both at 300°K and 77°K for the 700 sample.

This is despite the fact that Mössbauer spectra at 77°K shows a six-line magnetic hyperfine patterns (typical of a 'ferrimagnetic' material) and the ESR spectra at 77°K shows a 'ferrimagnetic' resonance for this 700 sample, as described later in the **sections 5.5 and 5.6.** The reason for this behaviour is that the relaxation time (τ) is much lower than the time (τ_{obs}) for the magnetization measurement at 77°K, so that the nano particles of magnetite 'appear' super-paramagnetic at 77°K, as pointed out above in this section showing the importance of different measurements.

However, no 'hysteresis loop' is observed at 77°K, indicating that the "blocking temperature" for the 700 sample is still lower than 77°K. It is noteworthy that the value of M_S at 4°K for the 700 and 900 samples are 6.73 and 5.97 emu/g respectively. This shows that the number of Fe ions in the 'ferrimagnetic' state is slightly higher in the 700 sample, as also revealed by the Mössbauer data at 4°K.

With a brief description of the method of preparation of the material in the next **section 5.3,** *i.e.* preparation of basalt glass ceramic with ultra-fine particles of magnetite, the magnetization data based on the above theoretical analysis will be presented in the **section 5.4.** Subsequently, both the Mössbauer and ESR spectroscopic data will be presented along with some experimental procedures in the **sections 5.5** and **5.6,** which is to be concluded in the **section 5.6,** in order to give a complete description of "nano super-paramagnetic particles" of magnetite in a diamagnetic glassy matrix.

5.3. MATERIAL PREPARATION

It is known that basaltic rocks are available in abundance in nature, *i.e.* on the earth's crust. Many years ago, the magmatologists of France were interested to work with the materials scientists like us to explore how the magmas were fast-cooled to a 'glassy' or 'crystalline' state onto the earth's crust. The idea was to know about the 'history' of our planet's creation through the creation of such materials available in nature. Since, this basaltic rocks contain a large amount of iron oxide, it must have some kind of magnetic property and it is definitely possible to extract some useful information on our planet through magnetization measurements on such materials. By this brief account, the readers would appreciate the importance of such materials in magmatology or, geology. However, no attempt is made here in this book to go towards this interesting subject, since we are mainly concerned with nano materials.

In India, basaltic rocks are found in the Deccan range of mountains, and the Deccan basalts also contain a large amount of iron oxides in order to justify experimental work on magnetization. The geologists can obviously talk more on this subject and can identify other basalts, which are available in India and which are also useful for various interesting studies. For the present work, the basaltic rocks were taken from the famous Holyoke Basalt Flow at Westfield in the state of Massachusetts (USA). Its chemical composition is : 52.0% SiO_2, 14.1% Al_2O_3, 12.8% Fe_2O_3, 9.3% CaO, 6.4% MgO, 3.2 Na_2O, 1.2% K_2O, 1% Ti_2O (wt.%).

The entire raw material processing and consequent glass fabrication were done at Corning Glass Works (USA) [1]. The basaltic rocks were crushed and ground to a fine powder, and then it was melted 'twice' to bring homogeneity in a large platinum crucible at 1500°C for 16 hours. After casting large plates of glass, they were annealed at 525°C. First of all, a dilatometric experiment was carried out at a heating rate of 3°C/min. From $\Delta L/L_0$ vs. T plot, the glass transition temperature (T_g) was estimated to be 635°C. The glasses were cut into small square (1 cm × 1cm) pieces and heat-treated at 600, 650, 700, 800 and 900°C respectively for 8 hours (5 samples) so that crystallization and other studies could be performed. There were mainly six samples including the as-annealed glass, which is called a 'blank glass'.

5.3.1. Nano Particles and X-ray Data

On all the above six samples, the X-ray diffraction measurements were carried out, which is based on the intensity of diffracted beam [I(θ)] at different angles of diffraction (2θ) following Bragg's diffraction Law : $n\lambda = 2d \sin\theta$, where λ is the wavelength of CuK_α radiation, and *d* is the inter-atomic distance. It was observed that there is no crystalline peak in the 'blank glass', 600 and 650 samples. The peaks due to "magnetite" (Fe_3O_4) having an inverse spinel structure appeared for the 700 sample. The strongest peak was at *d*-spacing of 2.50 Å.

For the 800 and 900 samples, there were peaks due to magnetite and also due to pyroxene ($CaMgSi_2O_6$) giving the strongest peak at *d*-spacing of 2.97 Å plus some other minor phases. However, the 'Transmission Electron Microscopy (TEM) showed that magnetite was present in the 650 sample also and obviously for 700 sample, while the last two samples contained a mixture of phases. From the X-ray data, no quantitative estimate of the amount of magnetite or pyroxene was made. However, from the well-known Scherrer equation, the particle sizes of magnetite were estimated from the respective magnetite peaks at 2.50 Å for 650, 700, 800 and 900 samples, which were found to be **4.5 nm, 5.5 nm, 6.4 nm** and **7.0 nm** respectively. A detailed study of 'Small Angle Neutron Scattering' (SANS) showed for the 700 sample that the particle size should be of the similar order as that found by X-ray data [2]. From these data, it can be said that the particles size distribution is quite narrow within the domain of this study.

5.4. MAGNETIZATION DATA OF NANO PARTICLES OF MAGNETITE

In the **section 5.2,** adequate description has been given on the theoretical side of the magnetization in terms of 'saturation magnetization' (M_S) and 'volume' (V) of the super-paramagnetic particles of magnetite, which is an interesting 'entity' for a detailed study. It is interesting since such glass-ceramic material with ultra-fine particles of magnetite could be used as many 'magnetic devices', such as 'high-density magnetic storage of information' in hard disc of the computers, where a new horizon is just unfolding (see later) for terabyte range. It has also applications in magnetic imaging, etc.

The magnetization measurements are definitely useful. As pointed out earlier in the **section - 5.2,** these measurements have to be performed at different temperatures and also as a function of magnetic field. For the 'blank glass', 700 **(5.5 nm)** and 900 **(7.0 nm)** samples, the magnetization (M) measurements were carried out in a well-standardised "magnetic balance" as a function of a wide range of magnetic field (H) between 0 and 60 KiloGauss (KG) at different temperatures between liquid helium temperature (4°K) and 800°K.

In order to estimate the values of saturation magnetization (M_S), the measurements were carried out at only one temperature, *i.e.* ≈ 270°K, for all the six samples. For the 'blank glass' and 600 samples, M_S increased 'linearly' with H, but for all the other four samples, M_S increased 'non-linearly' with H and then saturated near H = 50 KG, at which the 'spins' were blocked. None of the samples showed any 'hysteresis loop' at 300°K.

First of all, the M_S values (at ≈ 270°K) are plotted against the temperatures of heat-treatment of all the six samples in **Figure 5.3.** The time of heat-treatment was kept constant at 8 hours for each of these samples, which means that the particle size is unique in each sample, by assuming of course that there is no distribution of nano particles or, at least the distribution should be very narrow within each sample. This has the implication for processing nano-particles of magnetite within a glassy matrix, since the magnetic property is very sensitive to the particle size, even within this narrow range of nano-size as explained in the **section 5.2.**

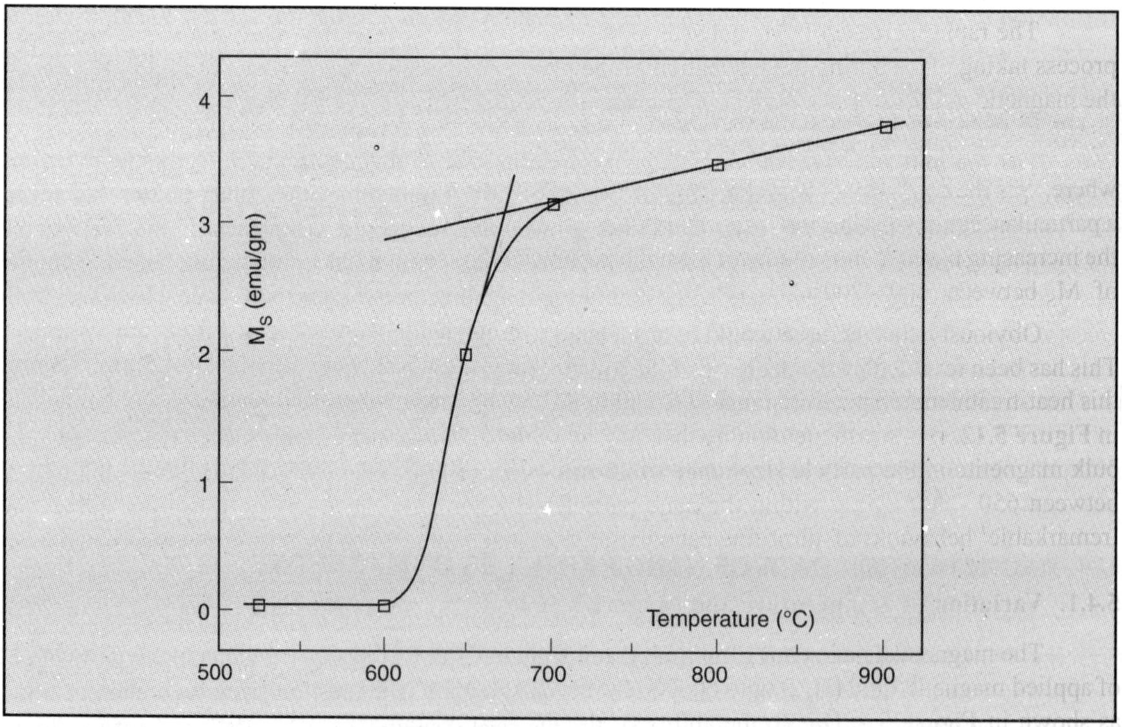

Figure 5.3 : Saturation magnetization (M_S) at 270°K against heat treatment temperature.

It is seen from **Figure 5.3** that M_S values for the 'blank glass' and 600 samples are low and quite negligible, since they mainly show paramagnetic behaviour with not so large values of magnetic moments. Then, M_S increases above about 625°C, which might indicate the nucleation of the magnetite phase, upto about 670°C, and this increase is quite rapid between 625 - 670°C as seen from **Figure 5.3.** Then, there is a change of slope, and M_S increases much less rapidly, which could be ascribed to a growth process of nano particles of magnetite.

Generally speaking, in most physical systems when one physical quantity (here, in this case, saturation magnetization), *i.e.* the independent variable shows a "distinct" change of slope in its behaviour against one dependent variable, *i.e.,* temperature in this case, it is common to conclude that there are "two" processes operative. Hence, in the interpretation of the magnetic data of **Figure 5.3,** it can be said that there is a 'nucleation' of nano-particles of magnetite between 625-670°C and there is a 'growth' process operating between 670 - 900°C. But, there are more interpretations in store, if we take a much closure look in these data.

Although M_S shows almost a linear relation with increasing particle size in the temperature range (600 - 900°C), it is interesting to note that the increase of M_S is much more pronounced between 600 - 700°C than that between 700 - 900°C. Between 700 - 900°C the particle size, and hence the particle volume increases, but there is a 'slow' increase of M_S, which could be interpreted as due to the increase of the 'ferrimagnetic phase' thereby reducing the contribution of the 'super-paramagnetic' particles. As shown later in the **section 5.5,** this has also been revealed by the Mössbauer data at 300°K. Furthermore, this could be due to the increasing dilution of the surface Fe ions on the nano particles of magnetite by other non-magnetic cations, such as Ca^{2+}, Mg^{2+}, etc. and it is also revealed by the Mössbauer spectra at 4°K with H_{ext} = 50 KG, wherein the 'spins' of the nano particles of magnetite seem to be blocked.

The rapid increase of M_S between 600 - 700°C could be ascribed to a 'cation re-distribution' process taking place during the nucleation of the nano particles of magnetite. According to Neel [15], the magnetic moment of one molecule of magnetite is written as

$$M_\beta = (4 + 2\gamma)\mu_B \qquad ...(5.6)$$

where, γ is the coefficient characterizing the 'inverse spinel' structure, and μ_B the Bohr magneton. For a particular 'cation distribution' other than 'normal', $0 < \gamma < 1$. The increasing value of γ associated with the increasing number and volume of the nano particle of magnetite could account for the rapid increase of M_S between 600 - 700°C.

Obviously, this increase would be associated with the increase of the 'symmetry' of the Fe ions. This has been revealed by the decrease of the 'quadrupole splitting' and the increase of 'isomer shift' in this heat-treatment temperature range (*i.e.* within this nano range of magnetite particles) as shown later in **Figure 5.12.** It is worth mentioning that M_S at ≈ 300°K increases by about 70% in the case of pure bulk magnetite in the particle size range of **20 nm - 75 nm** [16], whereas it increases by about 50% between 650 - 900°C, *i.e.* within the nano particles size range of **4.5 nm - 7.0 nm**. This shows the 'remarkable' behaviour of ultra-fine nano particles of magnetite embedded within a glassy matrix.

5.4.1. Variation of Temperature and Magnetic Field

The magnetic data in emu/g, measured at different temperatures (4, 77 and 300°K) as a function of applied magnetic field (H_{ext}) upto 60 KG, for the 'blank glass', 700 and 900 samples are interesting, as shown in **Figure 5.4.** Due to the ability of the material containing nano particles of magnetite to "remember" an earlier magnetic field (*i.e.* remanance), each sample was subjected to a decreasing alternating field at 300°K before each experiment. Then, the evolution of the magnetization (M) was studied as a function of temperature upto 800°K with an applied magnetic field (H_{ext}) equal to 9.56 KG. These data are shown in **Figure 5.5.**

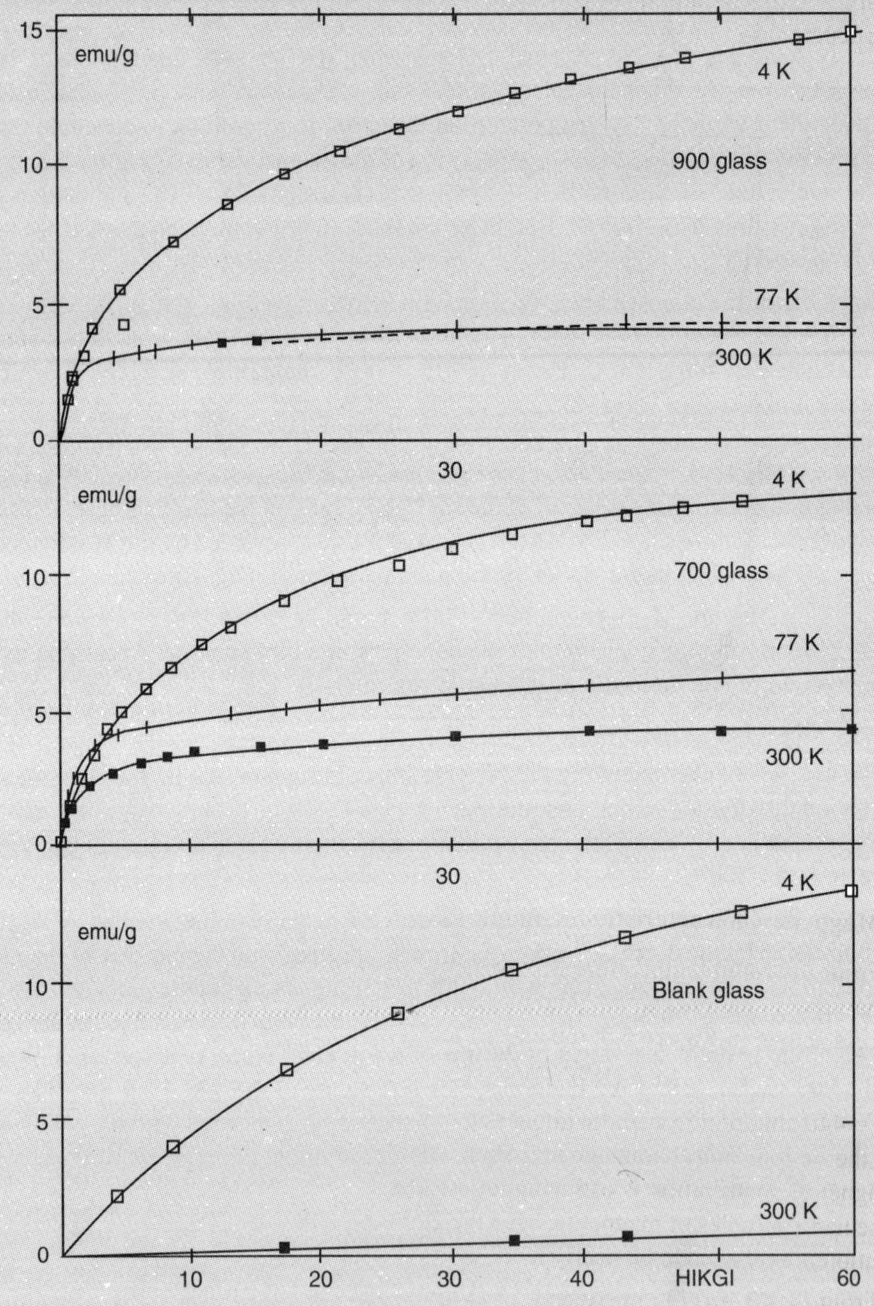

Figure 5.4 : Magnetization as a function of the magnetic field at different temperatures.

In the analysis of all the experimental magnetization curves as a function of magnetic field [M(H) in **Figure 5.4**] and as a function of temperature [M(T) in **Figure 5.5**], the paramagnetic part of the contribution of the Fe^{2+} ions, which are evaluated by the Lorentzian analysis of the Mössbauer data [6-7], is first deduced from each of the M(H) and M(T) values. It should be noted that this correction is necessary, and this is carried out with $\mu_{Fe^{2+}} = 540$ emu/g of iron and $N_{Fe^{2+}} \approx 3.2 \times 10^{-4}$ mol/g of glass in this particular case.

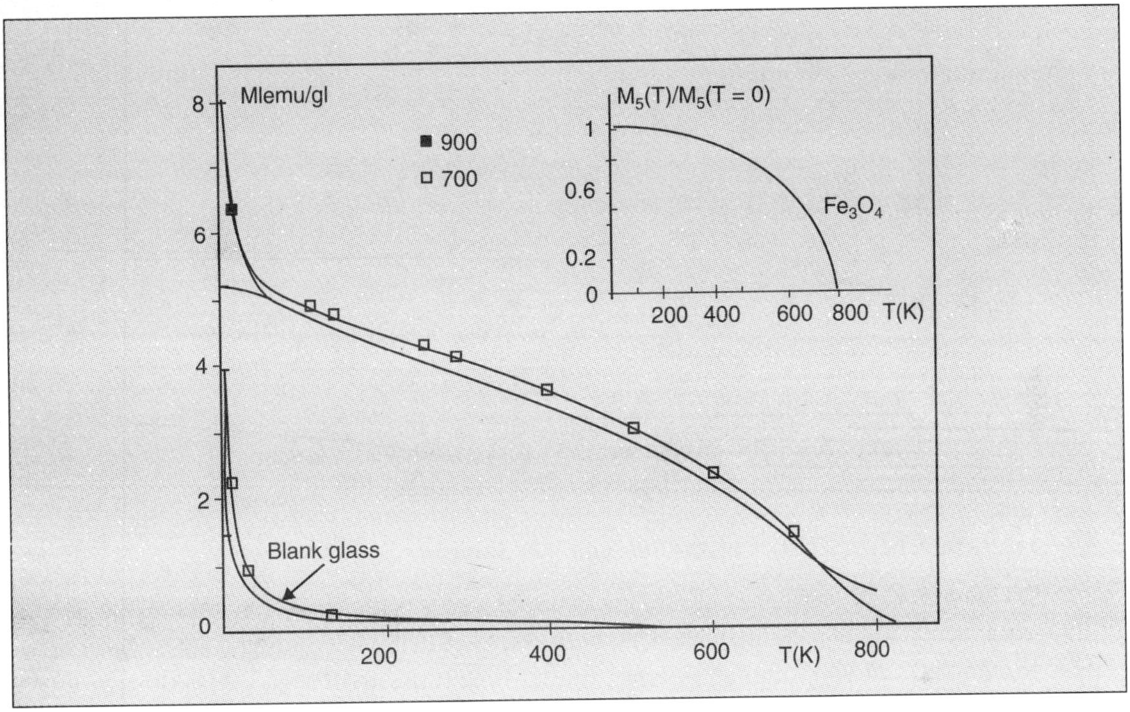

Figure 5.5 : Magnetization as a function temperature (H = 9.56 KG).

5.4.2. Magnetic Characteristics of Blank Glass

From the room temperature Mössbauer spectra of the 'blank glass' [6-7], it is found that this sample is simply 'paramagnetic'. The magnetic moment is measured as a function of magnetic field [M(H)]. As shown in **Figure 5.5,** the experimental curve at T = 300°K can be described as follows

$$M = (N'\mu'^2 H)/3k_B T \qquad ...(5.7)$$

with $\mu' = 440$ emu/g of iron and without any remanance and by assuming a N′ value corresponding to 80% of the Fe ions in a 'paramagnetic' state. At 4°K, from the remanance value (see **Table 5.1**), the 'ferrimagnetic' contribution is estimated by assuming a saturation magnetization value of 50 emu/g for these nano particles of magnetite, as in the 700 sample. At this low temperature, the 'ferrimagnetic' contribution is evaluated to be nearly 5%.

From the M = *f*(T) experiment, *i.e.* **Figure 5.5,** the dotted line means that the 'paramagnetic' contribution of the Fe^{2+} ions is already deducted. The values of $\chi^{-1} = H/M$ are plotted in **Figure 5.6**. In this analysis, the 'ferrimagnetic' contribution has been neglected in view of the small remanent magnetization value at 4°K. It is seen that χ^{-1} is a simple linear function of T : with $\chi^{-1} = 3k_B T/ N'\mu'$

with μ′ = 390 emu/g of iron. This value is somewhat lower than that estimated previously (*i.e.* 440 emu/g of iron), but is of the same order of magnitude. The difference could be due to approximate evaluation of the Fe^{2+} contribution.

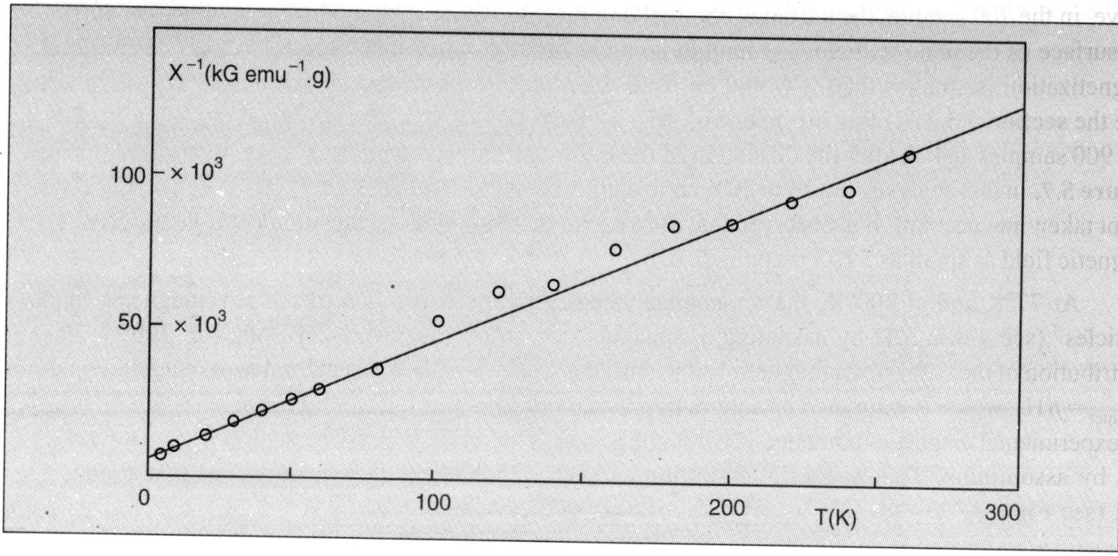

Figure 5.6 : The inverse magnetic susceptibility as a function of temperature for the blank glass.

5.4.3. Magnetic Characteristics of the 700 and 900 Samples

In **Figure 5.4,** only the behaviour for increasing H values is illustrated for 700 and 900 samples at different temperatures of measurement. It has been estimated that at 4°K, the remanent magnetization values for the 700 and 900 samples are very similar to half those of M_{60KG}-M_{Fe2+}, according to $\mathbf{M_R} = \frac{1}{2}\,\mathbf{M_S}$ [12]. This indicates that at this temperature, all the nano particles of magnetite in these two samples are surely below their 'blocking' temperature, and hence they naturally behave as 'ferrimagnetic' particles.

For the 700 and 900 samples, from these data at this low temperature for different magnetic fields after deducting the contribution of Fe^{2+} ions, the curves are drawn for $M = f(1/H)$ in order to be able to extrapolate their values of magnetic moment at $H = \infty$, which are

For the 700 sample, M ($T = 4$°K, $H = \infty$) = 4.95 emu/g of glass.

For the 900 sample, M ($T = 4$°K, $H = \infty$) = 5.85 emu/g of glass.

Obviously, it was assumed that at T = 4°K and $H = \infty$, all the 'spins' are aligned along the direction of the applied magnetic field. The proportion of magnetite in the 'ferrimagnetic' state at 4°K, without any applied magnetic field, is given by the ratio $2M_R/M$ ($T = 4$°K, $H = \infty$). These data agree well with those evaluated from the Mössbauer data [6-7].

From the values of M ($T = 4$°K, $H = \infty$), the values of 'saturation' magnetization of the nano particles in the 'ferrimagnetic' state are estimated as

For the 700 sample, M_S ($T = 4$°K, $H = \infty$) = 50 emu/g of magnetite [Particle Size = **5.5 nm**]

For the 900 sample, M_S ($T = 4$°K, $H = \infty$) = 59 emu/g of magnetite [Particle Size = **7.0 nm**]

These values are smaller than the magnetization value of the 'bulk' magnetite (*i.e.* ≈ 98 emu/g of magnetite).

The difference between these values can be explained by the difference in particle size. As noted above, in the 700 sample, the particles are smaller than that of the 900 sample. The canting of 'spins' at the surface of the nano particles is found to be more frequent in the 700 sample, so that the resulting magnetization is smaller than it would be if all the 'spins' were 'aligned' along the same direction (see the **section 5.5.2**). If this difference of M_S is taken into account, the $M = f(H)$ curves of the 700 and 900 samples at 4°K after the deduction of the Fe^{2+} contribution 'superpose' quite well, as shown in **Figure 5.7.** In this analysis, the 15 to 20% contribution of magnetite that is not in a 'ferrimagnetic' state is not taken into account. It is observed that 90% of the 'saturation' magnetization is obtained, even if a magnetic field as small as 5 KG is applied.

At 77°K and at 300°K, the remanence values give the proportion of the ferromagnetic nano particles (see **Table 5.1**) by assuming a constant 'saturation' magnetization from 4 to 300°K. The contribution of the 'super-paramagnetic' nano particles can be evaluated at each value of magnetic field: $M_{Super} = f(H)$, which is estimated by subtracting the contributions of Fe^{2+} and 'ferrimagnetic' state from the experimental magnetic moments. The last one is calculated from the remanence value as : $M_R = ½\, M_S$ and by assuming a 'ferrimagnetic' evolution as : $M_{Ferri} = f(H)$, which is similar to that found at 4°K (see **Figure 5.7**).

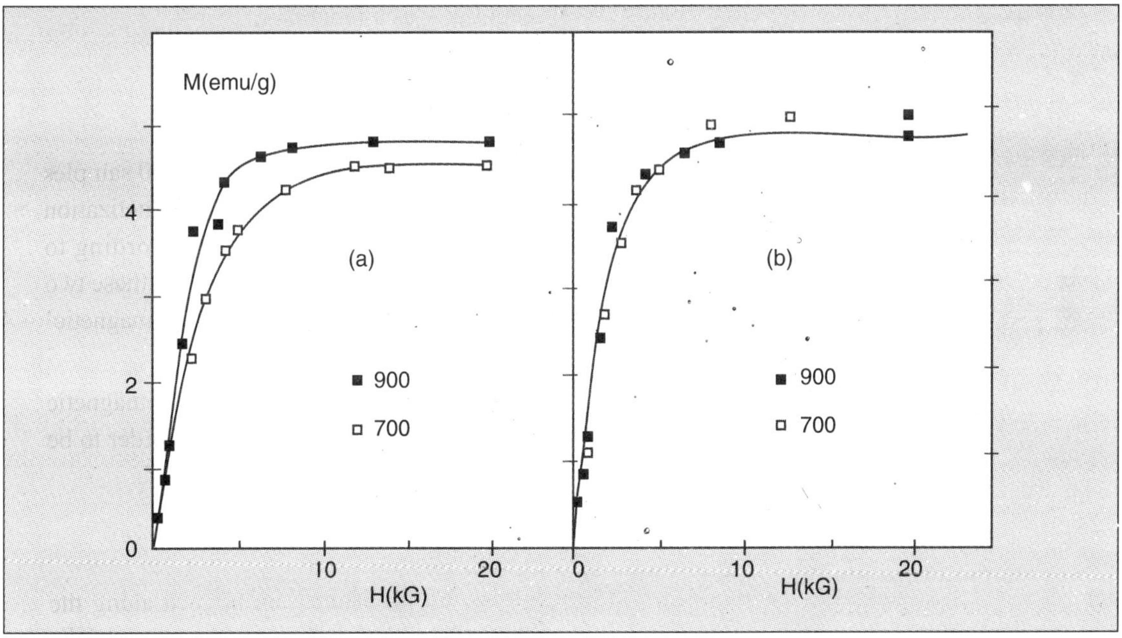

Figure 5.7 : Magnetization against magnetic field at 4°K (*a*) after deducting Fe^{2+} contribution, (*b*) M value multiplied by the ratio $M_{S\,900}/M_{S\,700}$.

In **Figure 5.8**, the values of M due to 'super-paramagnetic' contribution vs. H_{ext}/T are plotted for both the 700 and 900 samples. The magnetization curves superpose well both at 77 and 300°K, and agree well with the Langevin function (the full lines). From the high field part of the curves, the 'mean particle diameter' and the saturation moments (M_{Super}) of the 'super-paramagnetic' nano particles were estimated as per the equation (5.4). The nano particles in the 700 sample are smaller than those in the 900 sample. The low field part of the curves is not used, because the precision is poorer in this region. This typical 'super-paramagnetic' behaviour confirms the earlier 'hypothesis' that 20% of the iron ions are in a 'paramagnetic' state (Fe^{2+}), some in 'ferrimagnetic' state and the balance iron ions are in a 'super-paramagnetic' state.

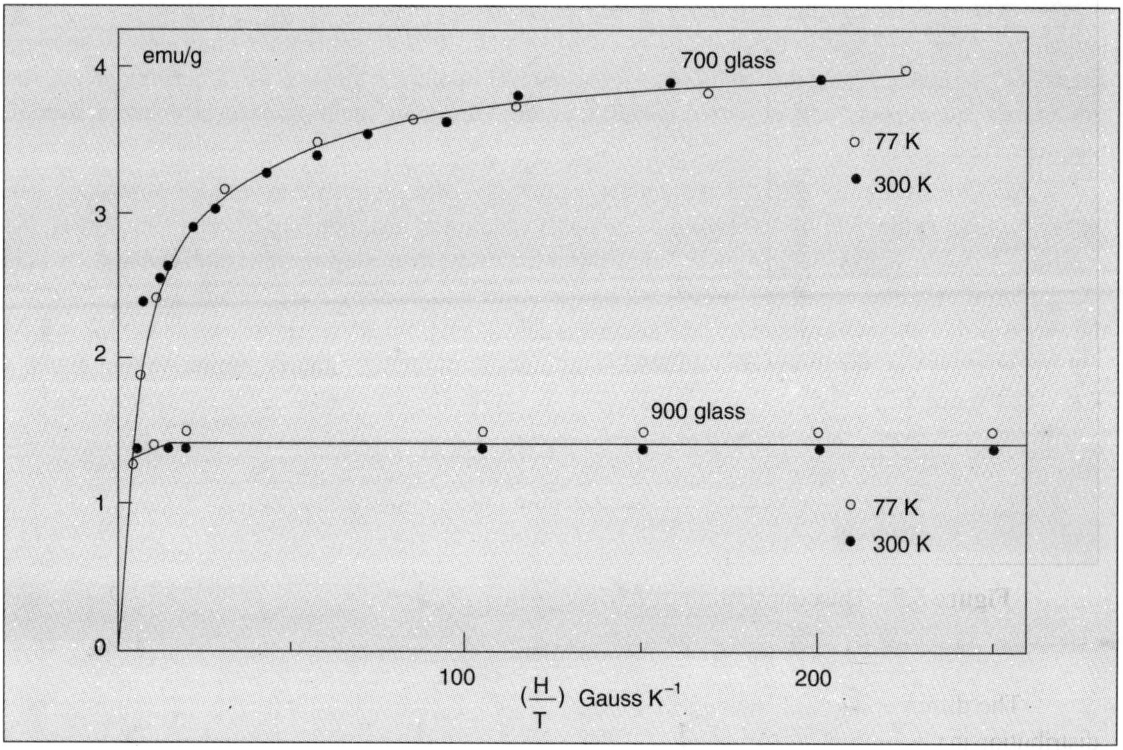

Figure 5.8 : Superposition of the magnetization curves of the superparamagnetic particles against H_{ext}/T , both at 77°K and 300°K.

The M = *f*(T) curves for the 700 and 900 samples can be described by assuming that all the remaining 80% of the iron ions are in a 'ferrimagnetic' or a 'super-paramagnetic' state of magnetite. This means that

$$N(V)_{Ferri} + N(V)_{Super} = \textbf{Total Magnetic Volume = Constant}$$

As the temperature increases, some 'ferrimagnetic' nano particles transform into 'super-paramagnetic' nano particles due to faster relaxation → the 'spins' in the same nano particles flip-flop too fast. Hence, we can write

$$M = M_{S\text{-}Ferri}\,[N(V)]_{Ferri} + M_{S\text{-}Super}\,[N(V)]_{Super}\,[\coth \alpha - 1/\alpha]$$

with $$\alpha = M_{S\text{-}Super} V_{Super} H_{ext} / k_B T.$$

Both $M_{S\text{-}Super}$ and $M_{S\text{-}Ferri}$ are supposed to be dependent on temperature in the same manner as 'bulk' Fe_3O_4 [17], as shown in **Figure 5.5,** whose values have already been estimated earlier. For each temperature, it is possible to calculate the **%ferrimagnetic fraction.** This is shown in **Figure 5.9** for the 700 and 900 samples as a function of temperature (H = 9.56 KG). As the temperature increases, the proportion of the 'ferrimagnetic' nano particles decreases. It is seen that all the nano particles are in a 'paramagnetic' state at T > 800°K. This temperature is lower than the "Curie" temperature of the bulk magnetite (851°K).

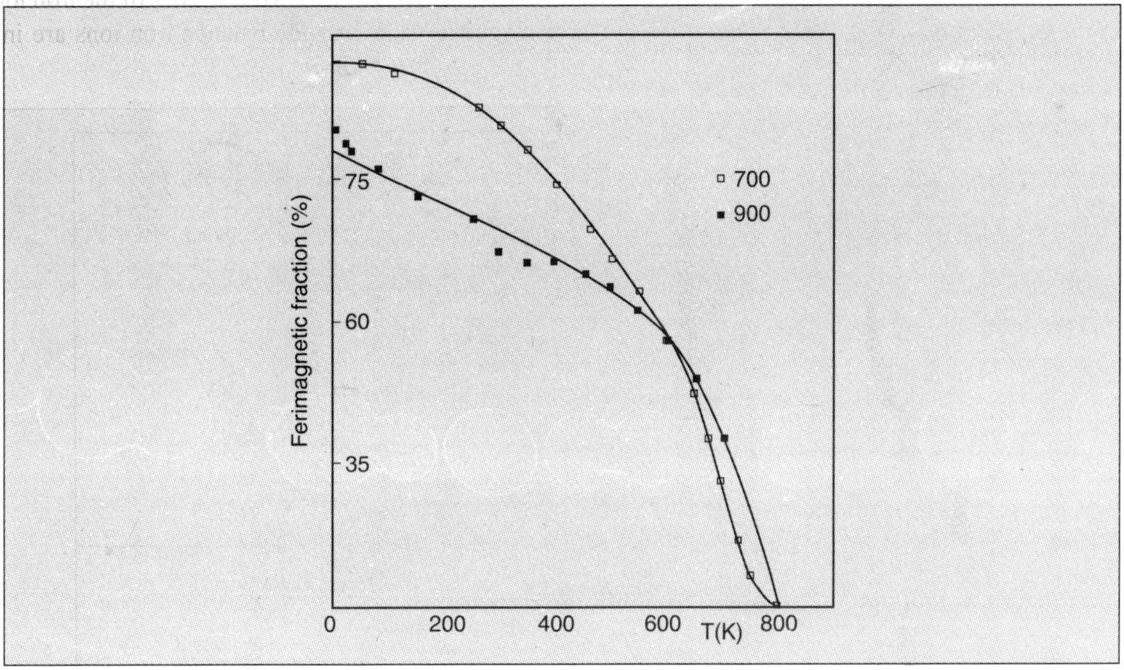

Figure 5.9 : The contribution of ferrimagnetic particles (% volume of magnetite) as a function of temperature (H = 9.56 KG).

The differentiation of the curves of **Figure 5.9** results in the 'volume weighted' particle size distribution in the 700 and 900 samples [18-19], as shown in **Figure 5.10.** The temperature is proportional to the volume of the particle. In the 700 sample, the particle size distribution is larger than that in the 900 sample, but the 'mean value' is smaller. Thus, there is a distinct change of magnetic behaviour between the 700 sample and the 900 sample, within such a narrow range of particle sizes between **5.5 nm** and **7.0 nm,** as found out by X-ray data in the **section 5.3,** or between **2.8 nm** and **3.6 nm** as found out by magnetization data.

In summary, the magnetization measurements on the basalt glass samples heat-treated at different temperatures show the evolution of crystallization and also the magnetic behaviour. The 'blank glass' has paramagnetic behaviour with M = 390 emu/g of iron. The 700 and 900 samples contain 20% of paramagnetic Fe^{2+} ions and 80% of nano particles of magnetite. The nano particles of magnetite behave like a super-paramagnetic or a ferrimagnetic material depending on the crystal size, even within the "nano range". The mean particle diameter has been evaluated for the super-paramagnetic particles of these two samples as **2.8 nm** to **3.6 nm.** The saturation magnetization of the nano particles is smaller (28 to 52 emu/g) than those of the bulk magnetite (~ 98 emu/g). A volume particle size distribution has been calculated for the nano particles of super-paramagnetic magnetite [8].

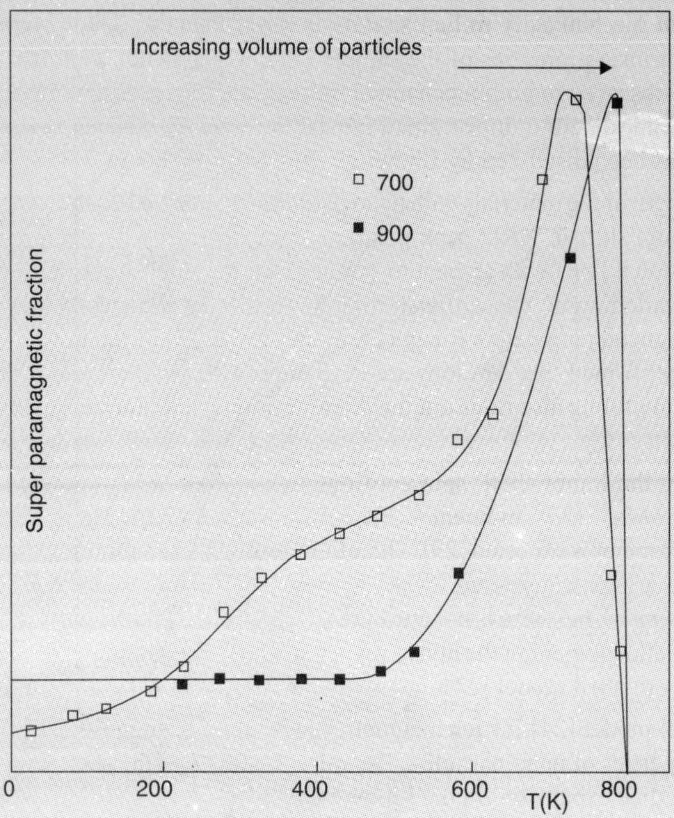

Figure 5.10 : Volume particle size distribution as a function of temperature (T is proportional to the volume of the particle).

5.4.4. Lattice Expansion in Ferrites with Nano Particles

The magnetic properties of bulk and nano particles of $MnCr_2O_4$ spinel oxide show that it is a ferrimagnetic insulator showing paramagnetic (PM) to collinear ferrimagnetic (FM) transition below $T_C \approx 45°K$ and collinear FM to non-collinear FM state below 18°K. The nano particle was prepared by employing the novel technique of mechanical milling. The measurement shows a few unusual magnetism in the nano particles in terms of a 'lattice expansion' with the decrease of particle size. A sharp magnetic transition at 18°K associated with the non-collinear spin structure is not seen in the nano-particles [20], but the non-collinearity of spin is shown in case of nano particles of magnetite, as described in the **section 5.5.2** [6]. The decrease of magnetization in the nano particles is interpreted by the "core-shell" model, and it is attributed to the increasing disorder effect. However, the increase of T_C, and systematic increase of Bloch exponent with decreasing particle size, seem to be unusual, which is highlight by a 'lattice expansion' of the nano particles [20].

Generally speaking, the decrease of particle size results in a decrease of 'lattice parameter' in nano materials. This is true in case of a wide class of nano materials, synthesized by various procedures as noted above. There are several explanations for the 'lattice expansion' in nano materials, such as change in oxygen coordination number with the cations and change of valence state of cations [21], and contribution of 'excess volume' of grain boundary spins [22]. There are some arguments that the 'lattice expansion' in mechanically milled samples may be related to the mechanical strain induced effect, rather than intrinsic properties of the sample [22], which is not well justified, since the 'lattice expansion' has been observed in both mechanical milled nano particles as well as in those prepared by the chemical route. Some details of the origin of the 'lattice expansion' in the nano particles of the above magnetic material has been discussed by Bhowmik [20], as given below.

The XRD pattern of the materials with nano particles are identical with the bulk samples. Although small, but a systematic shift in XRD peak positions are observed with decreasing particle size. The absence of any additional lines with respect to standard cubic spinel structure of bulk sample indicates that there is no crystallographic phase transformation in the materials containing nano particles. The agreement of the magnetic parameters (magnetic moment and T_C) of the bulk sample with previously reported value has confirmed that Mn ions are in divalent (Mn^{2+} : $3d^5$) state. The argon atmosphere during the mechanical milling also rules out the formation of significant amount of Mn ions with higher ionic (3+ or 4+) states.

By comparing the 'outer shell' spin configuration of Mn^{2+} ($3d^5$, moment = 5 μ_B), Mn^{3+} ($3d^4$, moment = 4 μ_B) and Mn^{4+} ($3d^3$, moment = 3 μ_B), it is evident that if Mn^{3+} or Mn^{4+} exists at all, the decrease of lattice parameter is expected [23]. Therefore, the change in valence state or the crystallographic phase transformation are not considered for the 'lattice expansions'. Hence, the reduction of magnetic moment with decreasing particle size is not attributed to the change in valence state of Mn ions. However, the decrease of magnetic moment in the above magnetic material containing nano particles seems to be consistent with the core-shell model [24].

The core-shell model [24] for ferrimagnetic nano particles suggests that shell contribution will dominate on the properties of nano particles. The microstructure of the shell with more disorder effect may influence the 'lattice expansion' in the following two ways :

1. Increasing the free 'excess volume' of the incoherent shell, *i.e.* grain boundary, spins in the interface structure, and
2. Lowering symmetry in oxygen coordination numbers with surface cations.

It has already been shown for $MnCr_2O_4$ that the increase of lattice volume is highly related to the change in oxygen coordination number with the cations [21]. Consequently, the lattice pressure on core spins may be reduced by the elastic coupling between the shell spin and the core spin lattices [22].

In fact, various factors such as breaking of long range crystallographic coherent length of bulk material and orientation between the shell spins exhibit inter-atomic spacing, which is different from the bulk lattice. Overall, the macroscopic average of inter-atomic spacing contributes larger lattice parameter in the nano particles.

Since the 'superexchange interaction' in spinel oxide strongly depends on both the bond angle and the bond length of A and B site spins, any change in the 'spin configuration' either in shell or core must be reflected in the change of magnetic properties in the nano particles. The gradual decrease of the non-collinear spin structure amongst the B site Cr^{3+} moments in the nano particles is reflected by the disappearance of sudden jump in magnetization at 18°K, as observed for the bulk sample [20].

This suggests an increase of B site Cr-Cr distance in the nano particles, which leads to the decrease of direct antiferromagnetic (J_{BB}) interactions between Cr moments, and increase of inter-sublattice superexchange interactions (J_{AB}). The strong J_{AB} with increasing B site Cr - Cr distance has been realized in many other spinels. If the 'lattice expansion' is related to the increase of Cr - Cr distance via O^{2-}, it is expected under the assumption $T_C \sim J_{AB}$, that T_C should be higher for the nano particle samples. The experimental data of *dc* magnetization show magnetic irreversibility and spontaneous magnetization above 45°K (*i.e.* T_C of the bulk samples) for the nano particle samples. This confirms the enhancement of ferrimagnetic order in $MnCr_2O_4$ nano particles. A similar increase of T_C with the 'lattice expansion' has been observed in other nano particle systems [25].

This work is very important in that the non-collinear ferrimagnetic order is associated with strong B site direct (antiferromagnetic) cation-cation interactions, which favour non-collinear B site spin structure below 18°K. The increase of T_C with decreasing particle size is a consequence of the decrease of non-collinear (B site) spin structure in the nano-particles. The experimental data indicate that the 'lattice expansion' has a direct effect on the decrease of non-collinear (B site) spin structure, and hence on enhanced ferrimagnetic order in the nano particles. The observed reduction of magnetic moment, large magnetic irreversibility between MZFC and MFC, and appearance of 'superparamagnetism' are some of the disorder effects in nano-particles [20]. The latter is explained by the data on a detailed Mössbauer experiment on the nano particles of magnetite in terms of a 'spin-cating' model in the **section - 5.5.2** [6].

5.5. MÖSSBAUER DATA OF NANO PARTICLES OF MAGNETITE

In the nano range, the ultrafine particles of magnetite display a super-paramagnetic behaviour below the magnetic transition (*i.e.* Curie) temperature of the bulk magnetite of 851°K, due to a 'fast' relaxation of the magnetization of such nano particles, which arises due to the fast flip-flops of the spins [8, 9]. This relaxation time is strongly dependent on the particle size and the temperature of measurement. Below the so-called blocking temperature (T_B), *i.e.* blocking of the fast flip-flops of the spins, these same nano particles show ferrimagnetic behaviour.

The Mössbauer technique is a very convenient tool, which can be used to investigate the iron containing compounds. In order to study the evolution of the magnetic behaviour as a function of temperature and particle size, a detailed Mössbauer spectroscopy was carried out on different heat-treated samples of the basalt glass at different temperatures with and without any applied magnetic field. This makes it the first such study of nano particles of magnetite within glassy matrix and also the most comprehensive, and hence it definitely merits a detailed reporting here. In the **section 1.6.1**, the theory of Mössbauer spectra is given in details.

The Mössbauer spectra of all the six samples, as mentioned in the **section 5.3,** were recorded at room temperature and at liquid helium temperature (4°K). Magnetic fields upto 50 KG were applied at 4°K. A source of ^{57}Co in chromium was used for the room temperature measurements, while a source of ^{57}Co in rhodium was used for other measurements. The 14.4 KeV γ-rays were detected by means of a 'proportional counter'. The source was moved by a conventional 'constant acceleration drive'. The velocity signal of the 'electro-mechanical transducer' was a 'symmetrical triangle'. The counts were stored in the 'time mode' of a 400 'channel analyser'.

The Mössbauer spectra taken at room temperature is shown in **Figure 5.11** for all the six samples. Each spectrum, except for the 800 and 900 samples, can be interpreted as resulting from the superposition of two doublets, as can be found out from a simple Lorentzian analysis. For the 800 and 900 samples, the spectra consists of the same 'two doublets' (as observed for the other four samples), superimposed to a diffuse six line magnetic 'hyperfine structure' **(HFS),** identified later as due to magnetite. The internal magnetic fields of these **HFS** are approximately 410 and 450 K respectively for these last two samples.

The isomer shift (δ) and quadrupole splittin (Δ) of the less intense doublets are almost the same for all the samples, *i.e.* δ = (1.05 ± 0.02) mm/sec, which is normally referred to iron metal, and Δ = (2.02 ± 0.02) mm/sec. These values are the 'characteristics' of the Fe^{2+} ions [26]. Here, it is worth mentioning that the above values are very close to those of 'orthopyroxene' [27, 28]. In fact, the crystals of this compound were detected by X-ray diffraction (see the **section 5.3**), but only in the 800 and 900 samples.

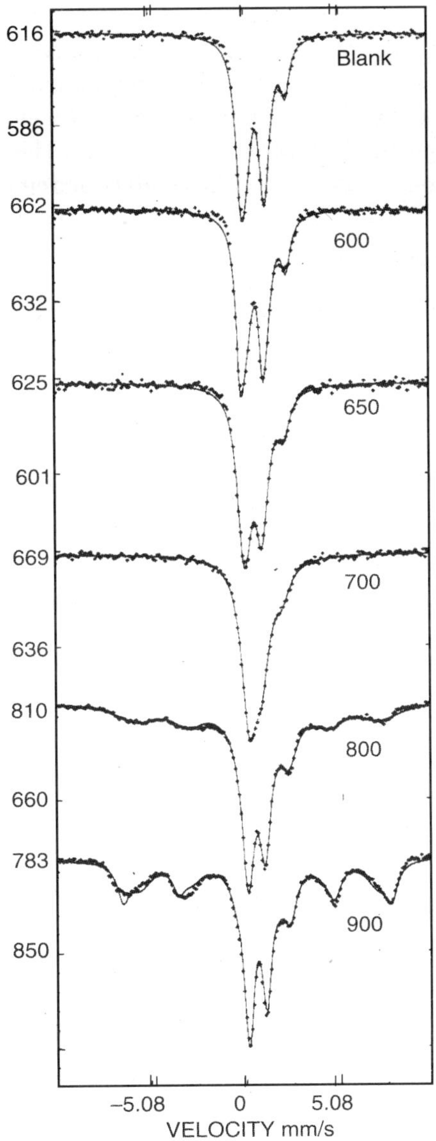

Figure 5.11 : The Mössbauer spectra at 300°K.

The values of δ and Δ of the 'second' doublet are plotted against heat-treatment temperature (*i.e.* increasing size of the nano particles) in **Figure 5.12.** These values are close to those of the Fe^{3+} ions [26]. However, the contribution of the Fe^{2+} ions cannot be completely ruled out. In fact, the 'second' doublet for the blank glass and 600 samples represents mainly isolated Fe ions, while for the last four samples, it represents mainly superparamagnetic Fe ions in the form of magnetite, as identified later by low temperature measurements.

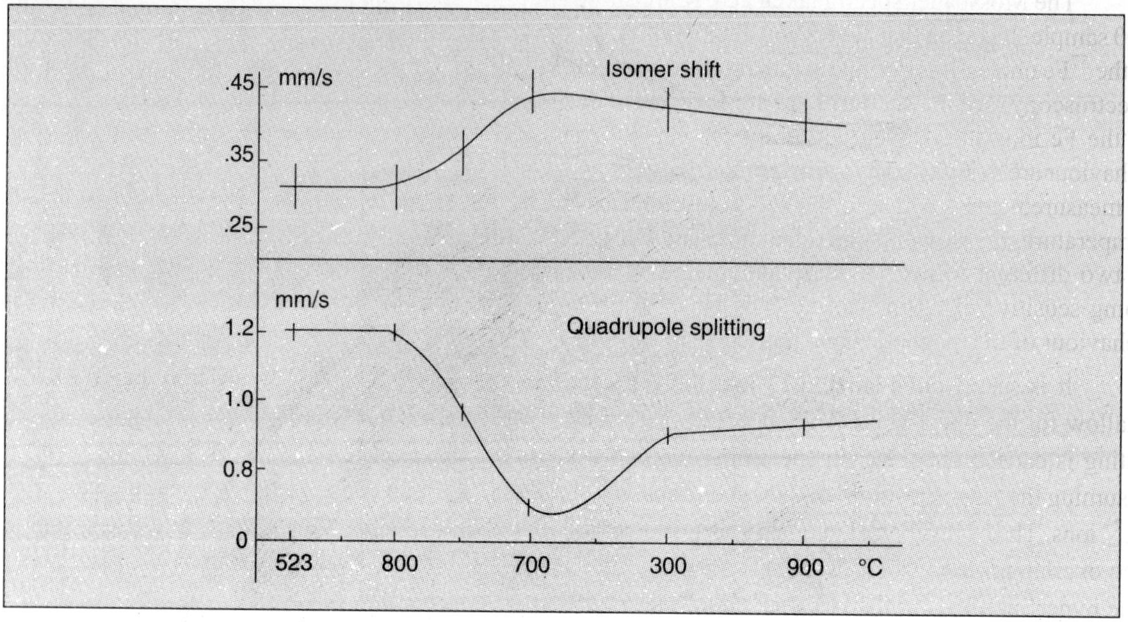

Figure 5.12 : The variation of the isomer shift and quadrupole splitting for Fe^{3+} measured at 300°K as a function of temperature of heat-treatment.

Let us assume here that the values of δ and Δ correspond to the predominant contribution of the Fe^{3+} ions and that the contribution due to the Fe^{2+} ions remain constant, as observed for the 'first' doublet throughout the range of heat-treatment temperature. It is noted that from **Figure 5.12** that the 'symmetry' of the Fe^{3+} ions increases from 600 to 700 sample with the nucleation and crystallization of the magnetite phase. After 700°C, the increase of Δ and the slight decrease of δ towards 900°C are brought about by the disturbance of the magnetic structure by other cations, *e.g.* Ca^{2+}, M^{2+}, etc., as discussed later. The maximum amount of magnetite is formed in the 700 sample (see later), which is thought to result in the most improved 'symmetry' of the Fe^{3+} ions, as shown by the smallest quadrupole splitting and the largest isomer shift for this sample.

It is a common practice to compute the 'areas' under the peaks in order to be able to do quantitative estimation of 'population' of different 'sites' in the samples, provided the "*f* - factors" are the same. These site populations are listed in **Table 5.1.** It is seen that the concentration of the Fe^{2+} ions, which are assumed to be "free" in all the samples corresponding to the less intense doublet (*i.e.* the 'first' doublet) ranges between 15 - 22%. The concentration of the rest of the Fe ions, which are assumed to be 'free' in the blank glass and 600 samples corresponding to the 'second' doublet, ranges between 78 - 85% for the first four samples. For the 650 and 700 samples, the concentration corresponds mainly to the 'assembly' of super-paramagnetic particles. For the 800 and 900 samples, the concentration or site population of the Fe ions in the super-paramagnetic state are much lower at 37% and 29% respectively, because there is a considerable number of the Fe ions in the 'bulk' state, *i.e.* 42.5% and 56% respectively. This makes their detection possible even at room temperature, as observed by a six-line magnetic hyperfine structure (HFS), which is typical of a bulk ferrimagnetic material [6].

5.5.1. Hyperfine Field in Nano Particles

The Mössbauer spectra taken at 4°K are shown in **Figure 5.13** for all the samples except for the 600 sample. It is seen that at very low temperature, all the samples give rise to a magnetic hyperfine field at the ^{57}Fe nuclei. It is definitely remarkable to note that by doing sensitive experiments like Mössbauer spectroscopy (see the **section 1.6.1** for theoretical details), which is based on the details of the nucleus of the Fe ions/atoms even within a very narrow particle size range of nano particles, two different behaviour are manifested by the same nano particles within the same glass matrix at different temperatures of measurements, *i.e.* due to relaxation mechanism of magnetization or spin flip-flops. The lower the temperature, the slower is the relaxation, and we are able to 'arrest' or 'detect' the same nano particles in two different forms, *i.e.* super-paramagnetic and ferrimagnetic. While it shows the usefulness of doing sensitive experiments like Mössbauer at different temperatures, it also indicates a remarkable behaviour of the magnetic nano particles.

It is seen from **Figure 5.13** that the **HFS** lines of the magnetic spectrum are very broad and shallow for the blank glass. They become sharper from the 650 to the 900 sample. A computer curve fitting is carried out for each spectrum, except for the blank glass wherein the error is too high, by assuming the 'superposition' of a magnetic spectrum, a central doublet and an additional doublet due to Fe^{2+} ions. The experimental magnetic spectrum is fitted (*i.e.* the solid lines of **Figure 5.13**) by assuming two overlapping of two six-line **HFS** spectra and by taking only the 'four external lines' into account. The hyperfine field for the last four samples is found to be around 510 KG, as for the Fe^{3+} ions in the tetrahedral site (*i.e.* A - site) of the bulk magnetite [29].

The Mössbauer data at 4°K are consistent with the presence of magnetite in all the samples, except in the blank glass and 600 samples. They confirm the results of XRD and TEM measurements, as described in the **section 5.3.1.** The areas under the concerned peaks are kept in the ratio **3 : 2 : 1 : 1 : 2 : 3.** The relative concentrations of various species are also shown in **Table 5.1.** The largest proportion of magnetite is found to be present in the 700 sample. The values of the $\%Fe^{2+}$ given by the experiments at 4°K agrees with those obtained from room temperature measurement within experimental error. By lowering the temperature does not significantly alter the values of the Mössbauer parameters of these ions.

Table 5.1. The Mössbauer Data for Nano Particles of Magnetite

Heat-Treatment Temperature	300°K $\%Fe^{2+}$	300°K %Superpara	300°K %Ferri	4°K $\%Fe^{2+}$	4°K %Superpara	4°K %Ferri
As-annealed	21.0	79.0	—	—	—	—
600	22.0	78.0	—	—	—	—
650	19.5	80.5	—	25.0	16.5	58.5
700	15.5	84.5	—	19.0	2.0	79.0
800	20.5	37.0	42.5	25.0	10.0	65.0
900	15.0	29.0	56.0	18.0	18.0	64.0

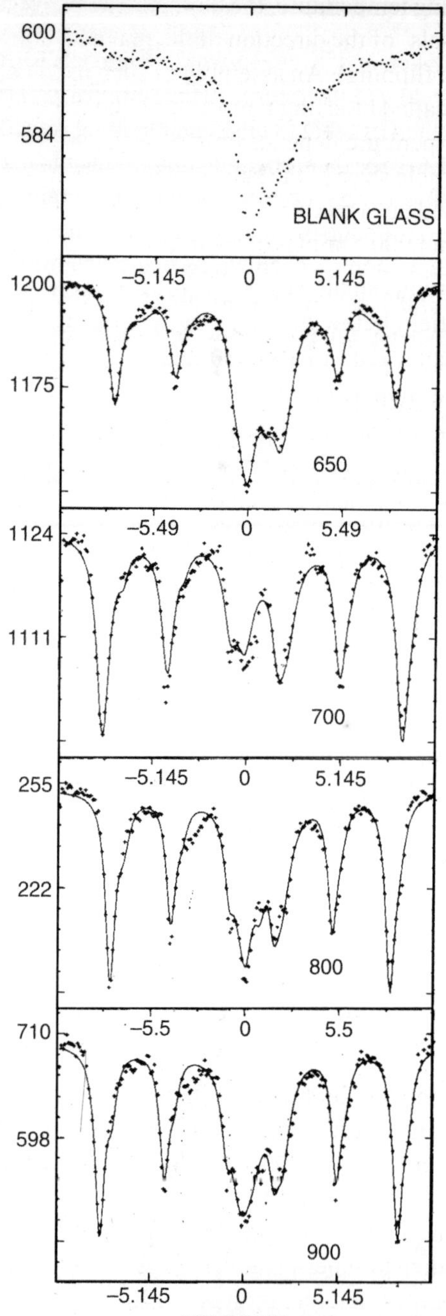

Figure 5.13 : The Mössbauer spectra at 4°K.

The magnetite present in the samples gives a 'magnetically split' spectra at 4°K. The magnetic splitting completely collapses at room temperature, except for the 800 and 900 samples, due to the faster flip-flops or faster relaxation above the so-called blocking temperature, which is a typical behaviour of small nano particles of magnetite. Below the Curie temperature of the bulk magnetite (*i.e.* 851°K) and above the so-called blocking temperature, the thermal energy or thermal vibrations of the magnetite lattice can cause 'many reversals' of the direction of the magnetization in each nano particle during the time of measurement (*i.e.* spin flipping). An assembly of such nano particles is in a super-paramagnetic state. However, below the so-called blocking temperature, the direction of magnetization is 'frozen-in'. The value of the blocking temperature depends on the particle size of the nano particles. Each sample contains a 'distribution' of particle sizes of the nano crystals of magnetite. At a given temperature, in one sample, it is then possible to find out a mixture of super-paramagnetic and ferrimagnetic particles, which is clearly evident in the 800 and 900 samples at room temperature (see **Figure 5.11**).

In several studies of various small particles of magnetite in the nano range, it is reported that the magnetic hyperfine splitting field below the blocking temperature is smaller than that found in larger crystals [30-32]. The Mössbauer data of the 800 and 900 samples clearly confirm this particular observation. The values of the 'hyperfine field' are found to be 410 KG and 450 KG respectively for these two samples. In the bulk magnetite, McNab et al found a value of 486 KG [29]. This interesting observation could be due to an "intrinsic size effect" [32] of the small nano particles of magnetite, or in other words, due to a "collective magnetic excitation" mechanism proposed by Morup and Topsoe [31].

It has already been said above that the small "nano" particle has a 'single domain' structure. The thermal energy or thermal vibration at the temperature of the experiment can be sufficient to equilibrate the magnetization in a time span which is "short" compared to that of the experiment, *i.e.* a super-paramagnetic state with a continuous spin flipping or faster relaxation that are difficult to be blocked above the so-called blocking temperature. In an assembly of single domain nano particles, below their Curie temperature, the frequency "f", at which the gross particle magnetic moment 'flips' amongst the easy direction of magnetization, is much greater than the larmour precession frequency (*i.e.* ν_L) of the Mössbauer nucleus ^{57}Fe. Since this precision time (*i.e.* $1/\nu_L$) is about 10^{-8} sec, very small particles with a size less than **10 nm** are required so that the 'anisotropy' energy per particle $KV \approx k_BT$. This criterion must be satisfied, since the anisotropy energy 'governs' the frequency of the 'spin flipping', *i.e.* relaxation, of the magnetization vector, when $H_{ext} = 0$. The relaxation time is written as follows :

$$\tau = \tau_0 \exp(KV/ k_BT) \qquad ...(5.8)$$

where, $\tau = 1/f$ is the time for the magnetic moment to flip through **180°** , the particle frequency factor (τ_0) is usually $10^{-10} - 10^{-9}$ sec, K is the anisotropy constant, V is the volume of the nano particle, and k_BT is the thermal energy.

In large volume particles, *i.e.* bulk magnetite, KV is very large, and consequently $\tau >> 1/\nu_L$. Hence, a six-line HFS pattern is observed, when the Mössbauer experiment is performed, *e.g.* for the 650 sample at 4°K. On the other hand, when $\tau << 1/\nu_L$, the nucleus can precess at a set frequency only for a small part of a period before H_{hf} can changes the direction. The "net result" of many such changes or "spin flips" in the direction of H_{hf} is no 'precession' at all, *i.e.* $H_{hf} = 0$, and consequently a central doublet is observed corresponding to super-paramagnetic Fe ions in the nano-crystalline magnetite phase, *e.g.* for the 650 sample at room temperature (compare **Figures 5.11** and **5.13**) [9].

For the 900 sample, there is a distribution of sizes of the nano particle, which gives rise to a 'superimposed' spectra. For example, for some nano particles $\tau >>1/\nu_L$ giving rise to six-line HFS pattern, and for some nano particles $\tau << 1/\nu_L$ giving rise to the central doublet, which is superimposed on the HFS pattern. The computer fitting work is done based on these premises, which also minimizes the error in the least square fitting procedure in the Lorentzian analysis..

As noted above, the value of H_{hf} in small nano particles of magnetite is smaller than the bulk magnetite. It is known that when the particles are small, then there are significant number of 'spins' on the surface of the nano particles. It can be argued that the surface ions have smaller hyperfine field than that of the ions in the interior of the nano particles giving rise to smaller average H_{hf} due to the surface effect [32, 33]. Here, it is quite pertinent to consider the model proposed by Morup and Topsoe [31], which states that the lower value of H_{hf} results from 'fluctuations' of magnetization direction around the energy minimum corresponding to an 'easy' direction of magnetization, *e.g.* <1 1 1> direction of magnetite. Morup and Topsoe could fit their experimental results of the Mössbauer spectra for small particles of magnetite without taking any 'surface effect' into account [31].

In this model, for $k_BT/KV << 1$, the hyperfine field for a particle of volume V at a temperature T can be given by the relation :

$$H_{hf}(V, T) = H_{hf}(\infty, T)[1 - (k_BT/2KV)] \qquad ...(5.9)$$

There is an important implication of equation (5.9), *i.e.* at T ≈ 0°K, H_{hf} will be simply independent of the particle size. In order to test this model, the Mössbauer experiment should be made at very low temperature (say, at 4°K) with the samples having different particle sizes, like in this present case from **4.5 nm** to **7.0 nm**. As shown in **Figure 5.13,** from the Mössbauer spectra at 4°K for the 650, 700, 800 and 900 samples, a value of H_{hf} is obtained as 510 KG, which is remarkably independent of particle size even within this narrow range of nano particles. McNab et al found a value of 511 KG for H_{hf} at 4°K for pure magnetite with particle size less than **10 nm.** Thus, this is the first experimental confirmation of 'collective magnetic excitation' or "magnon" model of Morup and Topsoe [31] for small nano particles of magnetite of varying sizes **(4.5 nm to 7.0 nm)** embedded in an inert/non-magnetic glassy or amorphous matrix. This has a tremendous implication for various applications in nano-magnetic devices.

The particle size of magnetite in the 900 sample is known from XRD as **7.0 nm.** Thus, the anisotropy constant (K) could be calculated from equation (5.9), by taking the value of H_{hf} from the 300°K spectra, *i.e.* 450 KG. The anisotropy constant (K) is found to be 0.98×10^6 erg/cm^3 . For the 800 sample, the value of K is found to be 0.77×10^6 erg/cm^3, by taking the estimated particle size of **6.4 nm** and the value of H_{hf} as 410 KG . Therefore, the anisotropy constant could be considered to be of the order of ~ 10^6 er/cm^3, which is very high for these small nano particles. McNab et al found a value of 8×10^4 erg/cm^3 in pure magnetite for particle size less than **10 nm,** by taking a distribution of particle sizes [29].

From the Mössbauer spectra at 4°K, *i.e.* hyperfine splitting, of the Blank glass and 600 sample, a rough estimate of the hyperfine field was made. By taking the anisotropy to be around ~ 10^6 erg/cm^3, an average particle size of the nano crystals of magnetite was evaluated for these samples, which were found to be **1.2 nm** and **1.5 nm** respectively. This shows that some sort of 'short-range' magnetic ordering was already present in the as-annealed glass. The SANS study later confirmed this observation (see the **section 5.7**) [5]. Therefore, the so-called 'blank glass' is not really 'blank', it also contains very small nano crystallites of magnetite of size **1.2 nm.** The 600 sample (which was heat-treated at 600°C, that is below T_g) contains slightly higher sized nano particles of **1.5 nm,** showing some interesting magnetic properties as discussed above.

In the present case, there is also a distribution of particle sizes, as determined from the magnetization data between 4°K and 800°K at H_{ext} = 9.56 KG for the 900 sample through a 'volume weighted particle size distribution' formalism [18, 19], but the distribution is found to be quite narrow, as evident from **Figure 5.10.**

5.5.2. Spin Canting in Nano Particles of Magnetite

There are some interesting Mössbauer results, when we apply a magnetic field (H_{ext}) parallel to the γ-ray direction. The Mössbauer spectra for 650 sample at 300°K, 4°K and 4°K + 50 KG are shown in **Figure 5.14,** and those for the 900 sample at 300°K, 4°K, 4°K + 10 KG and 4°K + 50 KG are shown in **Figure 5.15.** It is seen from the last figure that at 4°K and H_{ext} = 0, the 2nd and the 5th spectral line (*i.e.* $\Delta m_I = 0$ lines) are 'visible'. When H_{ext} = 10 KG, the 2nd and 5th lines are still visible with some intensities, and when H_{ext} = 30 KG (not shown here) still show some finite intensities. But, when H_{ext} = 50 KG, these lines are almost extinguished in the 'longitudinal configuration', when the γ-ray direction is parallel to the direction of the applied magnetic moment of the Fe ions. It should be noted that from 650 to 900 sample, the line-width becomes smaller, which could be ascribed to a decrease of disorder of the magnetite lattice.

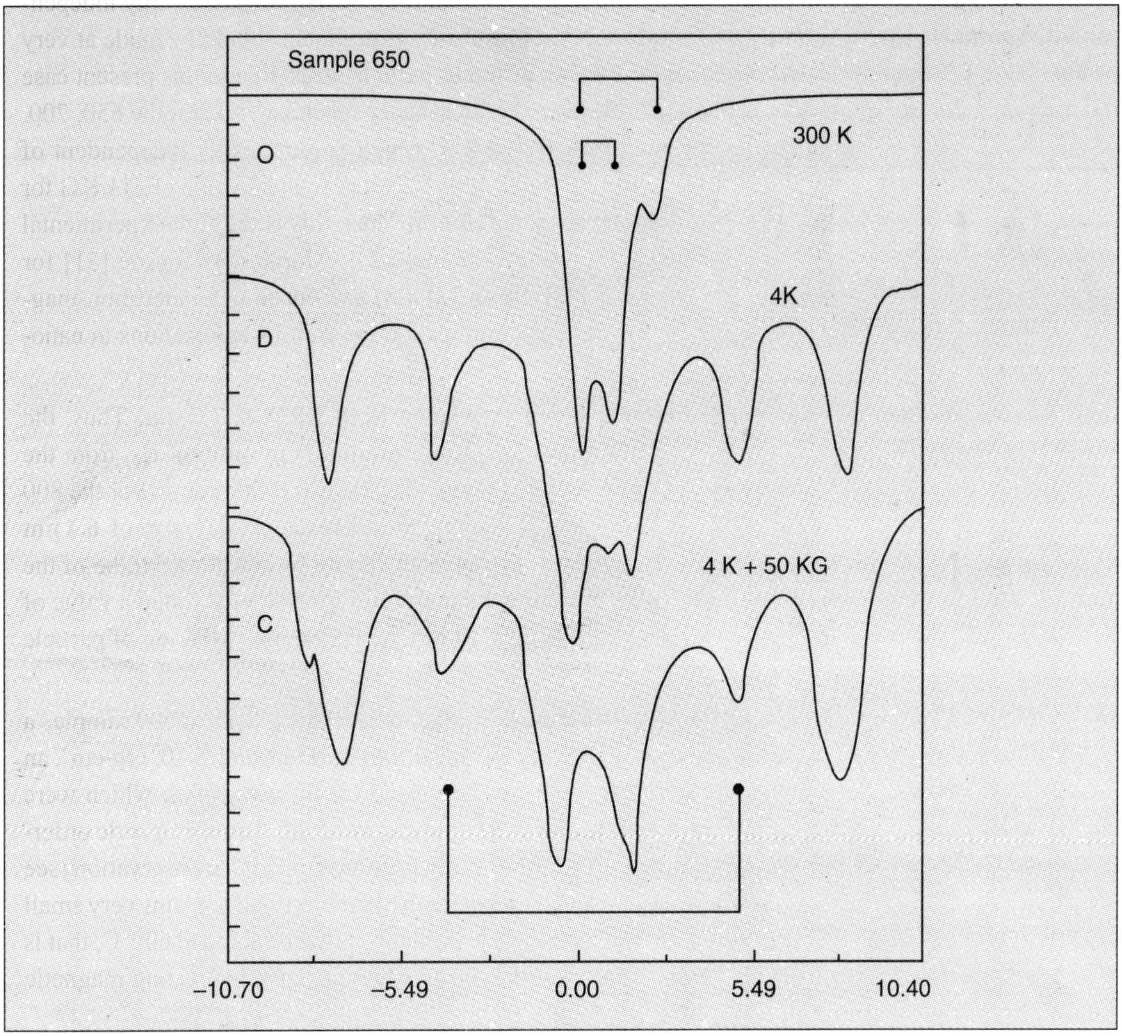

Figure 5.14: The Mössbauer spectra at 300°K, 4°K and 4°K + 50 KG for the basalt glass heat-treated at 650°C for 8 hours.

For the 650 sample, as shown in **Figure 5.14,** the 2nd and 5th lines do not vanish at all even at H_{ext} = 50 KG, showing a significant intensity, as marked by two arrows at the bottom of the figure. According to Neel's model of ferrimagnetism (as discussed in the **section 1.6.1**), antiparallel moments have unequal magnitude giving rise to a net resultant moment. When H_{ext} is applied parallel to the γ-ray direction in the Mössbauer experiment, the effective field on each sub-lattices A and B is given by the following relation :

$$H_{eff}(A, B) = H_{hf}(A, B) \pm H_{ext} \quad ...(5.10)$$

When H_{ext} is applied parallel to the γ-ray propagation, the $\Delta m_I = 0$ lines, *i.e.* the 2nd and 5th lines in the high field Mössbauer spectra, would vanish and the resultant spectrum consists of the superposition of the 4-line spectra with H_{ext} given by equation (5.10) [34]. While almost similar behaviour is observed in the 900 sample, the Neel theory is not satisfied at all for the 650 and 700 samples. Yafet anf Kittel extended the Neel theory by allowing the "Canted Spin" arrangements as [35] :

$$H_{eff}(A, B) = H^2_{ext} + H_{hf}^2(A, B) - 2\,H_{ext}\,H_{hf}(A, B)\,(\cos\theta_{A,B})^{1/2} \quad ...(5.11)$$

where, $\theta_{A,B}$ is the angles between the 'spin' on the sub-lattices (A, B) and H_{ext} .

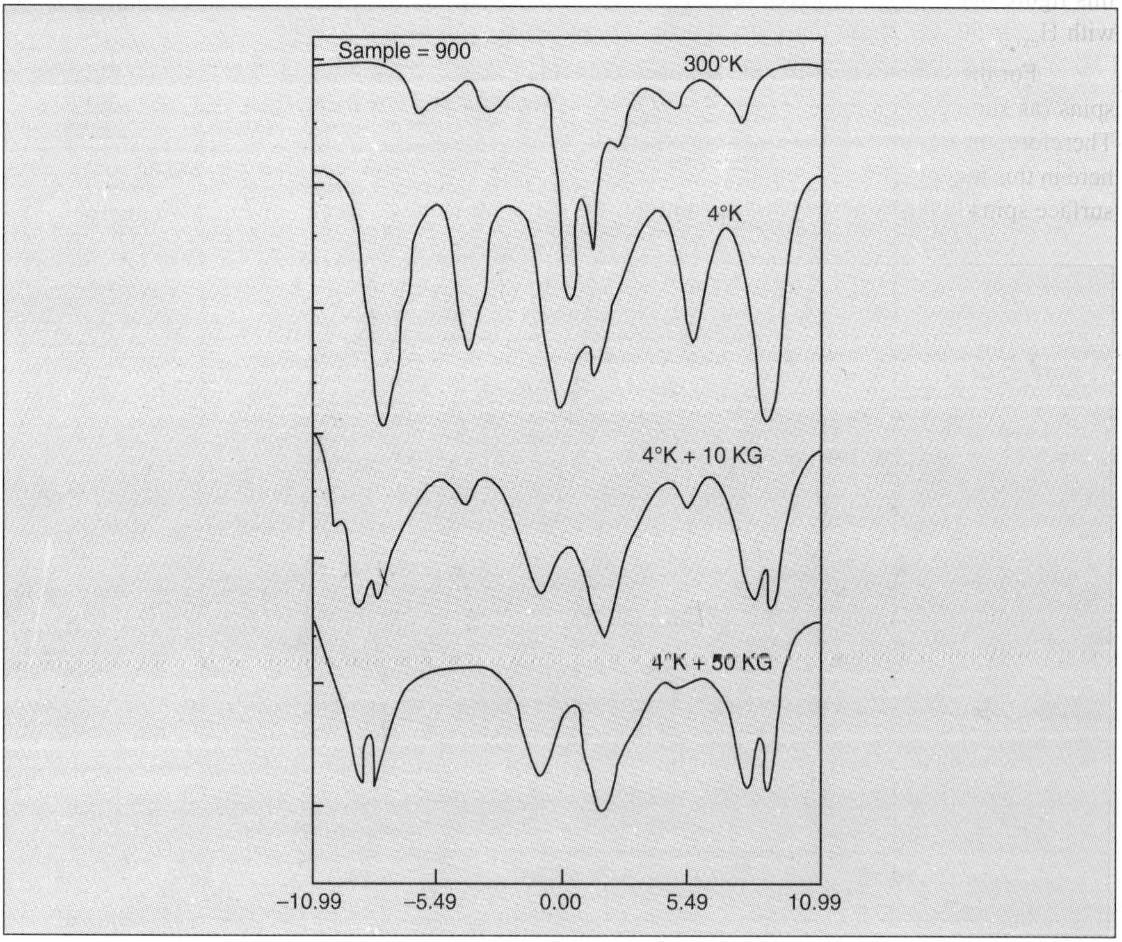

Figure 5.15 : The Mössbauer spectra at 300°K, 4°K, 4°K + 10 KG and 4°K + 50 KG for the basalt glass heat-treated at 900°C for 8 hours.

The appearance of $\Delta m_I = 0$ lines with $H_{ext} \neq 0$ has also been explained by Haneda and Morrish by assuming a "non-collinear" spin configuration of the surface Fe ions in small particles of γ-Fe_2O_3 following equation (5.11), *i.e.* spin canting. This is perhaps the 'first' experimental confirmation of 'spin canting' of the surface spins in small nano particles of magnetite within a glassy matrix. This also clearly shows that unless we go down to the nano level, we cannot understand the sensitivity of the arrangements of the surface spins changing the pattern of alignment from the 'canted mode' to more or less completely organised spin configuration within **5.5 nm** to **7.0 nm** of particle size.

The above situation can be described in terms of a model shown in **Figure - 5.16.** For the 900 sample, some surface spins may be canted even for H_{ext} = 50 KG, because the actual saturation of spin arrangements might occur at H_{ext} > 50 KG. This model is shown for the 700 sample with a particle size of around **5.5 nm,** although it is very much possible for the 650 sample, where more spins are present on the surface of the nano particles of magnetite, since the particle size is still smaller, *i.e.* **4.5 nm.** As said earlier, in the 900 sample, these surface ions are more or less aligned, *i.e.* less canted, which is clearly evident from **Figure - 5.15** as a function of magnetic field. It is noted from this figure that at H_{ext} = 10 KG, there is still some significant intensity of 2nd and 5th lines, whereas with H_{ext} = 50 KG, these lines are almost extinguished, indicating almost no 'spin canting' [9].

For the samples heat-treated at lower temperatures with still smaller particles size, these surface spins (as shown in the model figure) are quite difficult to be aligned even with high magnetic field. Therefore, the delicate role of smaller nano particles of having the surface spins aligned is emphasized here in this model. This also highlights the effective role of particle anisotropy for the alignment of the surface spins in terms of the 'fast relaxation' of the nano particles.

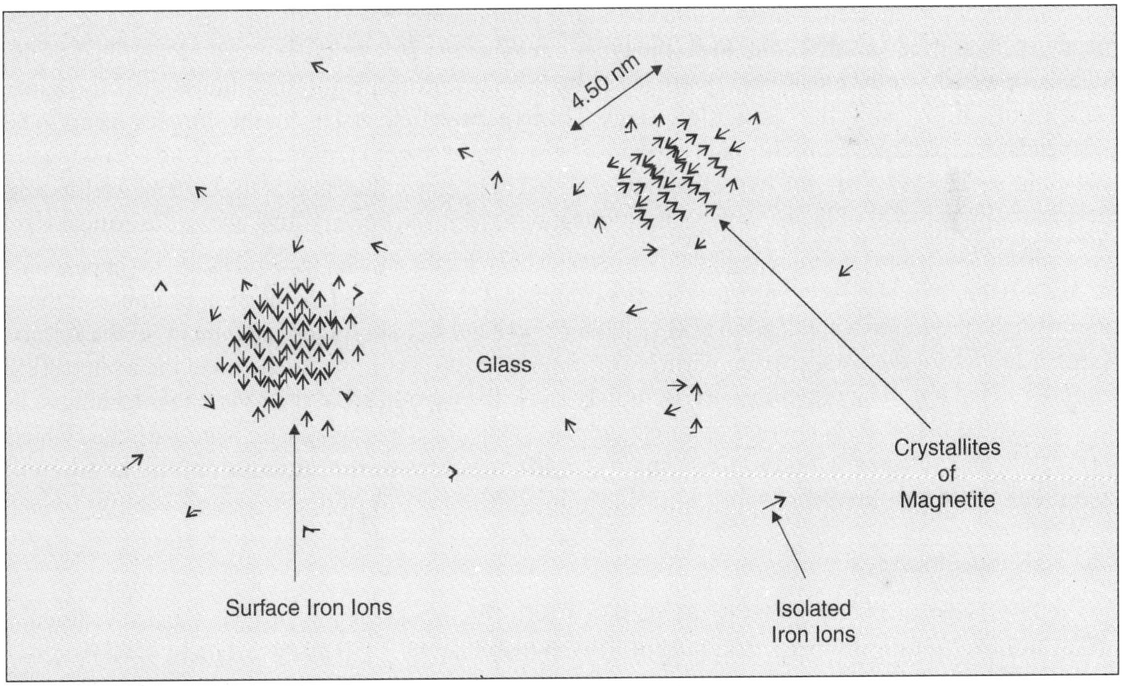

Figure 5.16 : Magnetic model of the basalt glass heat-treated at 650 or 700°C, where the spins appear to be canted.

As pointed out earlier, when the particles are small, the 'number of spins' on the surface is naturally more. Since the particle size of the 650 and 700 samples is less than that of the 900 sample, the surface spins are 'canted' in the former samples with smaller nano particles. It is quite remarkable to note that within such a narrow range of nano particle size, such a possibility can occur between **4.5 nm/ 5.5 nm,** *i.e.* for 600 or 700 sample showing spin canting, and **7.0 nm** *i.e.* for 900 sample not showing any appreciable canting of spins. So, there might be a sort of "critical size" even within this nano range, wherein the spins start getting aligned, say between **5.5 nm** and **7.0 nm.** Then, let us say that this critical particle size is **6.0 nm,** which is so sensitive in that this critical size determines whether or not a composite Mössbauer spectra of super-paramagnetic and ferrimagnetic contribution can be observed together, like in the case of 800 and 900 samples at 300°K. It also determines the extent of spin canting, *i.e.* how the surface canted spins observed below this critical size are progressively getting aligned, which means a 'saturation'. In the SANS data, some sort of criticality has also been observed in terms of re-dissolution of smaller nano particles when the larger nano particles grow, also within a narrow range (see the **section 5.7**).

Finally, it should be pointed out that the hyperfine field at saturation, say at $H_{ext} = 50$ KG, is found to be 528 KG, which is not significantly higher than that found at $H_{ext} = 0$. At $H_{ext} = 50$ KG, the hyperfine field is found to be 516 KG at $H_{ext} = 10$ KG. This again shows that the magnetic field necessary to align the surface canted spins arrangements in the 650 and 700 samples should be substantially higher than 50 KG, as also noted in several other magnetic materials [34].

In summary, it can be said that for the nano particles of magnetite within a basalt glass with an average partile size of **7.0 nm,** the Mössbauer spectra at 300°K show both super-paramagnetic and ferrimagnetic contributions, while below a critical size of about **6.0 nm,** only super-paramagnetic behaviour is observed. Below the so-called blocking temperature, the Mössbauer spectra at 4°K show magnetic 'hyperfine splitting' due to ferrimagnetic magnetite. This behaviour can be ascribed to the 'relaxation effect' of the small nano particles of magnetite with respect to the time of the Mössbauer experiment, *i.e.* inverse of the Larmor frequency. The hyperfine field at 4°K for the samples seems to be independent of particle size of the nano particles confirming a 'magnon' model. From the ferrimagnetic part of the spectra at 300°K, the hyperfine field is found to be lower than that of the bulk magnetite, and the anisotropy constant of the nano particles of magnetite seems to be very high above the critical size.

In the low temperature and high field Mössbauer data, for the smaller nano particles, the appearance of $\Delta m_I = 0$ lines with $H_{ext} \neq 0$ confirms the spin canting at the surface, while the disappearance of these lines with the progressive application of H_{ext} above the critical size shows the 'alignment' of the surface spins. A model for the smaller nano particles of magnetite is also given to elucidate their interesting magnetic behaviour. The present results definitely show the usefulness of the Mössbauer technique in elaborating the magnetic behaviour of single domain small super-paramagnetic nano particles, which were "uniquely" created within a glassy matrix at different heat-treatment temperatures, at different conditions of the measurement.

5.6. ESR Spectroscopy

The theory of ESR is given in the **section 1.6.2.** The measurements of ESR was carried out at 300°K in a Brucker spectrometer operating in X-band frequency (9.2 GHz), the magnetic field modulation frequency was 18 KHz. All the low temperature measurement were carried out using Varian Associate spectrometer at X-band frequency and the modulation frequency was 100 KHz. All precautions were taken to reduce the noise level to a minimum with a proper amplification and tuning to get the best possible signals. The ESR spectra were recorded in a standard plotter. Since the total intensity of ESR is dependent on the number of paramagnetic ions per unit volume, the weight of each sample was measured before inserting through a special quartz tube in the microwave cavity.

The typical ESR spectra, taken at 300°K, for some of the samples are shown in **Figure 5.17.** The peak-to-peak intensity (I), line-width (ΔH) and the integrated intensity ($I \times \Delta H^2$) for the $g = 2.0$ resonance are shown in **Table 5.2,** along with the respective weight of the samples. The spectra for the blank glass and 600 samples show both the resonances at $g = 4.3$ and $g = 2.0$ respectively, which are usually observed for paramagnetic materials containing Fe^{3+} ions.

Table 5.2. The ESR Data for Nano Particles of Magnetite.

Heat-Treatment Temperature	*Intensity, I (AU)*	*ΔH (G)*	*I . ΔH²*	*Sample Weight (g)*
As-annealed	2.71	750.0	1.524×10^6	250.0
600	0.59	883.3	4.603×10^5	250.0
650	25.41	166.7	7.061×10^5	10.0
700	35.88	666.7	1.595×10^7	10.0
800	19.57	1066.6	2.226×10^7	10.0
900	25.39	1150.0	3.358×10^7	10.0

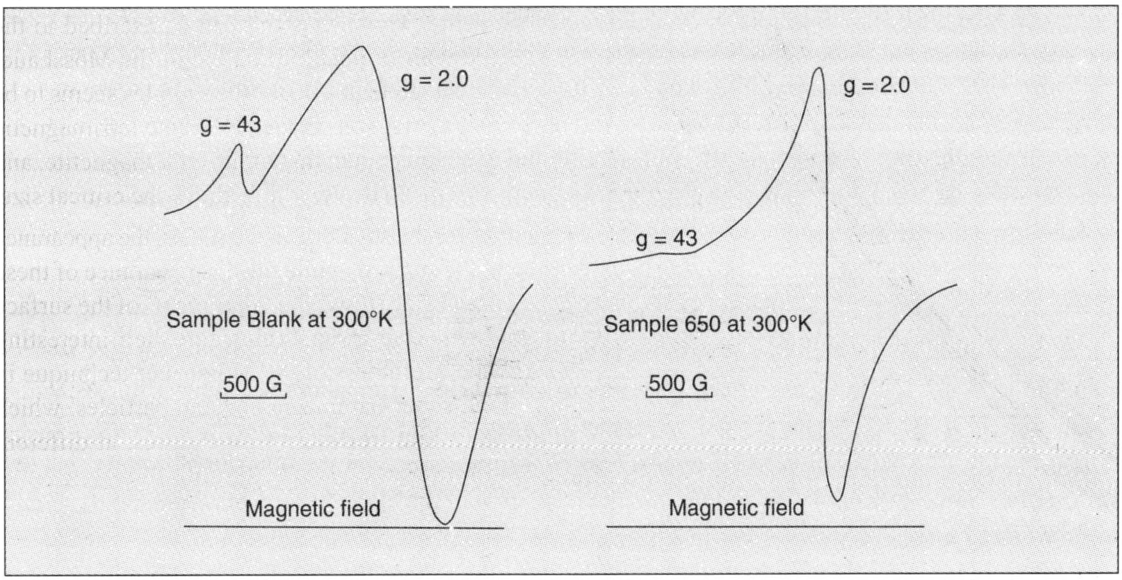

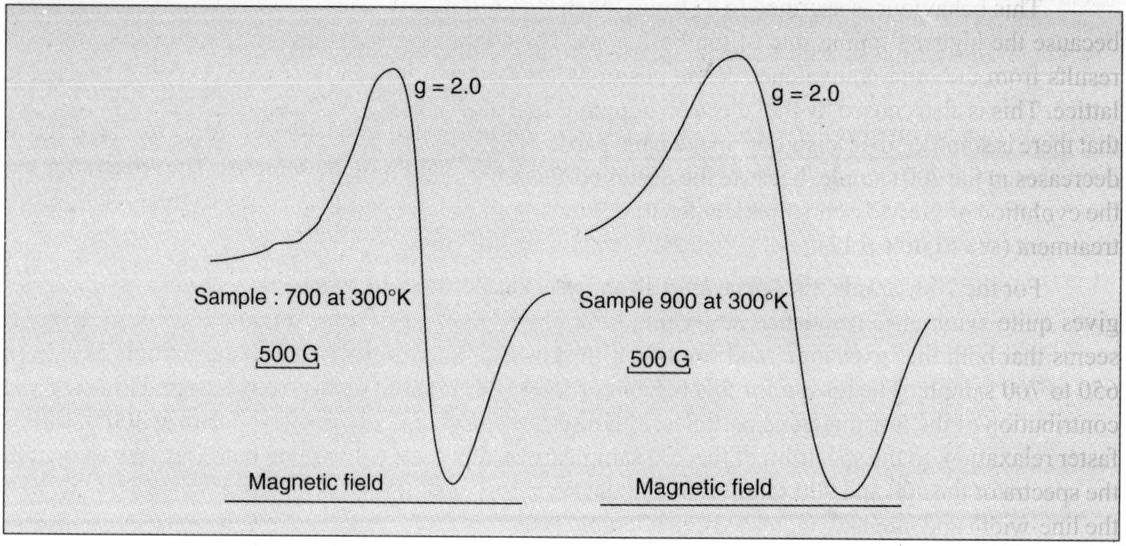

Figure 5.17 : The ESR spectra at 300°K for different samples.

The line-width for the $g = 2.0$ resonance is very high showing that there is a considerable 'exchange interaction' between the Fe^{3+} ions in these samples. For the blank glass, the magnetic ordering of the Fe^{3+} ions, as revealed by the Mössbauer spectra at 4°K (see **Figure 5.13**), could also contribute to the observed line-width. Similar inference can be drawn for the 600 sample as well.

At 4°K, the ESR spectra of the blank glass showed a strong $g = 4.3$ resonance with a very weak $g = 2.0$ resonance, but there was a strong absorption in the higher field region. From the analysis of the exchange Hamiltonian proposed by Owen [36], it can be suggested that the weakening of the $g = 2.0$ resonance at 4°K is due to 'antiferromagnetic interaction' between two Fe^{3+} ions, which are incorporated into small 'nano' clusters, as also revealed by the Mössbauer spectra at 4°K. If the interactions were between Fe^{2+} and Fe^{3+} ions, the $g = 2.0$ resonance should be stronger at 4°K than that at 300°K, which were not observed.

For the $g = 4.3$ resonance of the blank glass, it is found that the line-width increases from 108 G to 220 G between 4°K and 300°K. This is ascribed to the 'spin-spin' interaction rather than 'spin-lattice' interaction [37], which could cause an opposite temperature dependence of the line-width. Hence, even the Fe^{3+} ions in the glass which give rise to crystal field resonances are subject to 'exchange interaction' [38]. At 300°K, the $g = 4.3$ resonance progressively disappears from the blank glass to the other samples. The barely discernible appearance of this resonance for the 650 sample shows that the isolated Fe^{3+} ions have started migrating into the formation of the nano crystals of magnetite. It should be noted that for the 700 sample, this resonance almost disappears for the 700 sample.

The weight of the last four samples were 25 times lesser than those of the first two samples, but the observed intensities of the $g = 2.0$ resonance are still very high. This shows the presence of nano crystalline phases with super-paramagnetic behaviour in these samples. The spectrum for the 650 sample shows super-paramagnetic behaviour. This spectrum is very similar to that obtained for small nano particles of magnetite embedded in an inert carbonaceous matrix, giving the $g = 4.3$ resonance, but the line-width is very small [14].

This behaviour is ascribed to a strong 'exchange interaction' and also to a motional narrowing, because the higher flipping rate of the Fe^{3+} ions. The slight asymmetry in the shape of the spectrum results from the 'non-equivalence of the positions' of the Fe^{3+} ions in the nano crystalline magnetite lattice. This is also caused by the 'electron hopping' in the octahedral site in magnetite [14]. This shows that there is some kind of 'disorder' in the nano particles of magnetite for the 650 sample. This 'disorder' decreases in the 700 sample, because the observed spectrum is more symmetric. This is also reflected in the evolution of the Mössbauer parameters (*i.e.* δ and Δ) showing high symmetry around 700°C of heat-treatment (see **Figure 5.12**).

For the 700 sample, the spectrum still shows a super-paramagnetic behaviour, and the spectrum gives quite symmetric resonance at *g*-value little higher than 2.0, but the line-width is very large. It seems that both the 'exchange' and 'motional' narrowing decrease with increasing particle size from 650 to 700 sample. The reason for this behaviour is very difficult to justify at this stage. However, the contribution of the ferrimagnetic particles, although not detected in Mössbauer spectra at 300°K due to faster relaxation, to the spectrum of the 700 sample cannot be completely ruled out. This was evident in the spectra of the 800 and 900 samples in that the *g*-value was shifted to a still lower value, and the both the line-width and intensity increased considerably (see **Table 5.2**). All these results are quite consistent with the room temperature Mössbauer data, as described in the **section 5.5.**

For the 700 sample, the spectra showed a continuous change from 300°K down to 130°K in that the line-width increased rapidly with the decrease of intensity. The spectrum at liquid nitrogen temperature, *i.e.* at 77°K, shows a broad resonance at *g*-value much higher than 2.0, as shown in **Figure 5.18.** This is typical ferrimagnetic resonance [39, 40]. Although the magnetization measurements at 77°K show super-paramagnetic magnetite particles, but the Mössbauer spectra at 77°K show ferrimagnetic behaviour, the presence of ferrimagnetic particles gives such a strong resonance that the super-paramagnetic part of the resonance observed at 300°K is washed out. This was evidently manifested in the evolution of the spectra with decreasing temperature.. This shows that the interpretation of the ESR spectra is much more complicated than those of the other measurements, as discussed in the previous sections, because of the contribution of different 'magnetic interactions' to the observed line-width of a single resonance.

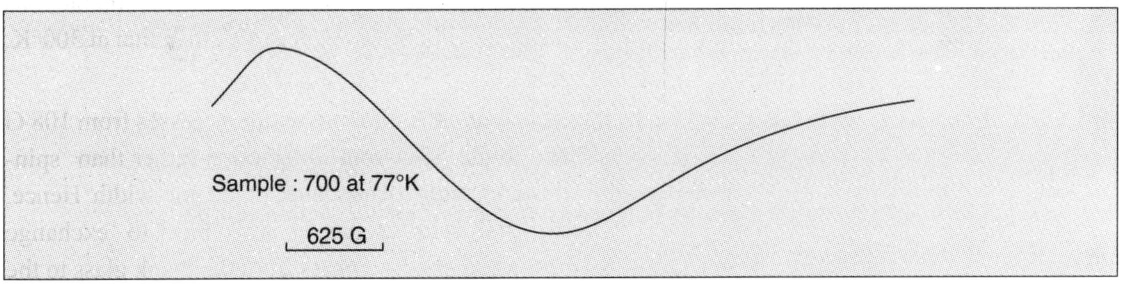

Figure 5.18 : The ESR spectra at 77°K for the 700 sample.

However, it seems more likely that the time required for the ESR measurement (τ_{obs}), which is almost comparable with that of Mössbauer measurement, is much less than the relaxation time (τ) of the nano particles at 77°K making the ferrimagnetic part of the resonance appear in the observed spectra. The importance of different measurements at different temperatures (300°K to 4°K) as highlighted before is again brought to the focus here.

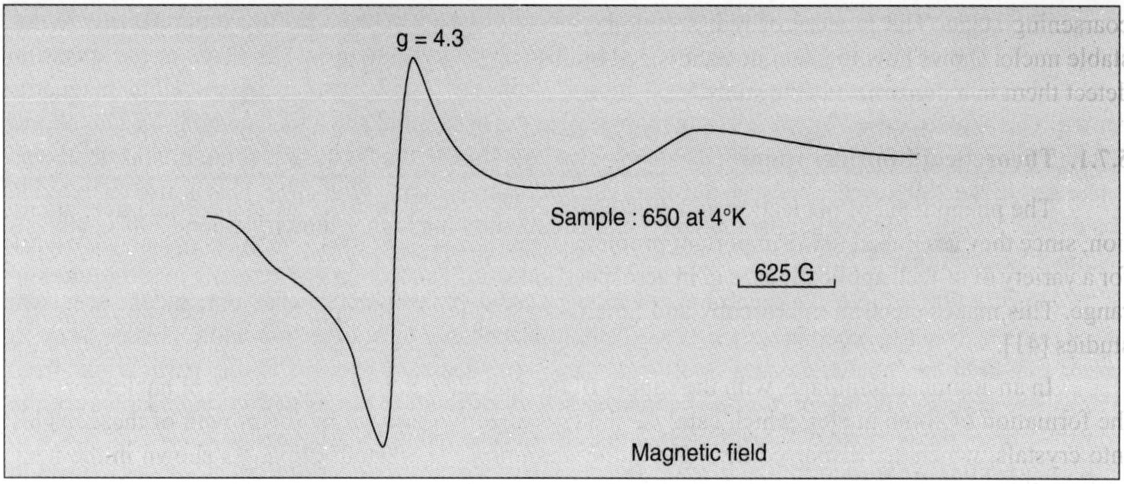

Figure 5.19 : The ESR spectra at 4°K for the 650 sample.

A very similar behaviour for the $g = 2.0$ resonance was observed for the 650 sample down to 135°K, but the $g = 4.3$ resonance remains the same as that at 300°K. No measurement was carried out at 77°K. However, the ESR spectra at 4°K show a strong g = 4.3 resonance with a broad ferrimagnetic component at a lower g-value, as shown in **Figure 5.19.** The appearance of the $g = 4.3$ resonance at 4°K with a significant intensity could be thought to be due to the increase of Boltzman's population between two energy levels, but it is not understood why this resonance remained stagnant down to 135°K. This behaviour could be mainly attributed to the 'spin-spin' relaxation for the 650 sample with a particle size of nano particles of magnetite of **4.5 nm**. There are still some Fe^{3+} ions, which are expected to remain outside the crystalline magnetite phase. The ESR spectra show that in addition to the magnetic interactions between the Fe^{3+} ions inside and outside the nano-crystalline magnetite particles, there could be magnetic interactions between the isolated Fe^{3+} ions and the Fe^{3+} ions on the surface of the nano-grains.

In summary, it can be said that although the time for ESR measurement is comparable to that of Mössbauer experiment, the contribution of different magnetic interactions complicates the analysis of the line-width. However, in general, the ESR data at 300°K support the Mössbauer data.

5.7. SMALL ANGLE NEUTRON SCATTERING

Preamble

A multi-component basalt glass is a unique system for studying the nucleation and crystallization behaviour of nano-sized magnetite particles, whose growth can be controlled to precipitate even **4 nm - 5 nm** size crystallites. The most sensitive technique for such a study is 'Small Angle Neutron Scattering' (SANS), which allows the determination of small nano-sized crystallites as a function of time at a given temperature, *i.e.* dynamic study of crystallization behaviour on nucleated samples at suitably designed temperatures to facilitate their growth.

The inter-particle interference effects show a pronounced maximum in the SANS spectra at the growth temperature of 710°C for different nucleated samples. The radius of such particles was estimated within the limits of Guinier approximation from the slope of the linear plots of I(Q) vs. Q^2. The number density of nano-sized particles decreases with the time of growth and saturates at longer time. This particular behaviour is interpreted as due to the redissolution of smaller nano particles, as the larger ones (still nano-sized) continue to grow with the stabilized number density in the Ostwald ripening or coarsening stage. The process of redissolution of the nano-sized particles during the growth of the stable nuclei shows how to generate nano-sized grains within a glass matrix and the power of SANS to detect them in a dynamic on-line study.

5.7.1. Theoretical Considerations

The phenomena of nucleation and crystallization in glasses have received a considerable attention, since they are linked to the important problem of the synthesis of glass-ceramics, which are useful for a variety of hi-tech applications, *e.g.* in aerospace industry. The fineness of structure is in the **100 nm** range. This makes electron microscopy and Small Angle X-ray Scattering (SAXS) well suited for the studies [41].

In an atomic assemblage, with the supply of thermal energy, there is a process of nucleation, *i.e.* the formation of some nuclei, which catalyse the crystalization process by the growth of these nuclei into crystals, which are dispersed throughout the matrix by the crystallization. As shown in the next section, this is the process of making glass-ceramics, or rather ceramization of a disordered structure like glass. The most readily observed physical property of nuclei is their ability to facilitate growth of a new phase under appropriate thermodynamic conditions. Therefore, the study of the growth process, immediately following the nucleation, is the most accessible though an indirect method of enquiry into the phenomenon of nucleation. However, it is often difficult to follow the very early stages of growth either because it occurs too rapidly, or because the size (or total mass) of the precipitates falls outside the conventional range of detection of the structural techniques. As will be evident later, small angle neutron scattering (SANS) is a very powerful technique to do the required job. However, first of all, we must understand the nucleation and crystallization phenomena, as detailed below.

5.7.2. Nucleation and Crystallization Behaviour

First of all, it is important to explain the nucleation and crystallization behaviour of glassy materials to understand their mechanism of transformation to glass-ceramics [1]. It is a quite involved subject and hence it is simply not possible to give a comprehensive review. However, a short description is attempted here in order to elucidate certain important points to be able to understand the study of nucleation and crystallization behaviour of the nano particles of magnetite by a very sensitive technique like 'Small Angle Neutron Scattering' (SANS) [2, 3].

When a material from a stable liquid state is cooled, it may start crystallizing at the crystallization temperature (T_C) by reducing its volume to a more stable state to be able to form the crystals. On the other hand, the material may now ignore this T_C, then it cools down further along the original path of cooling and at a certain temperature called the glass transition temperature (T_g), it forms a glass by changing the slope of this cooling curve at a viscosity of about 10^{13} poise. Hence, it is known that the glass is a thermodynamically metastable material and due to the blockage on the 'atomic rearrangements', it may not be transformed to a stable state. By this logic, a solid glass (which obviously has higher volume than the corresponding crystal) must be given sufficient energy, *i.e.* thermal energy, to make the 'atomic rearrangements' possible above T_g so that it can nucleate and crystallize to form small crystallites, *e.g.* in a basalt glass.

If the thermal treatment for the crystallization of a glass is carried out so that the 'precipitation' of the crystallites with special properties occurs throughout the material, the development of new material, called 'glass ceramics', is possible [1].

5.7.2.1. Homogeneous Nucleation

It should be remembered that the crystallization of a homogeneous phase (liquid or glass) does not occur in a material in the entire mass. First, it starts at a point and then it extends progressively from the discrete centers, and distributed in the entire mass. That's how the whole process takes place.

When a liquid is cooled below its fusion point, the crystallization occurs by the growth of crystals at a 'finite rate' from a 'finite number' of nuclei. The glass formation can be attributed to a low rate (*i.e.* almost nil rate) of crystal growth, or low rate of nuclei formation, or a combination of both the factors.

Here comes the question of 'stability'. The stability of a particle of the new phase in homogeneous nucleation will depend on two contributions :

1. One from a difference in free energy between the two phases, and
2. The other from the interfacial energy.

At the fusion point, the free energy of a given quantity of a material is the same in the crystalline and in the liquid forms. At lower temperatures, the crystalline form will always have lower free energy and the liquid will crystallize, if nuclei of correct sizes are available in sufficient numbers for germination.

In the nucleation state, very small particles (called embryos) become the starting point for the development of an 'ordered region'. These small particles, which are formed due to the 'structural fluctuations' produced by the thermal excitation, have different sizes - which fluctuates or varies constantly.

It will be seen later that this size attains a value of a 'critical size' so that it can serve as a 'starting point', *i.e.* a germ or a nucleus, for the formation of a new crystalline phase. The germination or nucleation across the system is called homogeneous - if all the elements of the parent phase are identical structurally, chemically and energetically with those of the nucleated volume. That's the necessary condition for homogeneous nucleation.

Evidently, this is only possible if the entire volume of material is chemically homogeneous and is free from any type of structural imperfections. In practice, this is quite difficult, since the surface itself is already an inevitable source of imperfections. It can further aggravate from the presence of some foreign impurities. In this case, the energy required to form a 'germ' or a 'nucleus' is found to be reduced for such impurity-laden sites, and the nucleation will be produced preferentially at their contacts. In that case, the nucleation is called heterogeneous nucleation.

In practice, it is quite rare that this alternative route of nucleation can be avoided, and we sometimes wonder whether the homogeneous nucleation is totally realizable. However, this case signifies an ideal situation.

The embryos becoming nucleus then increases in size by the successive addition of atoms taken out of the liquid phase, giving rise to the formation of a crystalline particle, which grows with a certain rate at the expense of the phase in its environment or the parent phase, *i.e.* the parent phase supplies the necessary atoms to the surface of the 'already formed' small embryos to help them grow further to complete the process of crystallization. This is called the "crystal growth stage".

5.7.2.2. General Conditions of Kinetics of Vitrification

The number I of nucleus produced per unit volume per unit time, *i.e.* the rate of nucleation, and the rate of crystal growth U both depend on the temperature, as shown in **Figure 5.20.** This figure needs an explanation.

For a liquid to form a glass, it must be cooled rapidly to avoid crystallization. Below the temperature of fusion (T_f), the liquid consists of a stable phase. As the liquid enters a stage of supercooling below (T_f), the growth can theoretically take place between T_f and T_3. However, the formation of the initial nucleus necessary before the growth can take place, occurs between T_2 and T_4.

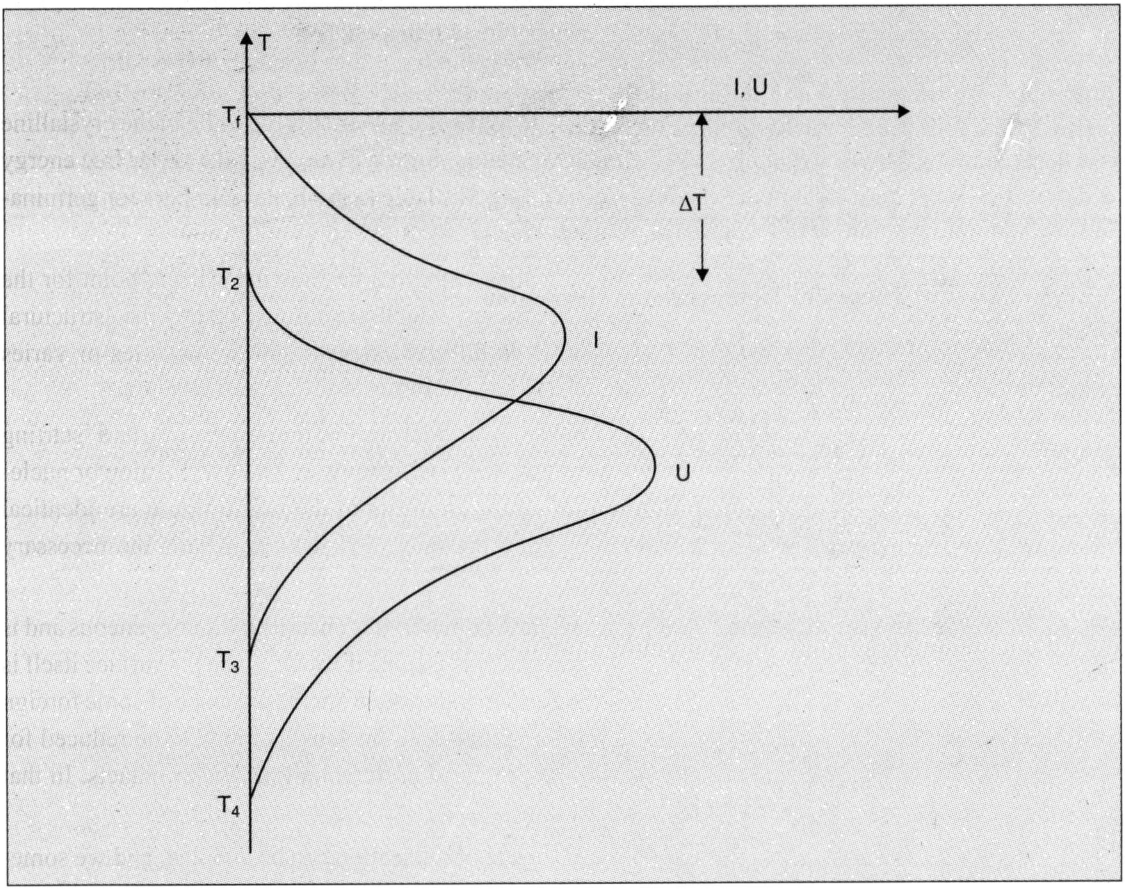

Figure 5.20 : The rate of nucleation and crystallization curves for a glass.

Thus, the critical region consists of a temperature between T_2 and T_3, and the possibility of crystallization will depend on the manner that one curve is imposed on the other (see **Figure 5.20**) as well as on the absolute values of the respective rates in the overlap region

If on the common interval $T_2 - T_3$, either I or U (or, I and U together) are too small, the crystallization is not detectable - and the system passes to a vitreous state. On the other hand, if I and U are quite important, *i.e.* strong overlap of the two curves, the total crystallization could not be avoided.

If in this interval $T_2 - T_3$, I is weak but U is strong, the crystallization could give rise to a small number crystals distributed in a glassy phase, whereas for the opposite case (*i.e.* I is strong, but U is weak), it will give rise to a partially crystallized material with finer grains.

Based on the above description, it can be said that for a solid glass like a 'basalt glass', both nucleation rate (I) and crystal growth rate (U) are dependent on the temperature, and hence this idea can be used to devise a strategy for 'heat-treatment schedule'. The maximum in I appears at a lower temperature than the maximum U. The most effective heat-treatment schedule firstly for nucleation is obviously attained at the temperature that gives the maximum nucleation rate.

Hence, a glass that has already cooled down (after forming) to room temperature is generally re-heated to this temperature, which is followed by a heat-treatment at a higher temperature for crystal growth. In a typical 'temperature-vs-time' curve, the temperature is first raised quickly to the nucleation zone and kept there for sometime, which is enough for enough number of nuclei to be formed in the glass. Then, the temperature is raised again to the crystal growth zone quickly and kept at this temperature for sufficient time for the nuclei to form stable crystal, and then the temperature is dropped rapidly to room temperature. This schedule depends on many factors including the composition of the glass. This is called 'two-stage' heat treatment schedule (see ref. [48, 50] later), which has also been used in case of basalt glass to precipitate nano particles of magnetite [2, 3].

5.7.2.3. Classical Theory of Homogeneous Nucleation

At the temperature where there is an appreciable mobility of the atoms, the atomic rearrangements continuously take place following the thermal agitation. If the phase is thermodynamically stable, this domain has a 'special' existence, since it is destroyed and replaced by the other phase. As the phase becomes metastable, certain fluctuation — which is the potential source for a more stable phase — can become permanent.

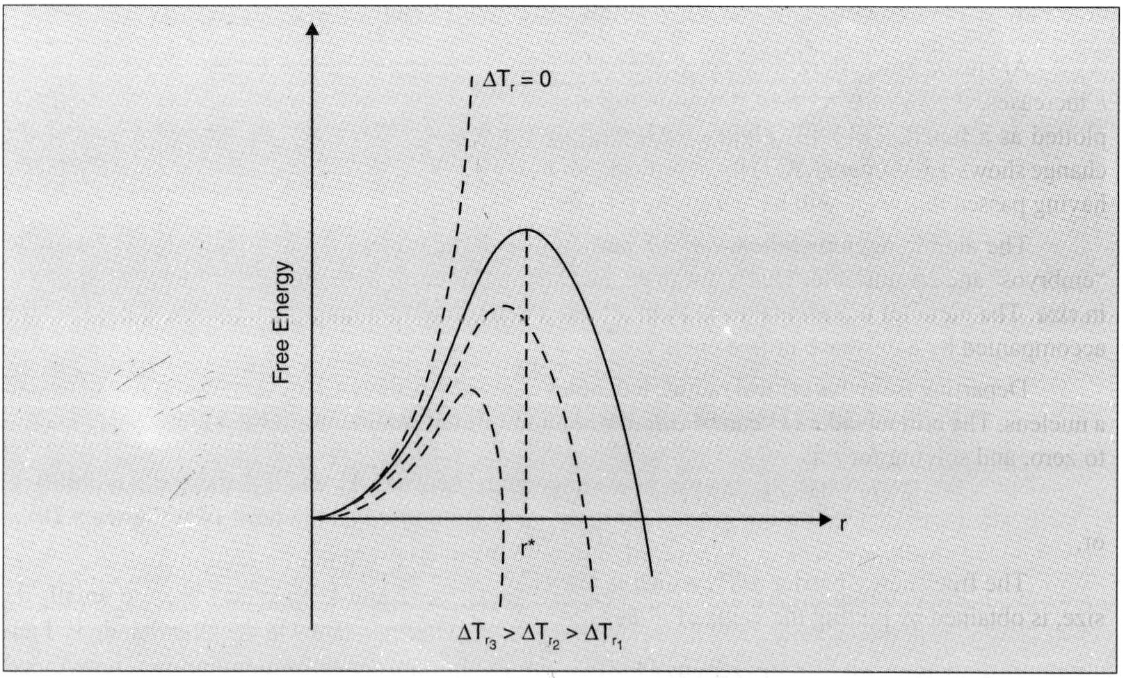

Figure 5.21 : Free energy curve against radius of the nuclei.

These fluctuations can have a variability of size, a form, a structure and a composition. In the most simple classical model, proposed by Volmer and Weber, and Becker and Doring, it is assumed that the embryos have a structure, a composition and uniform properties, which are identical to the 'future phases', and both these entities do not differ in their form and size (see ref. [48] later).

On the question of 'form', it is the 'form' which will make their energy of formation minimum. Thus, it is closely linked to the nature of the interface. If it is now assumed in the first approximation that the interfacial energy is independent of the crystallographic orientation and that the energy due to elastic deformation is negligible, the embryos will have a spherical form. The size of the embryos will result from the thermodynamic condition of stability (briefly mentioned above).

The molar free energy of the liquid and the crystalline phases is shown in **Figure 5.21** as a function of temperature. The curves grow at T_f, which represents the equilibrium between the phases. At this point, obviously, the difference in free energy ΔG between the two phases is zero. The portion of the curve representing the stable liquid or crystalline phase is denoted by continuous lines, and those of the metastable phases (*i.e.* supercooled liquid or superheated crystal) are represented by dashed lines.

For a temperature $T < T_f$ or a degree of undercooling $\Delta T = T_f - T$, the supercooled liquid is in metastable equilibrium with respect to the crystal. The transformation of the liquid to crystal is accompanied by a negative variation (ΔG_v) per unit volume. For a spherical nucleus of radius r, this corresponds to a liberation of a quantity of energy equal to $(4/3)\pi r^3 \Delta G_v$.

But this formation of the nuclei is accompanied by a modification of the interface, which contains a quantity of energy ΔG_S per unit surface area, which to a first approximation can be taken as equal to the macroscopic 'surface energy' (σ). For a nucleus of radius r, this requires an energy equal to $4\pi r^2 \Delta G_S$. Therefore, by taking stock of the whole situation, *i.e.* the bulk or volume effect and the surface effect, the free energy change for the nucleation ΔG_r is given by

$$\Delta G_r = (4/3)\pi r^3 \Delta G_v + 4\pi r^2 \Delta G_s$$

At small values of r, the surface term in r^2 will dominate and ΔG_r will be positive. However, as r increases, the volume term in r^3 will dominate and ΔG_r will become negative. The two terms are plotted as a function of r in **Figure 5.21,** with the summation as a solid line where the free energy change shows a maximum (ΔG^*) for a 'critical radius' (r^*). Above this critical radius (r^*), the fluctuation having passed this stage will have a strong possibility of growing.

The atomic agglomeration, *i.e.* the particles, with an average radius smaller than r^* is called "embryos" and are unstable. This is due to the increase in free energy, which accompanies their reduction in size. The particles of radius greater than r^* are called "nucleus" and are stable, since the growth is accompanied by a decrease in free energy.

Departing from this critical radius, it denotes a size of an embryo, which is susceptible to become a nucleus. The critical radius r^* can be calculated by setting the derivative of ΔG_r with respect to r equal to zero, and solving for r as

$$d(\Delta G_r)/dr = 4\pi r^2 \Delta G_v + 8\pi r \Delta G_s = 0$$

or,

$$r^* = -2\Delta G_s / \Delta G_v$$

The free energy barrier ΔG^*, which is associated with the formation of the embryos of critical size, is obtained by putting the value of r^* as :

$$\Delta G^* = \left(\frac{16\pi}{3}\right)\frac{\Delta G_s^{\ 3}}{\Delta G_v^{\ 2}}$$

Now, if v is the volume of one atom, the volume of a nucleus containing n atoms is nv. Hence, we can write

$$nv = \left(\frac{4}{3}\right)\pi\, r^3$$

or,

$$n = \left(\frac{4}{3}\right)\frac{\pi r^3}{v}$$

We can substitute for r^* in the above equation, the number of atoms in the critical-sized particle (n^*) is then obtained as

$$n^* = \left(\frac{32\pi}{3}\right)\left(\frac{\Delta G_s}{\Delta G_v}\right)^3$$

Since the formation of an 'embryo' involves a positive free energy change, the probability of such an occurrence would be quite small, *i.e.* it is not energetically favourable. However, the entropy of the solid-liquid system can be increased by the presence of a number of atom-clusters in equilibrium with the atoms of the liquid. It has to be noted here that the above description of the 'criticality' of the number and size of the nuclei is quite important to understand the nucleation and crystallization behaviour of any glass-ceramic system, including the basalt glass-ceramic, which has been studied dynamically by SANS [2, 3].

5.7.3. Small Angle Neutron Scattering

The 'Small Angle Neutron Scattering' (SANS) technique shows some advantages in following the 'growth process' of the precipitates in the bulk materials. Three particular advantages are generally exploited

1. The range in scattering vector (Q), which is possible to be studied, include the region around Q = 5 × 10^{-4}/nm, which usually falls between the scope of Small Angle X-ray Scattering (SAXS) and Light Scattering (LS) measurements.
2. The low absorption of neutrons even at long wavelengths **(~ 1 nm)** allows an easy investigation in the transmission of a bulk glass, which is heated directly in situ in the neutron beam, *i.e.* a dynamic on-line study.
3. The scattering contrast between the precipitates and the host glass matrix, particularly glasses containing transition metal TM oxides, like a basalt glass.

The striking effects have been revealed by SANS study in different glass-ceramics like basalt, cordierite etc. [2] during the early stages of the growth of the precipitates, *i.e.* the formation of just 'nano' particles. As a result of the continuous viscosity-temperature relation, the rate of crystal formation in glasses can be put on a 'favourable' time-scale to follow the 'kinetics of growth' simply by working at appropriate temperature.

The 'Small Angle Neutron Scattering' (SANS) technique was first used by Roth and Zarzycki [42] on samples heat-treated at different temperatures, and by Bandyopadhyay and Zarzycki [3] for the dynamic study of nucleation and crystallization behaviour at the high-flux reactor facilities created at Institut Laue Langevin (Grenoble, France).for such studies to be able to compare with those of SAXS data on various systems, and also for the early nucleation and crystallization studies, when the size of the nuclei or crystallites are in the very small nano range [2-4]. The readers with a knowledge of Fourier transform should appreciate the following part on the theoretical considerations for SANS study, where we take a real example of a material.

How Fourier transform comes into this picture?

The intensity of the scattered radiation in both cases of X-rays and Neutrons at small angles as a function of the scattering vector k is proportional to the square of the modulus of the Fourier transform of the 'fluctuations' of the local 'scattering power density', $P(r)$. As a real example, for a two component system like soda-silica glass, it is expressed as

$$P(r) = N(r)\{C(r)\, b_1 + (1 - C(r)b_2\} - \langle N(r)\{C(r)\, b_1 + (1 - C(r)b_2\}\rangle \quad ...(5.12)$$

where, b_1 and b_2 are scattering coefficients for stoichiometric silica and soda units respectively, $C(r)$ is the local mole concentration in silica, and $N(r)$ is the local value of the number of the stoichiometric units per unit volume. Since silica and soda units have the same molar mass ($M \approx 60$), $N(r)$ can be simply taken as proportional to the local density $\rho(r)$ as

$$N(r) = (N/M)\, \rho(r) \quad ...(5.13)$$

where, N = Avogadro's number.

The term in the bracket $\langle\,\rangle$ represents the 'average' value of the first term over the volume of the sample. In the case of X-rays, it can be readily shown that $b_1 = b_2$, since both units of silica and soda have the same number of electrons (*i.e.* 30). Hence, $P(r)$ is reduced and can be written as

$$P(r) = (N/60)\, [\rho(r) - \langle\rho(r)\rangle]\, b_2 \quad ...(5.14)$$

Thus, in the soda-silica glasses, the SAXS only measures the 'density fluctuations', and it is insensitive to the 'concentration fluctuations'. In the case of neutrons, the values of b_1 and b_2 are given as

$$b_1 = 1.576 \times 10^{-12} \text{ cm}$$

and,

$$b_2 = 1.280 \times 10^{-12} \text{ cm}$$

Hence, the SANS depends on both density and molar concentration fluctuations, which are expressed as

$$P(r) = (N/60)[\rho(r) - \langle\rho(r)\rangle]b_2 + (N/60)[\rho(r)c(r) - \langle\rho(r)c(r)\rangle](b_1 - b_2) \quad ...(5.15)$$

Thus, a comparison of SAXS and SANS data could separate the two effects. The solid states actually represent 'bounded atomic association', and the diffracted intensity is a consequence of the coherently scattered wave interference. The neutron scattering can be coherent (both elastic and inelastic) and incoherent. By neglecting the incoherent term, the differential cross section for neutron coherent scattering is given for a general case by the Fourier sine transform as :

$$(d\sigma/d\Omega)_{coh} = N\langle b\rangle^2 \{1 + \int_0^\infty 4\pi r^2\, [\rho(r) - \langle\rho(r)\rangle]\, \sin(Kr)/Kr\; dr\} \quad ...(5.16)$$

The term in the large bracket involves the structure factor incorporating the concentration and density terms of the individual scattering atoms. It is seen from this equation that the RDF of atomic density $[4\pi r^2 \rho(r)]$ is the Fourier transform of the diffraction intensity. Then, from the Fourier inverse transform, the density and concentration fluctuations (local) can be estimated in an amorphous material with an interface of phase separation. Hence, the importance of Fourier transform is again seen for SAXS and SANS problems.

5.7.4. Interpretation of the SANS Data

In a so-called homogeneous glass, the concentration fluctuations, which influence the spontaneous formation of 'clusters', are uncorrelated. Therefore, they should lead to spatially and temporally random appearance of the nucleation centres throughout the glass matrix. The growth of the new crystalline phase, however, modifies the solute 'super-saturation' ratio in the neighbourhood of growing particles by forming a 'depleted zone' extending from the particle surfaces [43]. The concentration at the surface maintains a dynamic equilibrium with the particle radius, but the gradient and the extent of the 'depleted zone' depend upon the diffusion rate of solute material and the age of the particle. It has been suggested that the presence of the 'depleted zones' can deform both the 'spatial and temporal randomness' of the nucleation process by lowering the probability of nucleation near existing nucleation sites, thereby creating short range order between the particles and eventually a saturation of the particle number density occurs [44].

When the diffusion rates are high, the 'depletion zones' can overlap to fill up the entire space available. All the particles then become influenced by the combined concentration profiles set up by the neighbouring particles. When the local concentration falls below a critical level, some particles will redissolve. The most affected will be the smaller particles close to or surrounded by the larger neighbours [5, 44]. This simply implies that, since the concentration of solute is greater around smaller particles than around larger ones, there is a diffusion of solute to the larger particles from the smaller ones, and these will eventually redissolve [5].

Let us mention very briefly about the theory of small angle neutron scattering (SANS), which is relevant for a study of nano particles, as described below :

By considering the fluctuations of electronic density, it is known that the local density [$\rho(r)$] of a material is not uniform in the microscopic scale, but it shows a small deviation with respect to the average density (ρ_0) as :

$$\rho(r) = \rho_0 + \Delta\rho \qquad ...(5.17)$$

It is convenient to decompose the fluctuations in their spatial Fourier components $\Delta\rho_x$ as :

$$\Delta\rho_x = \left(\frac{1}{V}\right)\int_v \Delta\rho(r) \exp(-ix.r)\, dr \qquad ...(5.18)$$

where V is the volume of the system.

For a dilute system of N identical particles of volume V, having a uniform electronic density ρ' contained in a matrix of density 'ρ', the intensity of the scattered radiation at small angles as a function of the scattering vector $Q = 4\pi \sin \theta/\lambda$ (where θ = scattering angle and λ = wavelength), is written as :

$$I(Q) = I_e (\Delta\rho)^2 NV^2 (1 - 1/3\, R_G^2.Q^2 + \ldots) \qquad ...(5.19)$$

where, I_e is the scattered radiation by an electron and $\Delta\rho = (\rho - \rho')$ is the difference of the electronic densities of the particles and the glass matrix.

The form of the scattering curve , *i.e.* I(Q) depends on the electronic gyration radius R_G, defined by the density $\rho(r)$ as follows

$$R_G^2 = \frac{\int_V r^2 \rho(r)\, dr}{\int_V \rho(r)\, dr} \qquad ...(5.20)$$

For a homogeneous particle of density $\rho(r) = \rho$ of radius R_S, it can be written as

$$R_G = R_S \sqrt{\frac{3}{5}} \qquad ...(5.21)$$

The intensity, I(Q), is represented by Guinier approximation as

$$I(Q) = I_e (\Delta\rho)^2 NV^2 \exp(-1/3\, R_G^{\,2}.Q^2) \qquad ...(5.22)$$

However, as shown later in this case, the main feature of the SANS spectrum is a well-defined peak, which is characteristic of the 'inter-particle interference' effects [2, 4, 45]. In the absence of the long-range order, the curve of I(Q) at $Q > Q_{max}$ is not strongly modified by the inter-particle interference, but it is characteristic of 'scattering' from the isolated particles for which the mean particle size can be determined within the limits of "Guinier" approximation, *i.e.* $Q R_G < 1.2$, from the slope of the linear plot of ln I(Q) vs. Q^2.

From the maximum of the 'scattering curve', a characteristic wavelength ($\Phi = 2\pi/Q_{max}$), which is interpreted as the 'mean distance' between the small precipitates, and hence a relative 'number density' of particles can be obtained ($N = \Phi^{-3}$) [2, 4, 45]. Therefore, the above three parameters (R_G, Φ, N) can be obtained from the simple evaluation of the scattering curves of the SANS spectrum. So, now we know : what we can get from SANS study on nano materials ?

5.7.5. Preparations for the SANS Study

This deals with the magnetic properties of a basalt glass contain in about 13 wt% iron oxide, which has been heat-treated at different temperatures, *i.e.* different particles of magnetite with a narrow range of nano-sized grains. On heat-treatment, the basalt glass transforms into a fine-grained glass-ceramic [1]. In order to control the quality of glass-ceramics, it is important to study the nucleation and crystallization behaviour of this system. Initially, X-Ray diffraction (XRD) and transmission electron microscopy (TEM) measurements were carried out on samples heat treated at 600, 650, 700, 800, 900°C for 2, 4 and 8h, and also on the as-annealed glass (*i.e.* the blank glass). As explained earlier, the XRD showed that there was no crystalline peaks for blank glass, 600 and 650 samples. The peaks due to magnetite appeared for the 700 sample ; the strongest peak being at the d-spacing of **0.250 nm.** For the 800 and 900 samples, there were also peaks due to the pyroxene along with those of the other minor phases. However, TEM showed the presence of magnetite in the 650 sample, and also in the 700 sample, while the last two samples showed the presence of a mixture of the phases. The particle sizes of the nano particles of magnetite were estimated to range between **4.5 nm** to **7.0 nm** between 650 and 900°C for 8 h only.

Since the particles are small, the supermagnetic behaviour of the particles is very interesting as explained above [6-9]. The Mössbauer measurements at 4°K showed that the maximum amount of magnetite was formed at 700°C, and the evolution of the Mössbauer parameters showed that the magnetite particles have the 'most improved symmetry' at this temperature [6, 9].

The study of growth process immediately following nucleation is the most accessible, though indirect, method of inquiry into the nucleation process [3]. In order to determine an optimum nucleation condition for magnetite in the basalt glass-ceramics, small angle neutron scattering (SANS) measurements were carried out dynamically at 710°C, as a function of time, on a series of samples (dimensions = 10 mm × 10 mm × 7 mm) heat treated (*i.e.* nucleated) between 550 and 665°C for different times. The importance of the SANS measurements for such a study has been discussed elsewhere [4]. The longer times upto 32 h were employed for lower temperatures and shorter times upto 8 h for higher temperatures of heat-treatment. Based on the results mentioned above, the growth of temperature was fixed at 710°C. This was also higher than the Curie temperature of the bulk magnetite (578°C), so that the effect of the 'magnetic scattering' on the SANS spectra could be neglected.

The SANS measurements were carried out in the spectrometer D-17 at the Institut Laue-Langevin (Grenoble); the details of this spectrometer are given elsewhere [46]. The detector distance was 2.85 m and λ_o = 1.282 nm. Absolute small-angle cross-sections were deduced from the experimental neutron intensity, measuring the incoherent scattering on a thick vanadium sample. The sample was put in a small furnace between the incident beam and the detector, and the temperature was raised to 710°C in about 10 to 120 min, except for one sample (nucleated at 577°C for 19 h) which was extended upto 360 min.

5.7.6. SANS Data for Nano Particles

The general form of the scattering curves, which were obtained on a sample nucleated by pre-treatment at 634°C for 2 h, is shown in **Figure 5.22.** The SANS intensity was almost negligible immediately after nucleation and corresponds to the zero time curve. On heating in situ at 710°C, SANS intensity increases with particle growth giving sufficient scattered intensity to allow the scattering curve to be followed every few minutes. This was almost the general feature for all the other samples nucleated at different temperatures. The important feature is that there is a pronounced maximum (Q_{max} ~ 0.005 nm at about 11 min) with the intensity falling toward zero at low Q [2]. A high density of the precipitating particles is responsible for an interference effect, which produces a maximum, Q_m, in the SANS spectrum [45]. The maximum is always apparent, even for the shortest times for which the particle size is very small, and remains stable during the growth period at higher crystallization times.

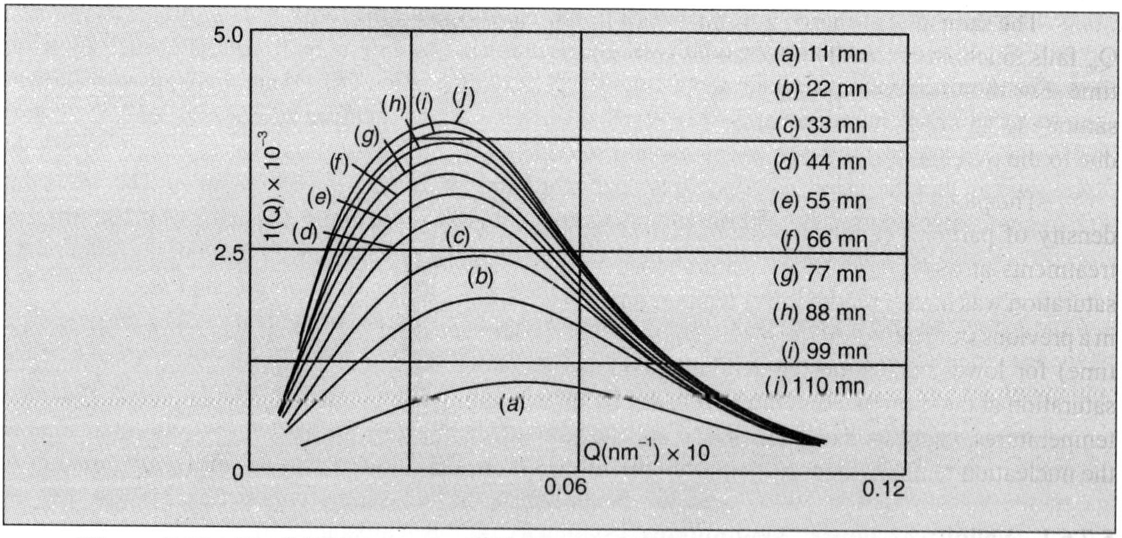

Figure 5.22 : The SANS spectra at different crystallization times at 710°C for a sample nucleated at 634°C for 2 *h*.

The values of Q_m are plotted against the time of crystallization for the samples nucleated at 634°C for 2, 4 and 8 h in **Figure 5.23.** It is seen that, for the 8 h sample, Q_m falls very rapidly during the early times of crystallization up to about 60 min, which is the characteristic of redissolution process with about a constant number of crystallites, as discussed later. A very similar feature is noted for the other two samples treated at 634°C for 2 h and 4 h, respectively.

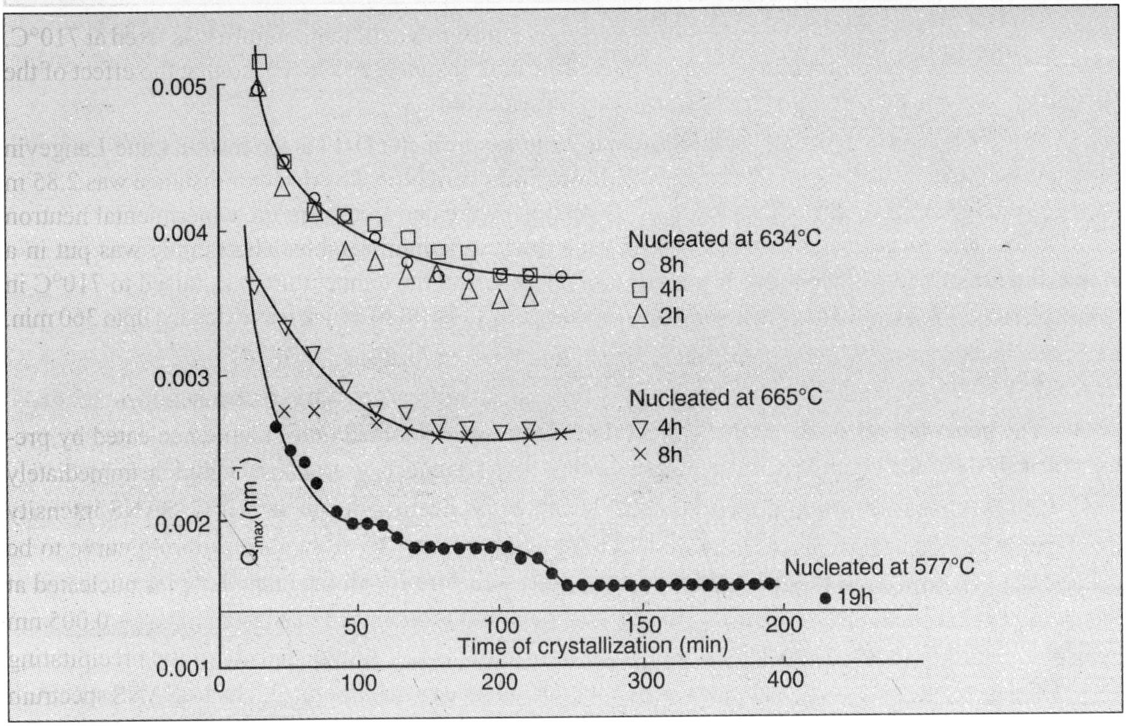

Figure 5.23. A plot of Q_{max} against crystallization time for different nucleated samples.

The same plot is shown in **Figure 5.23** for the sample nucleated at 577°C for 19 h. It is seen that Q_m falls much more rapidly up to about 250 min and then saturates to a constant value towards higher time. For the sample nucleated at 665°C (4 and 8 h), Q_m does not decrease so rapidly and tends to saturate to an almost constant value before 60 min, because the growth process starts earlier. This is due to the overlap of nucleation and growth processes at a nucleation temperature at 665°C [2].

The plots of a stabilized Q_m values as a function of nucleation condition showed that the number density of particles (directly proportional to Q_m) increases with nucleation time, but in the case of treatments at 634°C and 665°C saturation occurred after a few hours. The particle density at the saturation was higher for the lower temperature of nucleation [2].This behaviour has also been observed in a previous study on cordierite glass-ceramic [45]. The higher value of stabilized Q_m (at higher nucleation time) for lower nucleation temperatures, coupled with the observation that the time taken to reach saturation at the growth temperature (*i.e.* at 710°C) is longer for the sample treated at lower nucleation temperatures, suggests that there is a 'competition' between nucleation and growth processes even at the nucleation temperatures, particularly noticeable for the 665°C treatment, as mentioned above.

5.7.6.1. Validity of James' Assumptions

As said above, from the scattering curves, three important parameters can be calculated as

1. From the position of the maximum (Q_m) a rough estimate of the mean distance between the precipitates is obtained ($\Phi = 2\pi/Q_{max}$).
2. Hence the relative number density of the particles (*i.e.* the total number of particles per unit volume of the sample) is also obtained ($N = \Phi^{-3}$).

3. The size of the particles is obtained from the Guinier plots on the higher Q side of the maximum. Within the limit of validity $QR_G \le 1.2$, the scattering function S_Q is related to the radius of gyration (R_G) according to $S_Q = \exp\left(-\frac{Q^2}{R_G^{2/3}}\right)$ and $R_G = \sqrt{\left(\frac{3}{5}\right)}R_S$, where R_S is the radius of the "spherical" particle.

It has been mentioned before that a high density of precipitating nano particles is responsible for an interference effect, which produces a maximum (Q_m) in the SANS spectrum. From this value of Q_m, a characteristic wavelength Φ of density fluctuations has been calculated, which has been interpreted as the 'mean distance between the nano precipitates', and hence a relative density of nano precipitates is obtained. However, as pointed out by Guinier [47], this is only an approximate evaluation. For comparison purposes between the different samples, this relative number density can still be used as a parameter; this also gives an idea about the 'order of magnitude' of the number of nano particles [2].

The parameter Φ is not plotted, because it will show the same types of behaviour as Q_m. The parameter N plotted against crystallization time in **Figures 5.24** for the samples nucleated at 634°C for 2, 4 and 8 h. It is seen from **Figure 5.24** that N decreases rapidly upto 30 min, which can be considered due to larger nano particles or nuclei 'eating' the smaller ones before the growth process starts, *i.e.* a redissolution process for the smaller nano particles as the larger ones continue to grow at their expense. After 30 min, N does not change significantly upto about 60 min, and it then approaches a saturation value when there is a growth of 'stable nuclei'. For the sample nucleated at 665°C, this change is less remarkable, since the saturation arrives at a lower time due to earlier growth process. This is obviously due to the need of a shorter time for the growth process owing to the higher nucleation temperature.

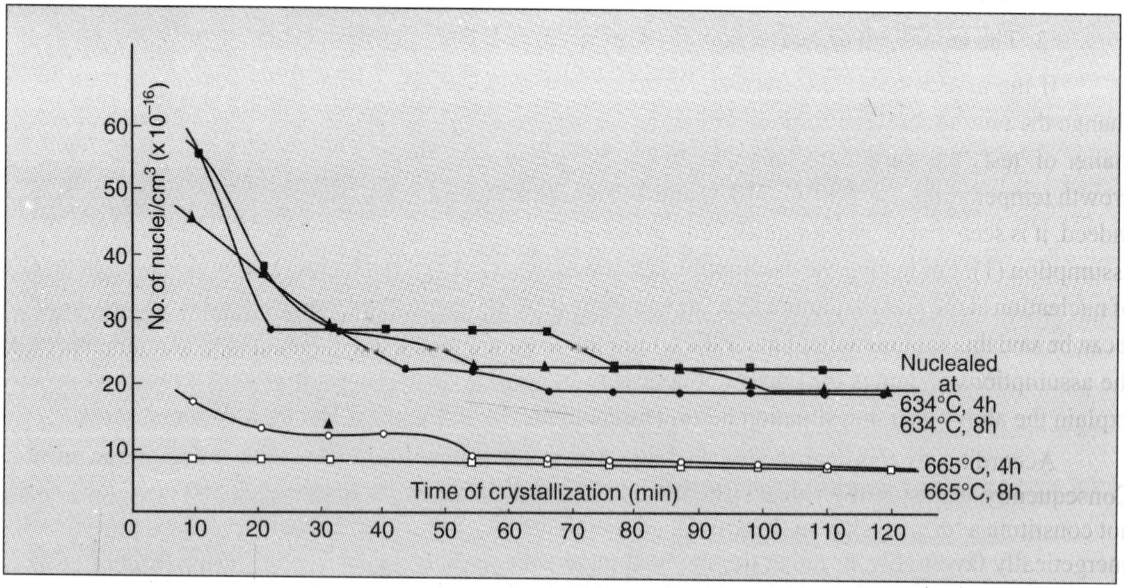

Figure 5.24 : Variation of the number of nuclei with crystallization time for different nucleated samples.

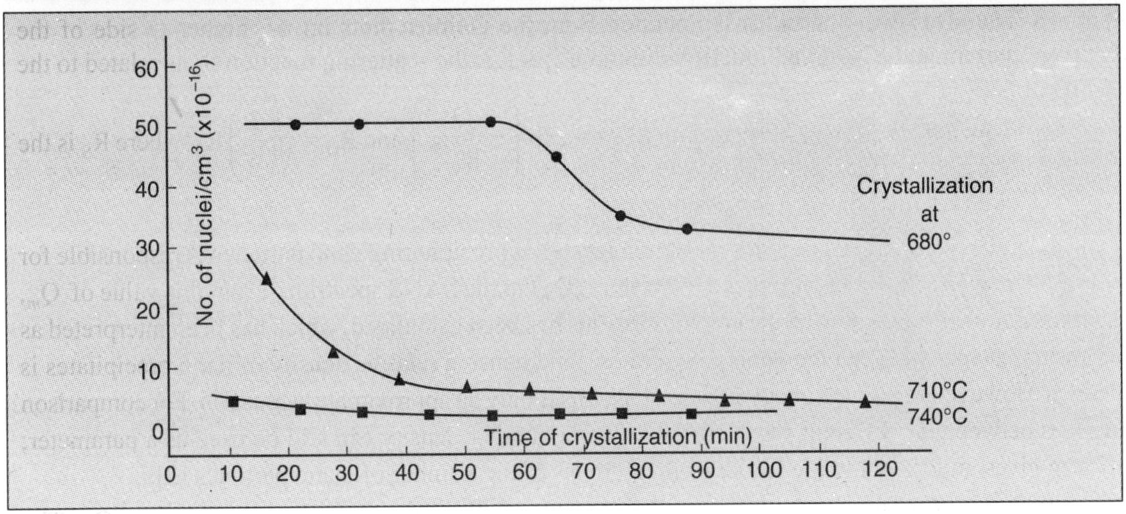

Figure 5.25 : The number of nuclei against crystallization time for the sample nucleated at 577°C for 19 h at three growth temperatures.

The above behaviour can be described according to the following model. The importance of two-stage heat-treatment for the study of nucleation of glass-ceramics has been emphasized by James [48], who made two assumptions as

1. After nucleation, the glass contains an assembly of nuclei, some of which will have grown into small nano crystals, the large majority of which do not redissolve on heating to the second stage of heat-treatment.
2. The nucleation rate at the growth temperature is negligible.

If the assumption (1) is correct, then by varying the growth temperature should not radically change the number of nano crystals, which was observed after a given nucleation heat-treatment. As a matter of 'test', the sample nucleated at 577°C for a very long time of 19 h was heat-treated at three growth temperatures, *i.e.* 680°C, 710°C and 740°C respectively. These data are shown in **Figure 5.25.** Indeed, it is seen that N varies significantly at these crystallization temperatures, which 'invalidates' the assumption (1). For testing the assumption (2), it was observed that the blank glass showed a high level of nucleation at the growth temperature, as shown later. This 'invalidates' the assumption (2). Therefore, it can be said that for the study of nucleation and crystallization behaviour of 'nano particles' of magnetite, the assumptions of James [48] does not appear to be valid for a two-stage heat-treatment. In order to explain the above data, the situation is shown schematically in **Figure 5.26.**

According to classical theory, the size of critical nucleus increases with rising temperature. Consequently, a cluster of 'critical size' at the lower nucleation temperature (say at 550 or 577°C) will not constitute a 'critical size' at the growth temperature and naturally will redissolve, since they are not energetically favourable, or rather thermodynamically unstable nuclei. However, during the nucleation heat-treatment, many of the nuclei that reach the 'critical size' will continue to grow and will attain a size larger than the 'corresponding critical size' at the growth temperature, which will make them 'stable' and also enable them to grow to larger particles at this growth temperature at longer time. Hence, the number of nano particles after a longer growth treatment would not be a good estimate of the number of nuclei formed at lower nucleation temperature, because of the 'redissolution' of many of these nuclei that were present in the 'original' nucleated sample.

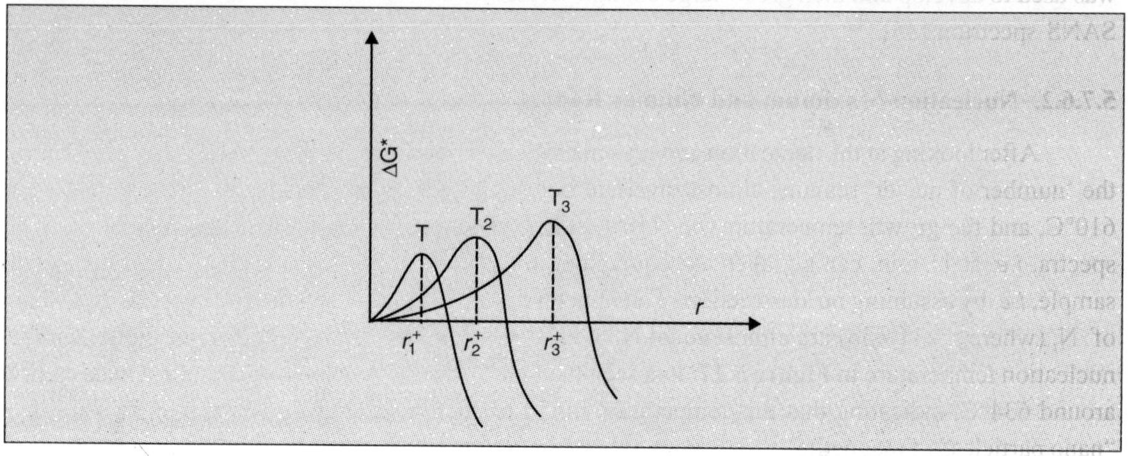

Figure 5.26. A schematic diagram of the free energy against radius of the nuclei at three different temperatures.

Now, let us say that T_1 is the nucleation temperature (T_N) and T_3 is the crystallization temperature (T_C). Therefore, according to **Figure 5.26,** at T_N the total number of nuclei (N*) with $r_1* \geq r$ is stable. If we increase the temperature to T_C and make them grow, the number of nuclei (N_1*) with $r_1* < r < r_3*$ are not stable, but the number of nuclei (N_2*), which are stable at T_C, will grow at this temperature with $r \geq r_3*$. This implies that at T_C, *i.e.* say at 710°C, $N* = N_1* + N_2*$ as $t \rightarrow 0$, but as $t \rightarrow \infty$, $N* = N_2*$, since N_1* number of nuclei have already redissolved during the initial time of crystallization at T_C [3].

Here, it should be mentioned that the nucleation rate of magnetite will depend on the level of super-saturation at the nucleation temperature, which is related to the amount of iron remaining dissolved in the glass matrix. As the precipitation occurs, the level of super-saturation will decrease with time, and hence the nucleation rate will itself decrease and gradually approaches zero. Thus, the number of nuclei (N) should reach a constant maximum or saturation value, as observed in this case of crystallization of nano particles of magnetite. Thereafter, Ostwald ripening effect may slowly take over. Thus, N should reach a 'maximum' simply due to all the available magnetite being eventually precipitated so that no new nuclei can be formed.

However, the saturation number of nuclei reached at a given temperature will depend on the rapidity of the overall precipitation process, which is governed by the nucleation and growth rates, and so will vary strongly with temperature. Hence, the use of two-stage heat-treatment might have resulted in the redissolution of the small nano particles of magnetite into the glass with the disappearance of certain number of particles and a partial dissolution of others. The process proposed here for the growth of nano particles is at least as 'probable' as the process of redissolution of sub-critical nuclei at the growth temperature. However, in practice, the situation might be more complex than the model proposed here, by considering the complex nature of the basalt glass [49].

It should be noted that a problems arises in applying a two-stage heat-treatment to the basalt glass for the study of nano crystallites. The two-stage method has been successfully applied to similar systems such as lithium disilicate, where the crystal and the liquid phases have the same composition, and the growth temperature used was not too high ; although not apparently applicable to a basalt glass, this method is still valid for many other systems. However, in basalt glass, the two-stage heat-treatment was used to develop and then grow 'large enough' crystallites (still in the nano range) in order to give a SANS spectrum [45].

5.7.6.2. Nucleation Maximum and Guinier Radius of Nano Particles

After looking at the nucleation and crystallization data on nano particles so far, if we assume that the 'number of nuclei' remains almost invariant between the respective nucleation temperatures, *e.g.* 610°C, and the growth temperature, *i.e.* 710°C, then the value of N from the first run of the SANS spectra, *i.e.* at 11 min, can be taken as "equivalent to the number of nuclei in the original nucleated" sample, *i.e.* by assuming no 'destruction of any nuclei during this 'short' heating-up process. The values of N_t (where $t = 11$ min) are almost equal $N^* (= N_1^* + N_2^*)_{t=0}$. The values of N_t are plotted against nucleation temperature in **Figure 5.27.** It is seen than the number of nuclei goes through a maximum at around 634°C, indicating that this temperature might be the 'optimum' temperature of nucleation of "nano particles" of magnetite.

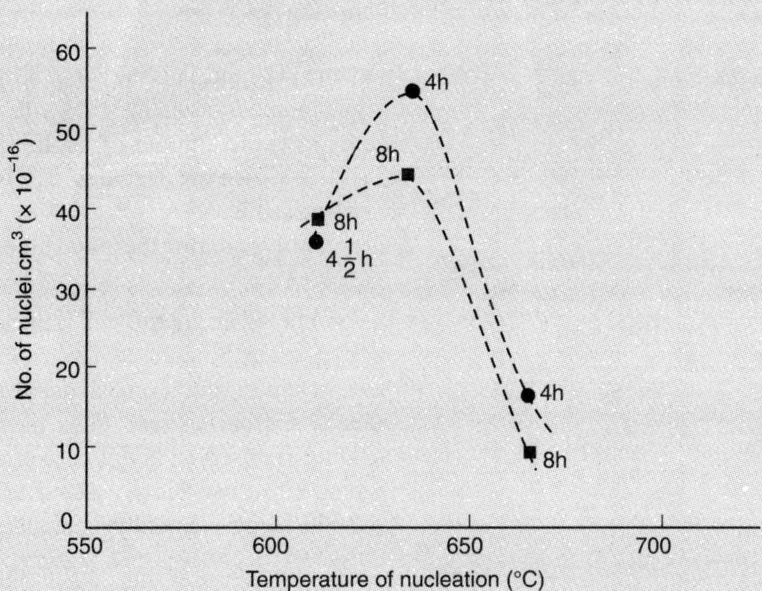

Figure 5.27 : A plot of the number of nuclei against the temperature of nucleation.

The Guinier radius of the nano particles is plotted against time of crystallization in **Figure 5.28** for different samples. It is seen that the sample nucleated at 634°C for 2 h, 4 h and 8 h, R_G increases quite rapidly up to about 30 min, then it slowly increases up to about 60 min, showing the termination of the redissolution process of smaller 'nano' particles. Finally, after 60 min, R_G reaches an almost constant value towards higher time indicating an Ostwald 'ripening process'. For the sample nucleated at 665°C for 4 h and 8 h, this change is less remarkable due to an earlier growth process, or rather due to an "earlier termination" of the redissolution process, owing to the higher nucleation temperature, thereby rendering the attainment of the stable nuclei for the growth process at T_C much easier.

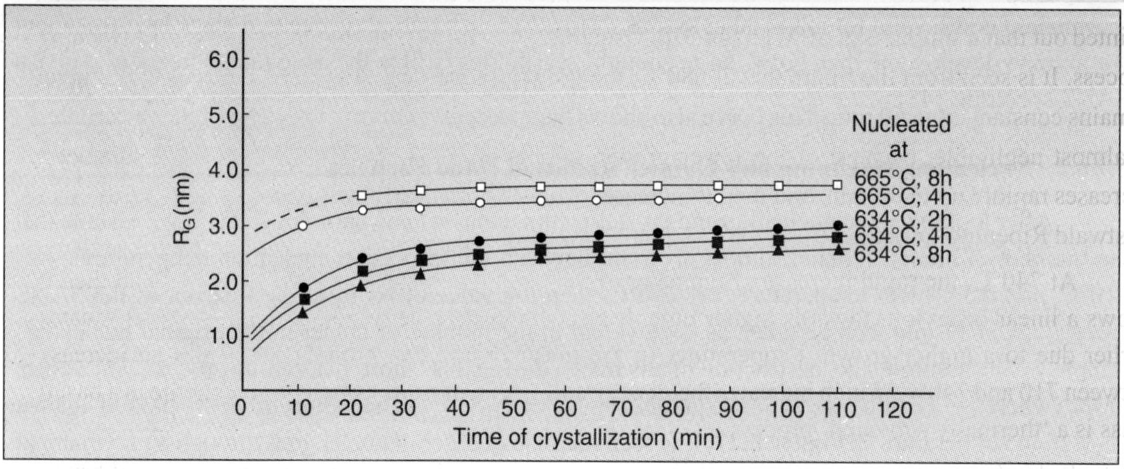

Figure 5.28 : The Guinier radius against crystallization time for different nucleated samples.

5.7.6.3. Ostwald Ripening for Nano Particles and the Growth

In order to examine the growth process, a sample nucleated at 577°C for 19 h was crystallized at three different temperatures, *i.e.* 680°C, 710°C and 740°C respectively. The plots of the cube of the diameter ($D^3 = 8R_S^3$) of the nano particles are shown against the crystallization time in **Figure 5.29.** It should be

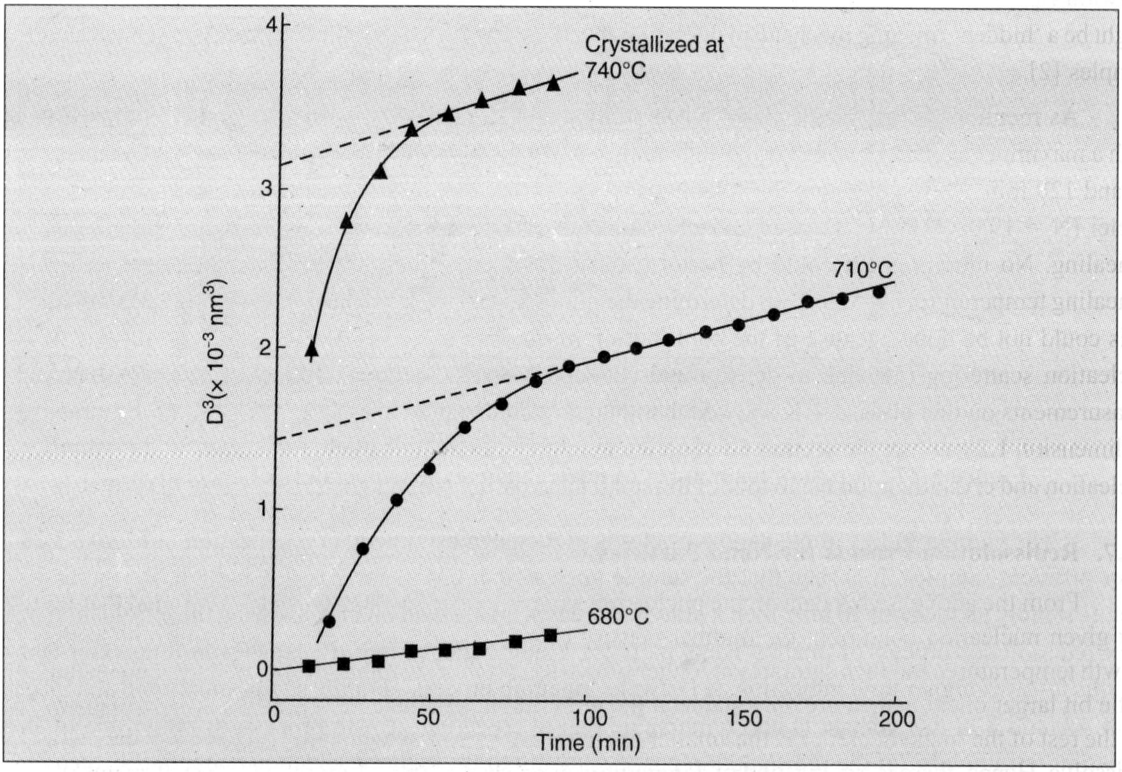

Figure 5.29. A plot of the cube of the diameter of the nano crystallites against the time of crystallization at three different growth temperatures.

pointed out that a similar plot of V [= $(4\pi/3)R_S^3$] against t_C could also be shown to indicate the ripening process. It is seen from the figure that at 680°C, the growth rate is very slow, and the number of nuclei remains constant up to 60 min. This curve also shows that the redissolution of the smaller nano particles is almost negligible, because of the lower growth temperature [3]. At 710°C, the particle diameter increases rapidly up to 80 min, and then it increases 'slowly but linearly' with time, which indicates an 'Ostwald Ripening' process.

At 740°C, the rapid increase of diameter of the nano particles is up to 40 min only, and then it shows a linear behaviour towards higher time. It also shows that the redissolution process terminates earlier due to a higher growth temperature. In the linear region, the growth rate shows an increase between 710 and 740°C, which indicates that the crystallization of nano particles of magnetite in a basalt glass is a 'thermally activated' process.

It should be noted that in the basalt glass, depending on the temperature, there is an 'equilibrium volume fraction' of magnetite of less than 10%. Actually, the Ostwald ripening only becomes predominant when the volume fraction is close to this limiting value. However, in some glass systems, *e.g.* the simpler system of lithium disilicate, the Ostwald ripening would occur for crystallinities approaching 100%, and hence would be negligible for the low crystal volume fraction applicable in most of the nucleation studies of that system [48]. In actual practice, the "impingement" of the crystals could occur at much lower crystallinities, rather than Ostwald ripening. Therefore, it should be kept in mind that Ostwald ripening is only important in certain circumstances [49, 50]. It may be construed that there might be a 'hidden' ripening mechanism operative at certain growth temperatures with different nucleated samples [2].

As mentioned, the 'blank glass' when studied at 710°C showed sufficient scattered intensity with a maximum even at 11 min. The particle radius (R_S) varied between **3.77 nm** and **6.13 nm** between 11 and 120 min. This shows that the 'blank glass' itself was already nucleated with a high density of nuclei (N = 19.0×10^{16} at 11 min), which could have occurred either during quenching or during annealing. No measurement could be performed on this glass either at room temperature or at the annealing temperature, *i.e.* 525°C, to determine the number and size of the nuclei in the original sample. This could not be done because of the contribution of the 'magnetic scattering' together with that of nucleation scattering (*i.e.* due to density and concentration fluctuations). However, the Mössbauer measurements on this glass at 4°K showed that there is a short-range 'magnetic ordering' of Fe atoms of dimension **1.2 nm** (see the **section 5.5**) [6]. It can be thought that this might have affected the overall nucleation and crystallization behaviour of the small nano particles of magnetite in a basalt glass matrix.

5.7.7. Redissolution Process for Nano Particles

From the above SANS data on the nucleation and growth of the nano particle, it is found that for any given nucleation condition, the number density (N_t) of 'nano nuclei' decreases with time at the growth temperature, and then saturates at longer time with a stabilized value. These are the 'nano nuclei' (little bit larger ones), which are only stable at the growth temperature and they will continue to grow, but the rest of the 'nano nuclei', *i.e.* the smaller ones, are not thermodynamically stable and so they will redissolve. Hence, there is a some kind of competition between the 'rate of growth of stable nano nuclei' and the 'rate of redissolution of unstable nano nuclei' at the growth temperature. In the thermodynamic evolution of the nano nuclei, the kinetic parameters of this 'evolution' should be considered as well.

Although the word 'redissolution' has been used in the previous sections several times, it is better to give it a concrete shape toward the actual meaning of this word by further analysis, which is given below.

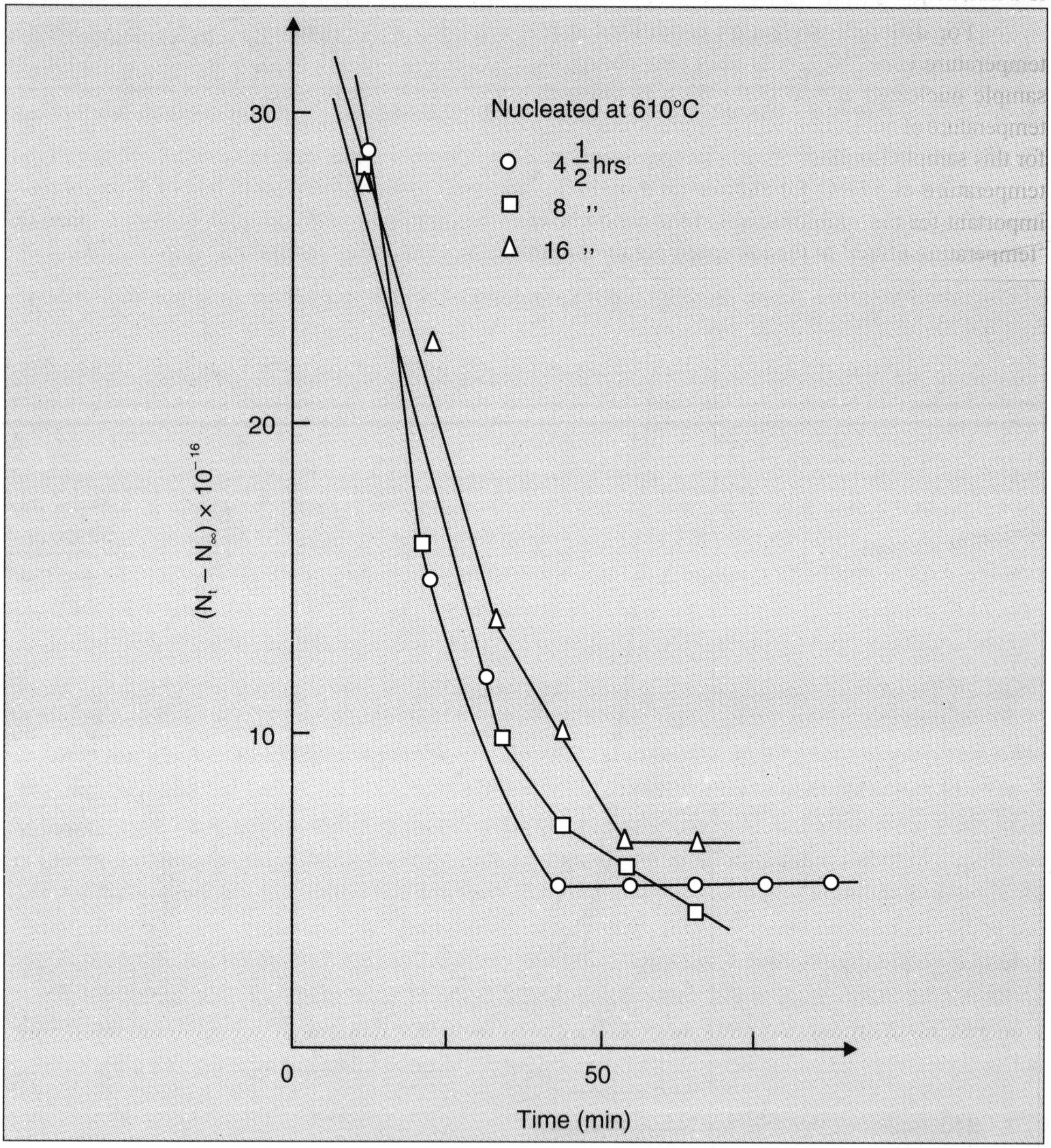

Figure 5.30 : The redissolution of nano particles against the time of growth at 710°C for the samples nucleated at 610°C.

At the growth temperature, if we take the number density at the longest time as N_∞, then at any time of growth, $(N_t - N_\infty)$ represents the number of nano nuclei which redissolve. The plots of $(N_t - N_\infty)$ as a function of time of growth are shown in **Figures 5.30** and **5.31** for all the nucleated samples. It is seen that almost all the curves show a more or less rapid decrease at the initial time of growth. This decrease is slowed down considerably up to certain time for lower temperatures of nucleation. After this stage, there is no more redissolution indicating a stable behaviour. This 'initial time' of rapid redissolution, *i.e.* a rapid decrease of $(N_t - N_\infty)$, can possibly be termed as an "incubation period", which corresponds to a kinetic phenomenon related to bulk atomic diffusion process in the basalt glass.

For different nucleation conditions at relatively lower temperatures, but the same growth temperature (*i.e.* 710°), it is seen that during the 'incubation period', the redissolution rate for the sample nucleated at 550°C for 32 h is lower than that of the 'blank glass'. This is due to higher temperature of nucleation, which seems to stabilize a certain number of nuclei. Since the time of nucleation for this sample is higher at 32 h, its redissolution rate is also lower than that nucleated even at a higher temperature at 577°C for shorter time of 19 h. Hence, it appears that the 'effect of time' is more important for the 'stabilization' of the nano nuclei - which means a lower redissolution rate - than the 'temperature effect' in the lower temperature region of nucleation, *i.e.* between 525 and 577°C.

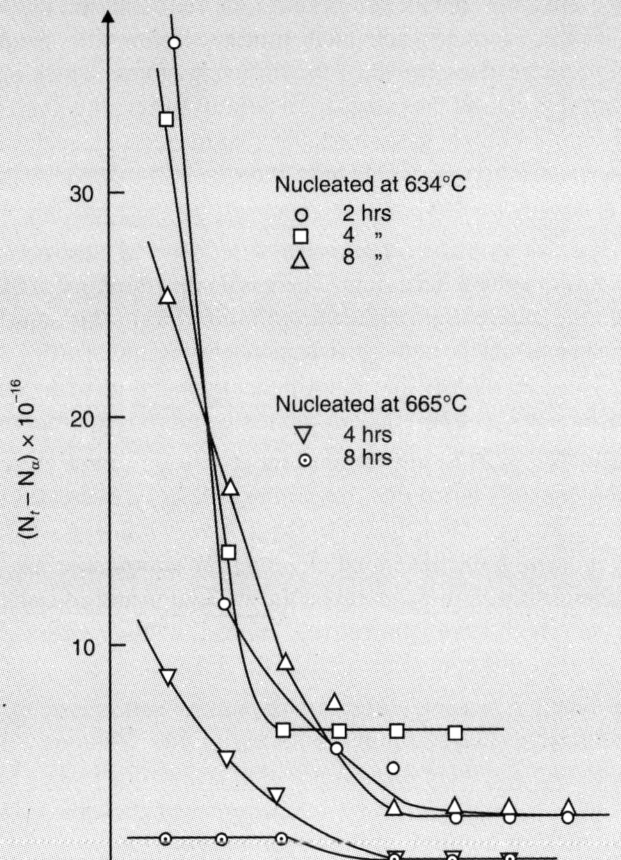

Figure 5.31 : The redissolution of nano particles against the time of growth at 710°C for the samples nucleated at 634°C and 665°C respectively.

For the same nucleation conditions, but different growth temperates, the observations for the redissolution of the nano nuclei are quite interesting. Thc 'incubation period' is longer for lower growth temperature due to a lower rate of atomic diffusion. During this 'incubation period', the redissolution rate is almost constant at 680°C, then it increases considerably at a higher growth temperature of 710°C, and then it decreases again at still higher growth temperature of 740°C.

The above data can be explained in the following manner :

At 680°C, the mobility of the atoms is much lower than that at higher temperatures up to a certain time, *i.e.* the 'incubation period', so that the redissolution rate is quite negligible, and the process is also quite long. At 710°C, the rate of redissolution increases considerably due to the increase of bulk atomic diffusion with temperature. However, at 740°C, this rate seems to be very much reduced, because many of the nano nuclei had already been redissolved during the increase of temperature, thereby making the attainment of the 'stable nano nuclei' much easier. The higher temperature represented the particles for which the Gibbs free energy would decrease with increasing size, favouring immediate growth once the temperature is raised in order to give enough 'thermal energy' for bulk atomic diffusion.

It should be noted after the 'incubation period', the redissolution rate decreased from 680 to 740°C, indicating this to be a 'thermally activated' process. Below this 'incubation period', there is some sort of 'blocking' in the rearrangement of the diffusing atoms, which not only kept the rate of redissolution constant, but also caused the 'incubation period' to be quite long, say at a temperature of 680°C. Therefore, the dynamic study of growth on different nucleated samples shows the 'advantages' in revealing some details, which otherwise would have been lost in the conventional technique of studying growth in two-stage heat-treatments [48].

For higher temperatures of nucleation, the number of 'redissolved' nano nuclei against time of growth (see **Figure 5.31**) shows a very similar behaviour. There is a rapid decrease of the 'redissolved' number density up to around 50 min, and then it reaches a saturation level. This saturation level is lower for higher temperatures of nucleation. At the highest nucleation temperature of 665°C for 8 h, the redissolution rate is almost negligible. This is obviously due to the higher temperature of nucleation. Therefore, in this temperature range of nucleation, *i.e.* 610 - 665°C, the temperature is an important factor.

By analyzing these curves, if the number density at $t \to 0$ is taken as the maximum number of 'redissolved' nano nuclei, then it is noted that this number [$N_{D(max)}$] decreases with nucleation time at any of these nucleation temperatures. While this 'decrease' is very little for 610°C, it is quite rapid at 634 and 665°C respectively, indicating the beneficial effect of nucleation at or just about T_g [45]. All the above data on redissolution clearly show its effect on the nucleation and crystallization behaviour of the nano particles of magnetite, which were 'innovatively created' within a basalt glass matrix by a simple procedure of heat-treatments at different temperatures for different time.

In summary, the dynamic 'on-line' study of growth of nano particles of magnetite shows a pronounced inter-particle interference effect in the SANS spectra. There is a decrease of the number density with the time of growth and thereafter, a saturation occurs due to the growth of the stable nano nuclei. This saturation level is higher for the lower temperatures of nucleation. This is interpreted as due to redissolution of the smaller nano nuclei, as the larger and stable nano nuclei continue to grow. The crystallization behaviour tends to follow an Ostwald ripening mechanism at higher time of growth. For lower nucleation temperature, the time of nucleation is an important factor, whereas for the higher nucleation temperature near T_g, the temperature seems to be an important parameter. The rapid decrease of the number density of the redissolved nano nuclei in the initial time is noted for all the nucleation conditions. At longer time of growth, the redissolution terminates progressively for lower nucleation temperatures. But for higher nucleation temperatures, there is a saturation effect before the termination of the redissolution process. These data on 'nano particles' within such a 'narrow range of sizes' is quite revealing and definitely merit further attention for our future endeavour in the search of 'newer nano materials'.

REFERENCES

1. G. H. Beall and H. L. Rittler, Bull. Am. Ceram. Soc., 55 (1976) 579.
2. P. Labarbe, A. K. Bandyopadhyay, J. Zarzycki and A. Wright, 'Stabilisation of crystal densities during nucleation of glass ceramics : a hidden ripening process', J. Non-Crystalline Solids, 43 (1981) 433.
3. A. K. Bandyopadhyay and J. Zarzycki, 'A redissolution process on the nucleation behaviour of a basalt glass ceramic studied by SANS', Proceedings of the Symposium on 'Phase Transformations in Vitreous Systems' by Society of Glass Technoloy, Warwick (UK), Ed. P. W. McMillan (1981).
4. A. K. Bandyopadhyay, P. Labarbe, J. Zarzycki and A. F. Wright, 'Nucleation and crystallization studies of a basalt glass ceramics by small angle neutron scattering', J. Materials Sc., 18 (1983) 709.
5. A. K. Bandyopadhyay and A. F. Wright, 'A redissolution process during particle growth of a glass-ceramic studied by SANS', Trans. Ind. Ceram. Soc., 46 (1987) 102.
6. A. K. Bandyopadhyay, J. Zarzycki, P. Auric and J. Chappert, 'Magnetic property of a basaltglass and glass ceramics', J. Non-Crystalline Solids, 40 (1980) 353.

   [Procedings of the 5th Univ. Conf. on glass Sc., Troy (USA), Eds. M. Tomozawa, R. A. Levy, R. K. MacCrone and R. H. Doremus, North Holland, Amsterdam, (1980)].
7. P. Auric, J. Chappert, A. K. Bandyopadhyay and J. Zarzycki, 'Mössbauer study of crystallite formation in basalt glass ceramics', J. de Physique, C1-41 (1981) 277.

   [Proceedings of the Intl. Conf. on Mössbauer Spectroscopy, Portoroz (Yugloslavia), (1980)].
8. P. Auric, J. Chappert, A. K. Bandyopadhyay and J. Zarzycki, 'Superparamagnetism and ferrimagnetism of the small particles of magnetite in a silicate matrix', J. Non-Crystalline Solids, 50 (1982) 97.
9. A. K. Bandyopadhyay and P. Auric, 'Hyperfine splitting and anisotropy of magnetite particles within a glass matrix by Mössbauer spectroscopy', Trans. Ind. Ceram. Soc., 46 (1987) 136.
10. L. Neel, C. R. Acad. Sci., 229 (1949) 664.
11. E. C. Stoner, Magnetism and Matter, Methuen, London (UK), (1934) pp. 312.
12. T. Nagata, Rock Magnetism, Merizen Comp, Tokyo (Japan), (1961) pp. 80.
14. I. S. Jacob and C. P. Bean, Manetism I I I, Eds. G. Rado and H. Suhl, Academic Press, New York, (1963) pp. 271.
14. S. M. Aharoni and M. H. Litt, J. appl. Phys., 48 (1977) 4298.
15. L. Neel, Ann. Phys., 3 (1948) 137.
16. Y. F. Krupyanski and I. P. Suzdalev, J. Phys., 35 (1974) 407.
17. J. Smit and H. P. J. Wijn, The Ferrites, Tech. Library, Philips, (1961).
18. P. Roggwiller and W. Kundig, Solid State Comm., 12 (1973) 901.
19. R. E. Vandenberghe, R. Vanleerberghe, E. de Grave and G. Robbrecht, J. Magn. Magn. Mat., 15-18 (1980) 1117.
20. R.N. Bhowmik, 'Lattice expansion and magnetic order in $MnCr_2O_4$ spinel oxide', Proc. Natl. Conf. on 'Nano-Science and Technology, J. U., Kolkata (India), January, (2005).
21. F.F. Fava et al., J. Phys.: Condens. Matter, 9 (1997) 10715.

22. R. Banerjee, E.A. Sperling, G.B. Thompson, H.L. Fraser, S. Bose and P. Ayyub, Appl. Phys. Lett. 82 (2003) 4250.
23. S. Krupika and P. Novak, 'Ferromagnetic Materials', Ed. E.P. Wolfarth, North-Holland, Amsterdam, Vol. 3, (1982) pp. 189.
24. R.H. Kodama et al., Phys. Rev. Lett., 77 (1996) 394.
25. A. Fukunaga, S. Chu and M.E. McHenry, J. Mater. Res., 13 (1998) 2465.
26. J. M. D. Coey, J. de Phys., 35 (1974) 89.
27. G. M. Bancroft e al, Geochim. Cosmochim. Acta, 32 (1968) 547.
28. G. M. Bancroft et al, Geochim. Cosmochim. Acta, 41 (1976) 2219.
29. T. K. McNab, R. A. Fox and J. F. Boyl, J. Appl. Phys., 39 (1968) 5703.
30. S. Morup, H. Topsoe and J. Lipka, J. de Physique, 37 (1976) C6-287.
31. S. Morup and H. Topsoe, Appl. Phys., 11 (1976) 63.
32. K. Haneda and A. H. Morrish, Phys. Letters, 64 (1977) 259.
33. A. M. Van der Kraan, Phys. Stat. Solidi (a), 18 (1973) 215.
34. J. Chappert and R. B. Frankel, Phys. Rev. Letters, 19 (1967) 570.
35. Y. Yafet and C. Kittel, Phys. Rev., 87 (1952) 290.
36. J. Owen, J. Appl. Phys. Suppl., 32 (1961) 213 S.
37. A. K. Bandyopadhyay, J. Materials Sci., 16 (1981) 189.
38. E. J. Friebele et al, Phys. Staus Solidi (b), 45 (1971) 323.
39. D. L. Griscom and C. L. Marquardt, Amorphous Magnetism, Eds. H. O. Hooper and A. M. Van de Graaf , Plenum Press, (New York), (1973) pp. 95.
40. V. K. Sharma and F. Waldner, J. Appl. Phys., 48 (1977) 4298.
41. J. Zarzycki and F. Naudin, J. Non-Cryst. Solids, 1 (1969) 215.
42. M. Roth and J. Zarzycki, 'A small angle neurron scattering study of SiO_2 - Na_2O glasses', J. Non-Cerysta. Solids, 16 (1974) 93.
43. J. Zarzycki, 'Heterogeneities in glasses and small angle scattering methods', J. Appl. Cryst., 7 (1974) 200.
44. A. F. Wright, P. W. McMillan and N. Bret, Proc. Intl. Conf. on Structure of Non-Ceryst. Solids, Cambridge (UK), July, (1982).
45. A. F. Wright, J. Talbot and B. E. F. Fender, 'Nucleation and growth studies by small angle neutron scattering and results for a glass-ceramic', Nature, 277 (1979) 366.
46. "Neutron Beam Facilities at I. L. L., High Flux Reactor", Institut Laue - Langevin, Grenoble, France, (1974).
47. A. Guinier, 'Small angle scattering of X-rays', Wiley, New York, (1955) Chap 4.
48. P. F. James, 'Kinetics of crystal nucleation in lithium silicate glasses', Phys. Chem. Glasses, 15 (1974) 95.
49. P. F. James and J. O. Isard, Private Communication.
50. J.Zarzycki, 'Les Verres et L'Etat Vitreux', Mason, Paris, (1982) pp. 143.

Chapter 6

Electrical Properties

6.1. SWITCHING GLASSES WITH NANO PARTICLES

PREAMBLE

In the **chapter 5** on "Magnetic Properties" of glass-ceramics, we talked about a simple technique of creating nano-crystalline particles of magnetite within a glassy matrix in terms of heat-treatment at different temperatures to precipitate small nano crystals by a mechanism of nucleation and crystallization. Here, in this chapter on electrical and electronic properties of nano crystals within a glassy matrix, the process is not so simple, but it is a quite interesting system to study nano materials [1]. Without going too far on switching behaviour of semiconding glasses, i. e. the Voltage (V) Current (I) characteristics, which passes a high current through the system with a slight change of voltage, we can say that the glasses are interesting systems to study such 'switching' properties [1, 2].

6.1.1. Introduction

The surface layers of certain silicate glasses containing Bi_2O_3 have been reported to show "memory switching" after they were subjected to sodium $\Leftrightarrow$ silver ion-exchange, which was followed by a reduction treatment in hydrogen [2]. Such glasses before ion-exchange and reduction stages have a micro-structure consisting of metallic bismuth particles with diameters in the nano range : **5 nm** to **25 nm** dispersed in an amorphous matrix. After ion-exchange and reduction treatments, the nano particles of silver of maximum diameter **100 nm** are also precipitated. The measurements carried out on 'thick films' of these glasses, which have not been subjected to any ion-exchange and reduction treatments, show that the switching action is the characteristic feature of the base glass itself. Here, the role of nano silver particles being only to lower the switching voltage [3]. This is some sort of an achievement, since we want a flow of current always at a lower voltage to make the 'switching device' efficient.

The off-state resistance of these glass films was tentatively explained as arising due to electron hopping between the metallic islands of nano particles of bismuth. Therefore, a more detailed study of both DC and AC electrical properties was necessary to be undertaken and the results had to be analysed. Since the electrical properties of glasses containing certain nano-metallic particles are quite interesting, the efforts were also made to study the electrical behaviour of some borosilicate glasses containing nano-crystalline selenium particles, because their properties are also found to be similar to those obtained for bismuth-containing glass systems.

6.1.2. Preparation of Glasses with Nano Particles

For bismuth-containing glasses, most of the work was done on a composition : 10 Na_2O, 18 B_2O_3, 64 SiO_2, 8 Bi_2O_3 (mole%). These glasses were identified as showing 'memory switching' [1, 3].

Some work was also carried out on samples of composition : 25 Na_2O, 10 CaO, 55 SiO_2, 10 Bi_2O_3 (mole%) to show that the electrical behaviour due to nano metallic bismuth is independent of the chemical composition of the glass matrix. Finally, in order to show the effect of nano-bismuth particles more clearly, some work was also done on the 'base glass' compositions : 10 Na_2O, 26 B_2O_3, 64 SiO_2 and 25 Na_2O, 20 CaO, 55 SiO_2 (mole%) respectively. The compositions of different glasses, which were used to study the effect of nano-bismuth particles on their electrical properties are summarized in **Table 6.1,** and which are numbered as 1 to 4. The compositions of the glasses, which were used to elucidate the effect of the dispersion of nano-crystalline selenium particles on the resulting composite matrix, are summarized in **Table 6.2,** and which are numbered as 5 to 8.

Table 6.1 : Compositions of Glasses containing Bismuth.

Glass No.	*Mole% SiO_2*	*Mole% B_2O_3*	*Mole% CaO*	*Mole% Na_2O*	*Mole% Bi_2O_3*
1	64	18	--	10	8
2	64	26	--	10	--
3	55	--	10	25	10
4	55	--	20	25	--

Table 6.2 : Compositions of Glasses containing Selenium.

Glass No.	*Mole% SiO_2*	*Mole% B_2O_3*	*Mole% Na_2O*	*Wt.% Se*
5	53.3	32.0	14.7	0.0
6	53.5	26.6	19.7	0.2
7	53.8	21.1	24.8	0.3
8	48.0	19.2	28.8	4.0

The glasses were prepared by melting reagent-grade chemicals in alumina crucibles in an electrically heated furnace at temperatures ranging from 1200 to 1400°C under normal atmospheric conditions. The selenium purity was 99.9%, which was introduced as powder in the mixture. The glasses were quenched by pouring the melts onto the aluminium moulds. These were then annealed at 500°C for an hour and then cooled slowly to room temperature within the furnace after being switched off.

A substantial amount of selenium was lost during melting, because of the low melting temperature of these selenium containing glasses (*i.e.* nos. 5 to 8) [4]. Therefore, the final compositions of these samples were determined after estimating their selenium contents by standard chemical methods [5]. For preparing glass 8, which contained the largest amount of selenium, the oxide glass was first made with adequate molar percentages of different components. This glass was then powdered and mixed with the appropriate amount of selenium powder in order to get the required composition. The mixture was taken in a quartz tube, which was evacuated and then sealed. The sealed quartz ampoule was heated to a temperature to 1000°C for 24 hours with gentle rocking for homogenization purposes. The quartz ampoule was taken out and allowed to cool at room temperature. The optical absorption spectra of the glasses 5 to 8 confirmed that selenium was present in its elemental form and not as an oxide [4].

The micro-structure of the glasses was studied with a Philips EM-301 Transmission Electron Microscope (TEM) as per the procedure described in **chapter 2.** The TEM micrographs show darker regions, which are ascribed to the presence of metallic bismuth grains, which are of non-spherical geometry and of different nano-sizes varying between **5 nm** and **100 nm.** In the TEM micrographs of the selenium containing glasses also, there are darker regions, which are ascribed to the presence of metallic nano particles of metallic selenium with size varying from **5 nm** to **200 nm.** The electron diffraction pattern of these glasses show that the metallic granules of selenium are crystalline [1, 2].

For electrical measurements, circular metallic electrodes (both aluminium or gold giving the same results) of diameters varying between 1 and 2 cm were evaporated onto two faces of the samples, which were ground and polished to a thickness in the range 1 to 2 mm. For glasses 5 to 8, silver paint was used as electrodes. The DC resistivity was measured by plotting their voltage-current characteristics over a decade of voltage by using an 'Electrometer' (Type 1230A, General Radio, U. K.). The linearity of the I-V curves was obtained for all the samples in the temperature range – 160°C to +200°C.

For AC resistivity measurements, a guard-ring was also deposited on one of the sample faces, and three-terminal measurements were carried out over a frequency range 100 Hz – 100 KHz with a 'Capacitor Bridge' (Type 716C General Radio, U.K.). For measurements at higher frequency, *i.e.* between 100 KHz and 30 MHz , a 'Boonton Q-Meter' (Type 260A) was used.

The possibilities of switching phenomena in selenium containing glasses were explored by subjecting them to a $Na^+ \Leftrightarrow Ag^+$ ion exchange, which was followed by a reduction treatment in hydrogen. The electrical measurements on these samples were carried out by the method described elsewhere [2]. The I-V characteristics of these samples were carried out by a 'Transistor Curve Tracer' (Type 575, Tektronix, USA). For all the electrical measurements on selenium containing glasses, the samples were kept in a dark chamber so that 'photo-conductive' effects could be avoided.

6.1.3. Electrical Data of Nano Particles of Bismuth and Selenium

6.1.3.1. Electrical Conduction in Bismuth Glasses

The plot of log ρ vs. 1/T is shown in **Figure 6.1** for the glasses 1 to 4. It is seen that for glasses 1 and 3 containing nano particles of bismuth, there are two linear curves, whereas for the glasses 2 and 4 containing no bismuth, there are simple linear plots over the entire temperature range. These results indicate that for the glasses 1 and 3, there are two conduction mechanisms operative in two different temperature ranges. By assuming an Arrhenius type of variation of resistivity as a function of temperature, which is written as

$$\rho = \rho_0 \exp(E/kT) \qquad ...(6.1)$$

the activation energies of these glasses are estimated in the two different temperation ranges, *i.e.* between 20° - 200°C and 200° - 500°C, which are shown in **Table - 6.3.** It is evident from the

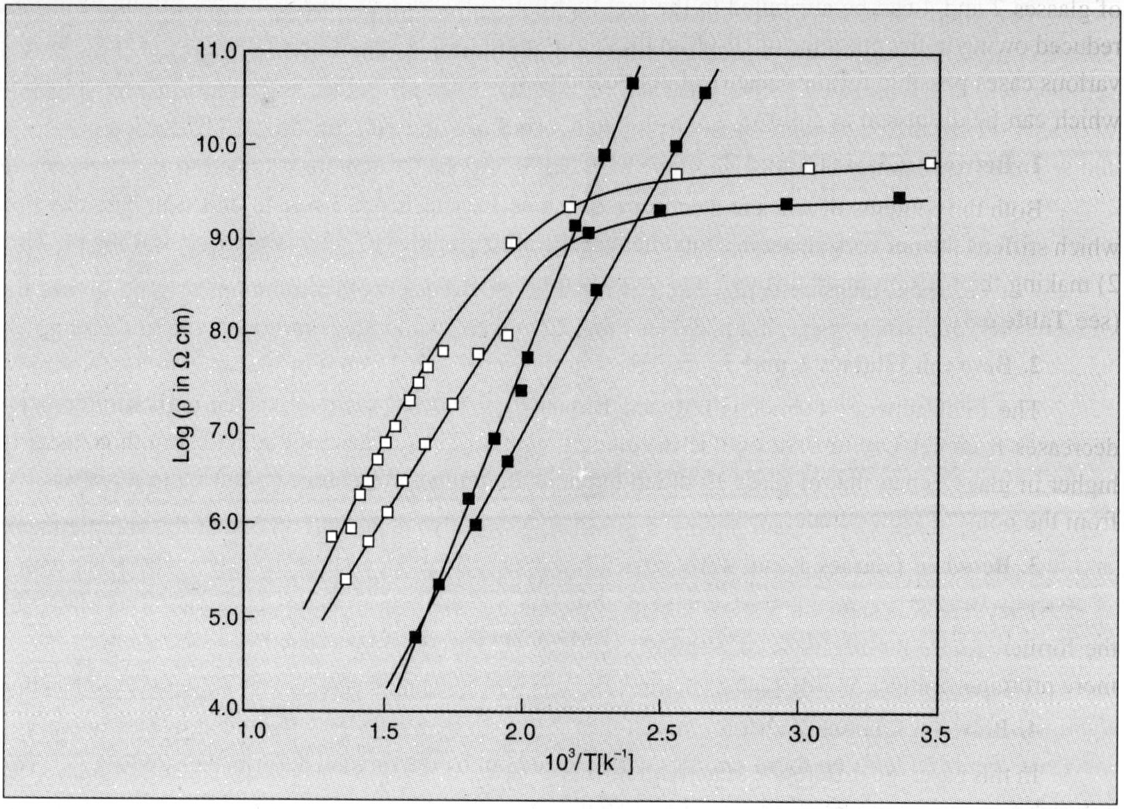

Figure 6.1 : Log ρ against 1/T for glasses 1-4 at *f* = 1 kHz.

table that the resistivity of the glasses 1 and 3 in the higher temperature range arises due to the movement of Na^+ ions through the glass matrix, which obviously requires a higher activation energy due to the 'diffusional' jump of Na^+ ions in the glass matrix [6].

Table 6.3 : Activation energies of conduction in glasses 1 to 4.

Glass No.	*Activation Energy (eV) : 20°C - 200°C*	*Activation Energy (eV) : 200°C - 500°C*
1	0.03 ± 0.00	1.10 ± 0.01
2	1.42 ± 0.01	1.42 ± 0.01
3	0.03 ± 0.00	0.90 ± 0.01
4	1.10 ± 0.02	1.10 ± 0.01

In oxide glasses, it is known that if the ratio of the network former (*i.e.* SiO_2, B_2O_3, etc.) to network modifier (*i.e.* Na_2O, CaO, Bi_2O_3, etc.) is reduced, the concentration of non-bonding oxygen ions is increased. This loosens the structure that makes it congenial for Na+ ion jump, which involves lower activation energy. The slightly lower activation energies of the glasses 1 and 3 compared to those of glasses 2 and 4 can be attributed to the fact that the 'coherence of silica network' in the former is reduced owing to the presence of Bi_2O_3 in these glasses, which act as network modifier [7]. There are various cases possible for the 'electrical conduction' due to Na^+ ions in the higher temperature range, which can be discussed as follows

1. Between Glasses 1 and 2

Both the contents of silica and soda are constant, but borax is increasing from 18 to 26 mole%, which stiffens the network structure (total network former = 64 + 18 = 82% in glass 1 and 90% in glass 2) making the Na^+ ion jump difficult and thereby the activation energy increases from 1.10 to 1.42 ev (see **Table 6.3**)

2. Between Glasses 1 and 3

The total network former is reduced from 82% to 55%, and thereby the activation energy decreases from 1.10 ev to 0.90 ev due to less structural cohesion. Moreover, the bismuth content is higher in glass 3 than that of glass 1, which explains the effect of bismuth in loosening the structure from the point of view of energy.

3. Between Glasses 1 and 4

They have the same activation energy, even if the ratio of network former/modifier is higher in the former, where the presence of bismuth weakens the structure more. So, the effect of bismuth is more prominently seen in this case.

4. Between Glasses 2 and 3

In this case, the difference of activation energy is the highest (from 1.42 ev to 0.90 ev), since the network former is substantially reduced as well as there is bismuth present in glass 3, which makes the 'weakening' of the structure maximum and hence the lowest activation energy.

5. Between Glasses 2 and 4

Here, the reduction in activation energy is from 1.42 ev to 1.10 ev due to the substantial reduction of network former from glass 2 to 4, but the absence of bismuth in glass 4 does not allow a reduction of activation energy further. So, the effect is clear.

Therefore, it is seen that the above glasses 1 to 4 containing bismuth are quite interesting for a detailed analysis for activation energy. In the lower temperature region, the conduction in glasses 1 and 3 is via 'electron hopping' between isolated islands of nano-metallic bismuth particles, and hence the activation energy is extremely low at 0.03 ev.

Both the DC and AC resistivity data obtained at different frequencies for glass 1 are shown in **Figure 6.2.** A change in slope for the DC resistivity curve is also observed at a temperature of 120°C.

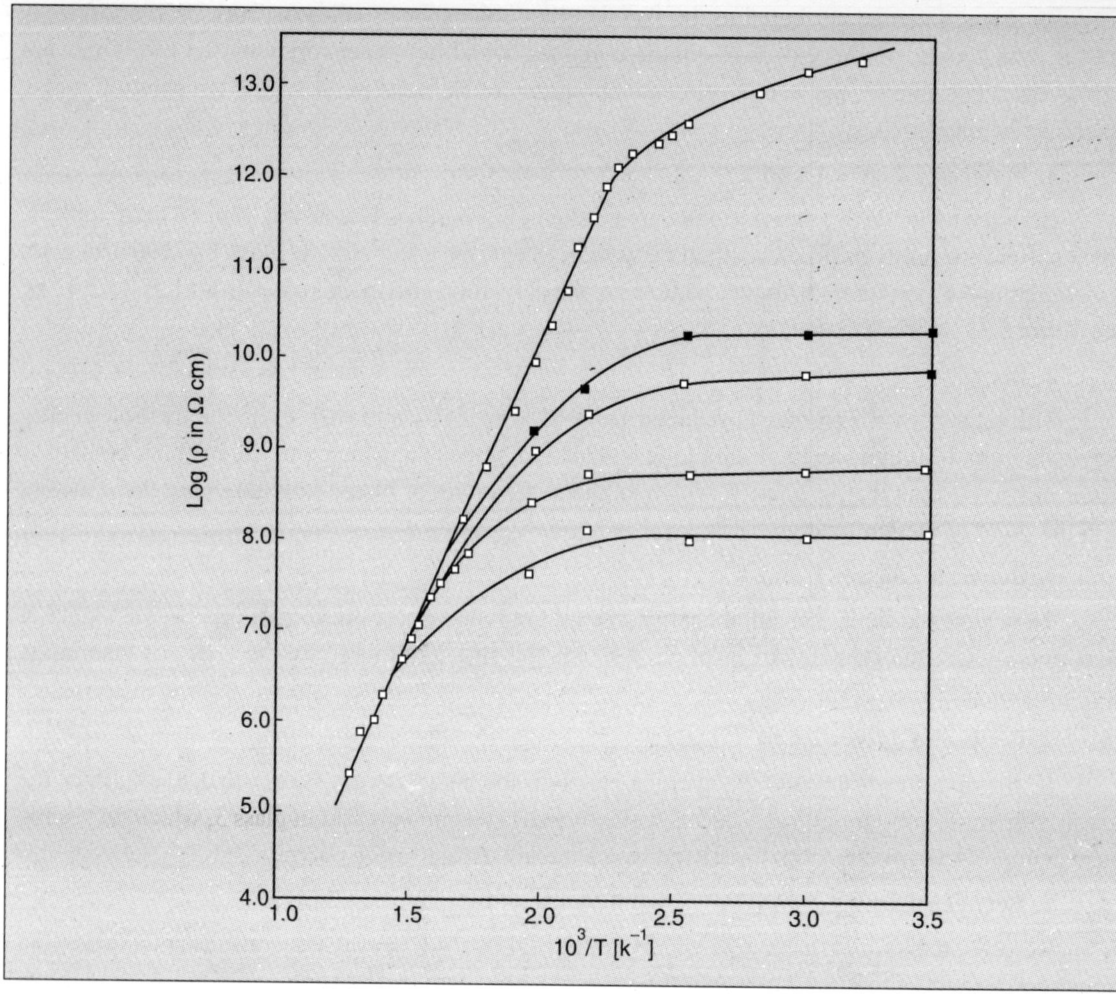

Figure 6.2 : Log ρ against 1/T for glass 1 at different frequencies : Higher Curve = DC, Other Curves = AC.

6.1.3.2. Electrical Conduction in Selenium Glasses

The DC resistivity data for the glasses 5 to 8 containing nano-crystalline selenium grains are shown in **Figure 6.3,** which shows a sharp decrease of the resistivity as a function of selenium concentration at low temperature. For these selenium containing glasses, it is seen that the slope of the resistivity curve changes in the temperature range – 60°C to 120°C, depending on the composition. The temperature at which, this break of slope occurs increases with the increasing selenium content. The activation energies for conduction is calculated for these glasses from the slopes of the curves and are shown in **Table 6.4.**

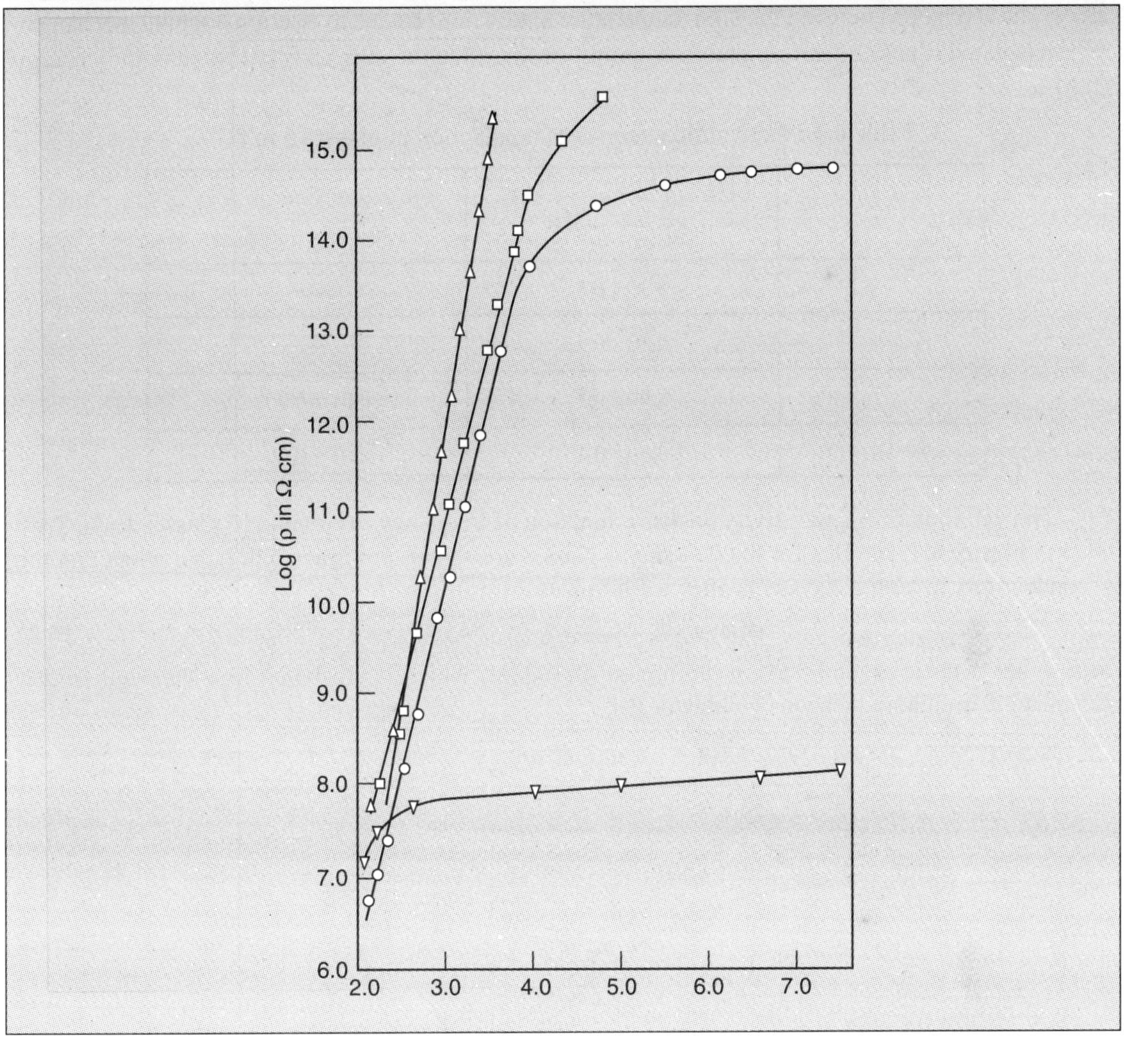

Figure 6.3 : Log ρ against 1/T for glasses 5 - 8 : Triangle = glass 5, Square = glass 6, Circle = glass 7, Inverted triangle = glass 8.

The resistivity in the higher temperature region is contributed by the diffusional movement of the sodium ions through the glass matrix [6]. The decrease in the activation energy for the glasses 6, 7 and 8 as compared to that glass 5 is attributed to the higher concentration of sodium ions in the former [8]. Moreover, the 'network coherence' is reduced due to the progressive inclusion of a higher amount of selenium in the glass matrix.

Here, the situation is little bit different than bismuth containing glasses. While the network modifier (soda) is progressively increasing from glass 5 to 8, the network former (silica) remains almost the same, and thereby the ratio decreases substantially from 5.80 to 2.33. This could be the reason for a reduction in the activation energy from 1.30 ev to 0.50 ev from glass 5 to 8. Since the selenium content goes up from 0 to 4% from glass 5 to 8, the effect of selenium in weakening the structure cannot be completely ruled out. The reduction of activation energy is relatively more pronounced in case of selenium glasses. The activation energy in the high temperature region must be due to electron hopping mechanism between isolated islands of nano-metallic selenium grains, which requires a very low activation energy of 0.01 ev.

Table 6.4 : Activation energies of conduction in glasses 5 to 8.

Glass No.	*Activation Energy (eV) 20°C - 200°C*	*Activation Energy (eV) 200°C - 500°C*
5	1.30±0.1	--
6	0.80±0.1	--
7	0.90±0.1	0.01±0.00
8	0.50±0.1	0.01±0.00

The variation of AC resistivity data as a function of frequency for (bismuth) glasses 1 and 3 are shown in **Figure 6.4,** and that for the (selenium) glass 8 are shown in **Figure 6.5.** It is evident that the AC conductivity in both the systems obey a relation as

$$\sigma(f) \propto f^{n} \qquad ...(6.2)$$

where, n has a value around 0.90 in the region of 10^6 Hz, which is attributed to a Maxwell-Garnet mechanism in an inhomogeneous conductor [9].

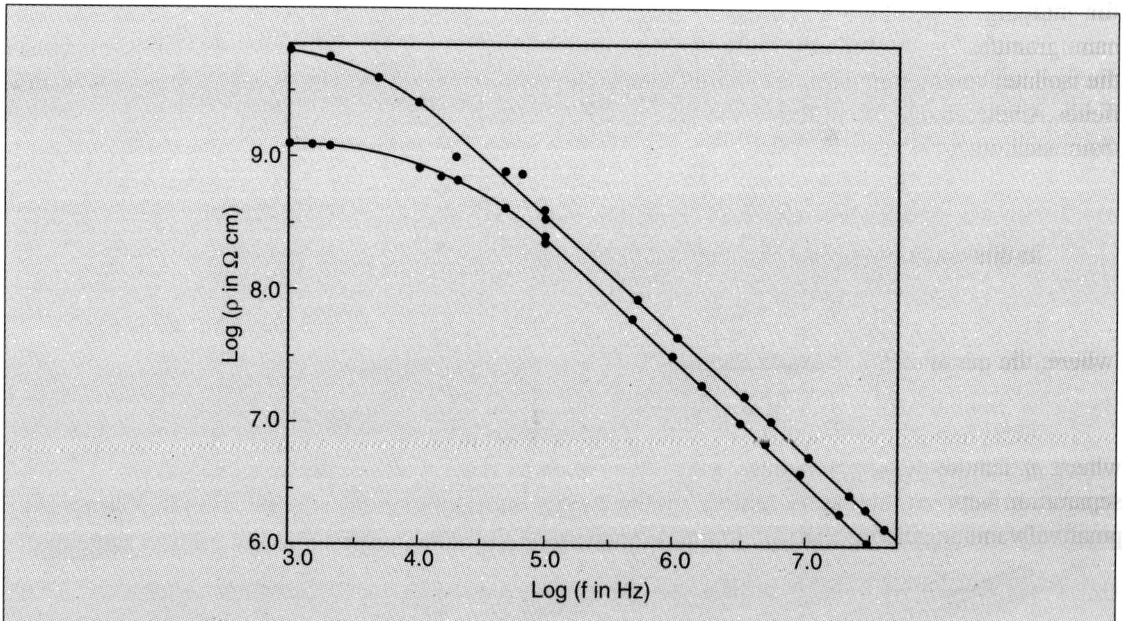

Figure 6.4 : Log ρ against log f for glasses 1 and 3 at 200°C : Filled circle = glass 1, n = 0.90, Circle = glass 3, n = 0.92.

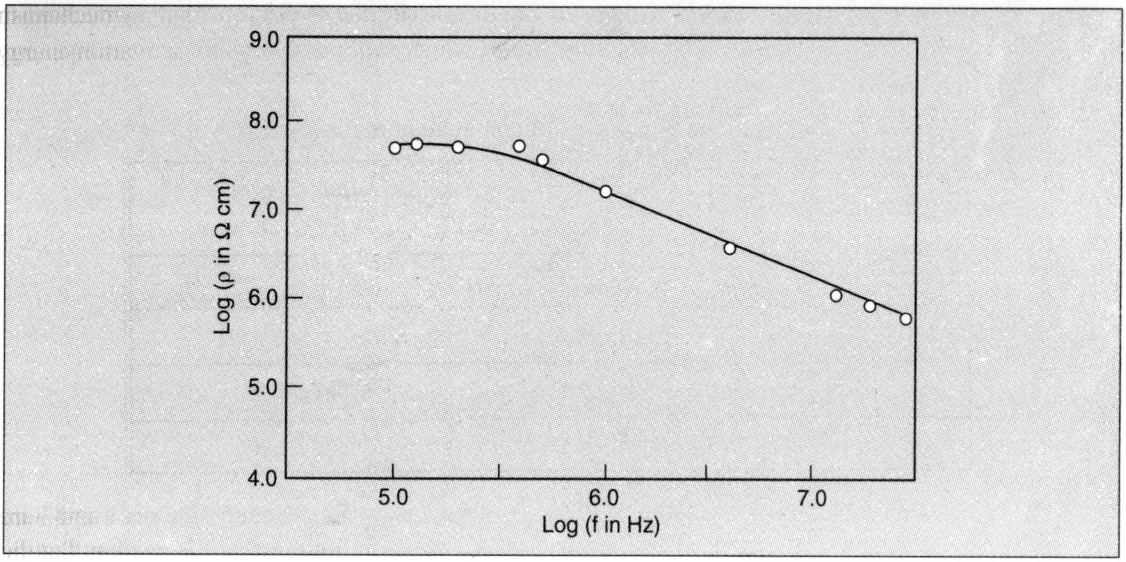

Figure 6.5 : Log ρ against 1/log *f* for glass 8 at 25°C, *n* = 0.96

6.1.3.3. Tunneling Conduction in Nano Particles

The present materials are very similar to 'granular metals', which represent a physical system for studying 'percolation conductivity' [9]. The microstructural features of these "glass-conducting nano granules" system clearly show that the electrical transport due to tunneling mechanism between the isolated conducting particles will influence their DC conductivity behaviour [10]. For low electric fields, Abeles et al [9] have shown that the resistivity of granular metals, when the particles are isolated from each other, is given by

$$\rho = \rho_0 \exp[2(C/kT)^{1/2}] \qquad ...(6.3)$$

In this equation, ρ_0 is a constant and the value of the other constant is given by :

$$C = \chi S E_c^0 \qquad ...(6.4)$$

where, the parameter χ is expressed as

$$\chi = (2m\phi/h^2)^{1/2} \qquad ...(6.5)$$

where, m denotes the electron mass, ϕ the effective barrier height and h is the Planck's constant, S is the separation between the grains, and E_c^0 is the energy required to generate a pair of fully dissociated positively and negatively charged grains. This energy is given by :

$$E_c^0 = 2e^2/Kd \qquad ...(6.6)$$

where, K is expressed as

$$K = \varepsilon[1 + (d/2S)] \qquad ...(6.7)$$

where, ε is the dielectric constant of the insulating medium, and d is the size of the nano particles.

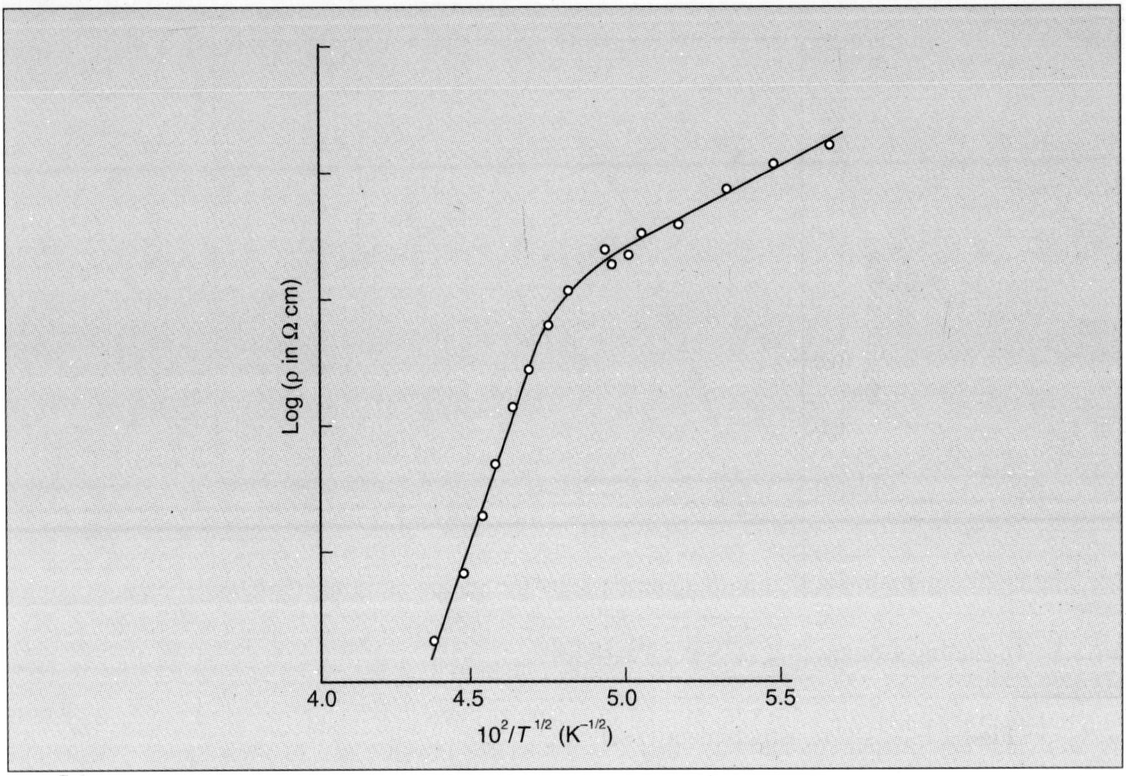

Figure 6.6 : DC resistivity data of glass 1 plotted as log ρ against $1/T^{1/2}$

The DC resistivity data of bismuth and selenium containing glasses are plotted as log ρ against $1/T^{1/2}$ in **Figures 6.6** and **6.7.** The linearity of the curves for temperatures below 120°C give us information on the 'tunneling' mechanism between conducting metallic islands.

The value of C belonging to different glasses are calculated from the slope of these curves. By taking $\chi = 10/\text{nm}$ [10] and S = **5 nm**, the values of E_c^0 are estimated from equation (6.4) for different compositions. The results are tabulated in **Table 6.5.** The values of E_c^0 are also calculated from equation (6.7) by taking $\varepsilon = 4$ [6] being the dielectric constant of the base glass and taking suitable values of d as found from the TEM micrographs. All these values are also given in **Table 6.5.** The calculated values of the energy are in reasonable agreement with those deduced from log ρ against $1/T^{1/2}$ plots, except for glass 8. This discrepancy is thought to be due to an 'uncertainty' in the value of S, which is taken approximately as **5 nm** for the calculation. The inter-particle separation in this glass composition is expected to be less than **5 nm,** which was the 'resolution limit' of the electron microscope used in this work [1, 2].

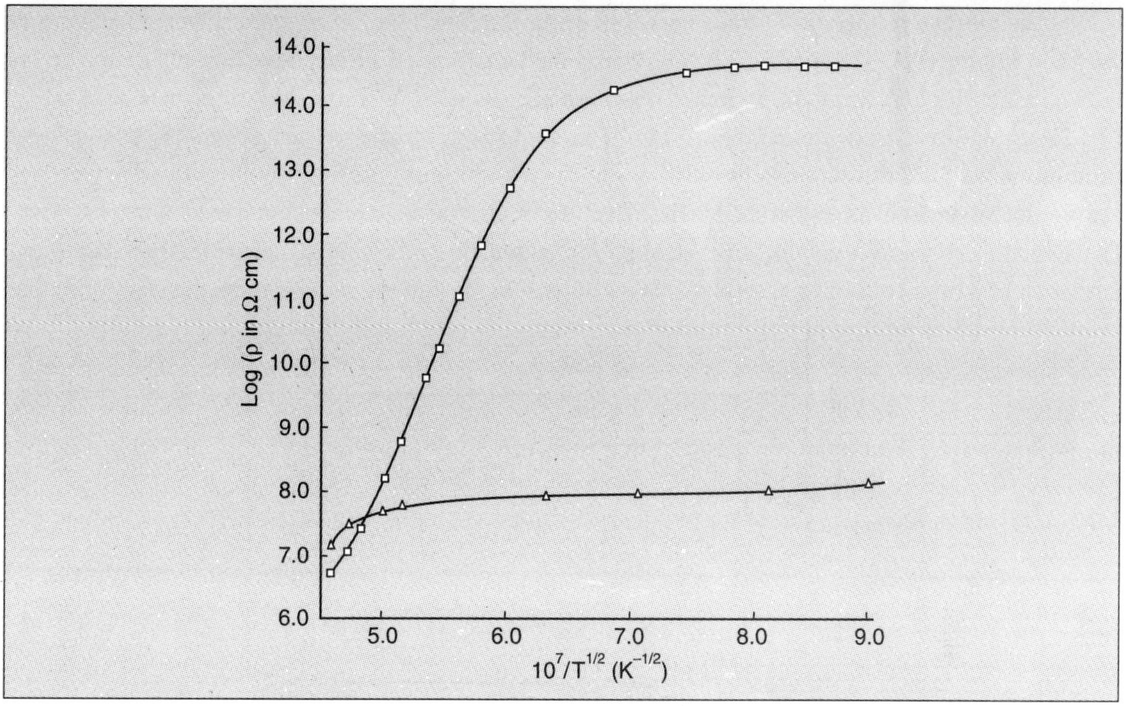

Figure 6.7 : DC resistivity data of selenium glasses plotted as log ρ against $1/T^{1/2}$.

Table 6.5 : Values of C and E_c^0 for different glasses.

Glass No.	*C (eV)*	*E_c^0 (eV)*	*d (nm)*	*E_c^0 (eV) Calculated*
1	1.90	4.00×10^{-2}	5	9.00×10^{-2}
7	2.00×10^{-2}	3.20×10^{-4}	100	6.50×10^{-4}
8	1.00×10^{-2}	2.00×10^{-4}	50	2.40×10^{-3}

Both the real and imaginary parts of the dielectric permittivity of glass 1 are plotted against frequency in **Figure 6.8** in the temperature range, where the 'tunneling' mechanism has been shown to be operative. In **Figure 6.9,** these data are represented by the Cole-Cole diagrams [11]. It is evident from these plots that a 'distribution' of relaxation mechanism is present in this glass 1 containing bismuth. The activation energy for dielectric relaxation is obtained by plotting log of the frequencies at which the 'loss maximum' occurs as a function of 1/T. The results are shown in **Figure 6.10.** From the slope of the linear plots, an activation energy value of 0.14 ev is calculated. This is also in reasonable agreement with the values of E_c^0, as shown in **Table 6.5.** Therefore, the observed dielectric relaxation is attributed to the 'tunneling' of charge carriers between the 'islands' of nano particles of bismuth within the glass matrix.

The current-voltage (I-V) trace on a Tektronix Oscilloscope for an 'ion-exchanged' glass 8 is shown in **Figure 6.11.** The width of the electrodes on the specimen surface and their separation are as shown in the figure. A threshold switch is observed at around 1.1 V from a low-resistance state (*i.e.* ≈ 12.5 Ω) to high-resistance state (*i.e.* ≈ 100 Ω). This switching is found to occur reversibly over several hundred cycles. Such effects are believed to be due to the presence of nano-crystalline selenium particles between the silver droplets precipitated in the glass matrix, after an ion-exchange and reduction treatments. The switching to a lower conductance-state by the application of a voltage seems to arise due to the formation of a high-resistance amorphous layer normal to the current path between the electrodes due to the 'localised' melting of crystalline selenium.

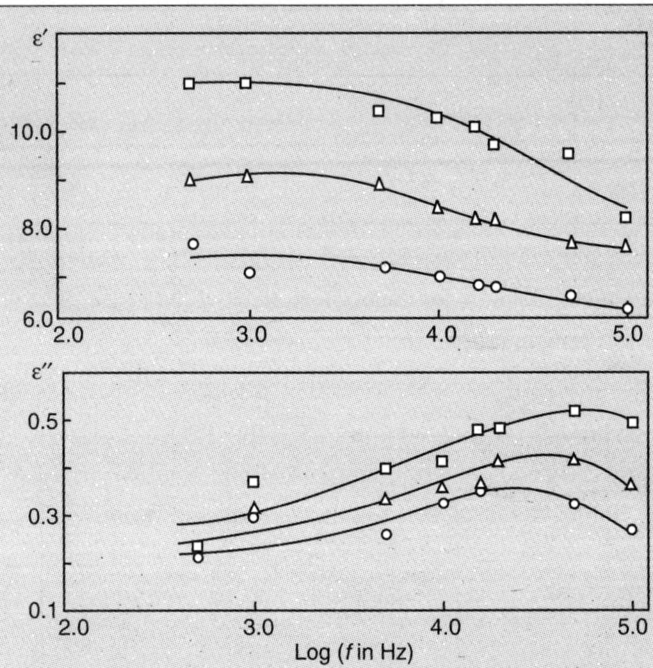

Figure 6.8 : Real and imaginary parts of dielectric permittivity for glass 1 at different temperatures against frequency : Circle = 16°C, Triangle = 59°C, Square = 118°C.

Such type of behaviour is termed as 'reversible switching', which has also been reported in the bulk crystalline selenium near the melting point [12]. The very low switching field of ≈ 1.1 V/cm observed in this study is ascribed to the presence of silver particles [2] in the system.

It has been shown that 'nano granular' metals represent an interesting system for studying the 'percolation' conductivity [9]. Such type of work was restricted either to computer simulation experiments [13, 14] or to simplified systems such as conducting papers with punched holes [15]. It is thought that the materials of the type described above could also be used for the same purpose [1].

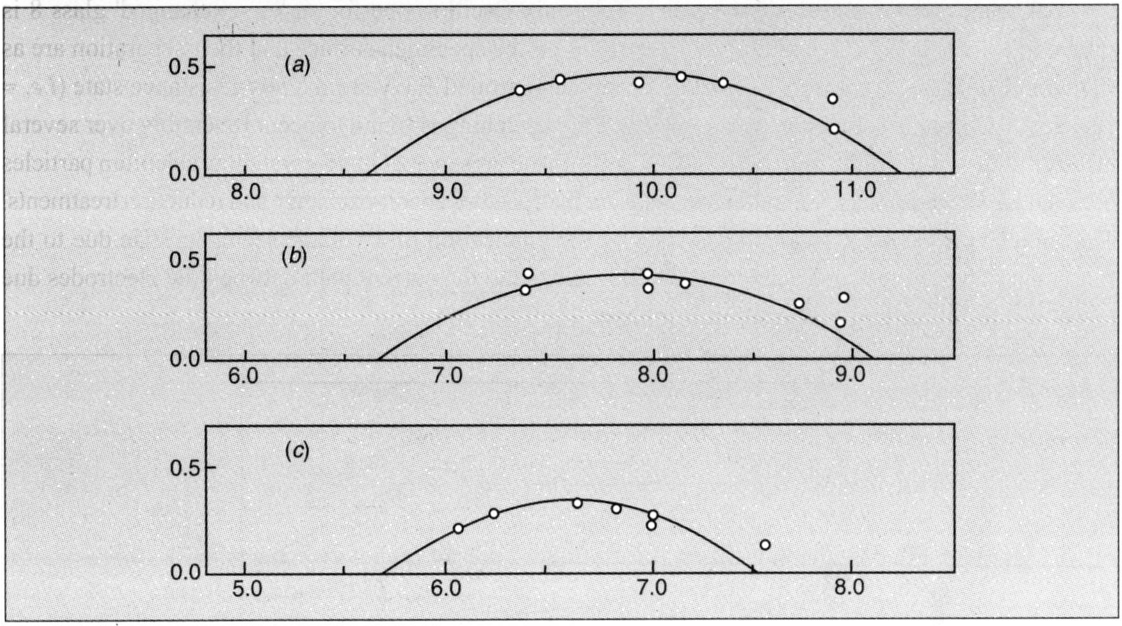

Figure 6.9 : Cole-Cole plots for dielectric permittivity data for glass 1 at different temperatures : (*a*) 16°C, (*b*) 59°C, (*c*) 118°C.

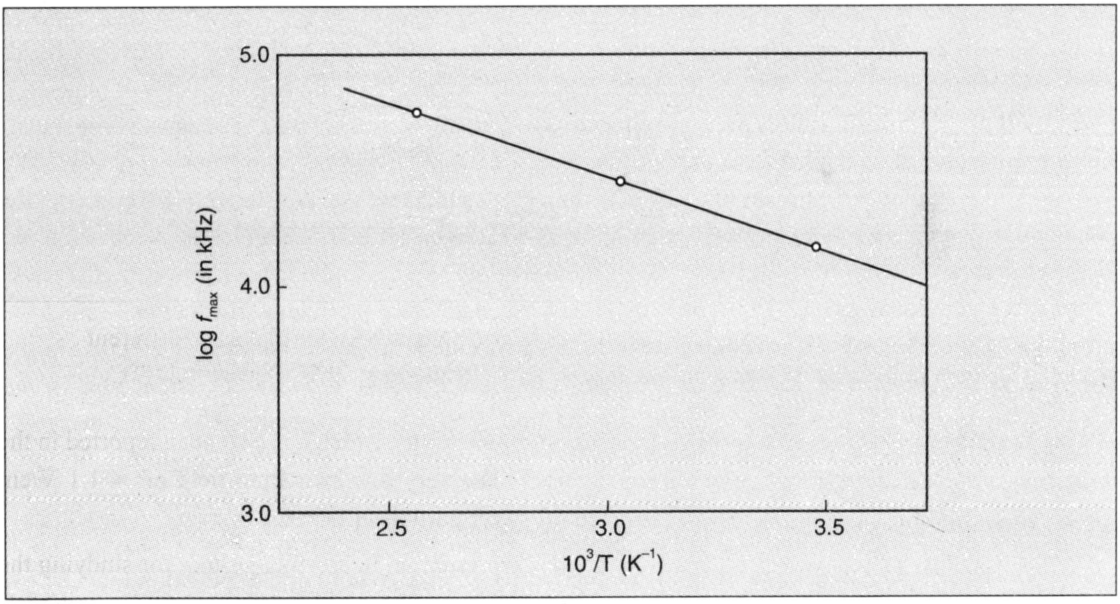

Figure 6.10 : Log of f_{max} against 1/T for glass 1 giving activation energy.

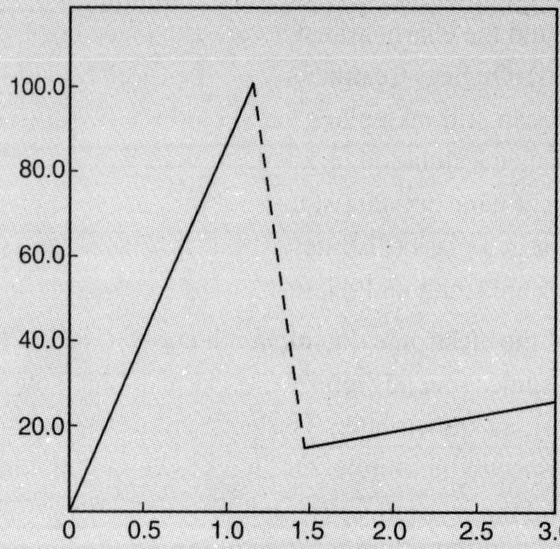

Figure 6.11 : V-I characteristics of the surface layer of an ion-exchanged and reduced sample of glass 8 : electrode width = 3 mm, electrode separation = 1 mm.

6.2. ELECTRONIC CONDUCTION WITH NANO PARTICLES

Preamble

The DC conductivity measurements have been made in a wide range of temperature on a basalt glass, which has been heat-treated at different temperatures in order to form nano particles of magnetite (see the details of formation of magnetite in **chapter 5**). A small polaron hopping mechanism is operative in the as-annealed glass (called a blank glass) between Fe^{2+} and Fe^{3+} ion sites in the isolated state. The sites are either similar or dissimilar in terms of symmetry, which is reflected on the activation energy for conduction. This mechanism is operative in the heat-treated basalt glass-ceramics between these ion sites in the nano particles of magnetite. The DC conductivity at 473°K and the activation energy show significant change at around 700°C of heat-treatment, when the nano particles of magnetite forms in the basalt glass, which has been correlated with the magnetic data on these samples, as detailed in the **chapter 5.** The TSPC and TSDC data are interesting, but the data analysis is complicated by the possible occurrence of the Verwey transition in the small particles of magnetite. The preliminary AC conductivity data support the small polaron hopping mechanism.

6.2.1. Introduction

The electronic conductivity of semiconducting oxide glasses containing transition metal (TM) ions is very interesting [16, 17]. These types of glasses show polaron hopping conduction mechanism between two different valence states of TM ions [18-20]. Hence, the redox equilibrium of different redox states of the TM ions is also interesting [21]. However, very little work has been reported on such glasses, which have been crystallized to form nano particles of magnetite dispersed in the glassy matrix [22, 23]. As described in the **chapter 5,** the magnetic property of these materials are also very interesting in terms different sensitive magnetic and crystallization measurements.

In a basalt glass containing about 13 wt% iron oxide, the iron ions in different redox states are dispersed in the glass matrix, and the charge transfer can take place between Fe^{2+} and Fe^{3+} ion sites in the small polaron states [18-20]. On heat-treatment, with the formation of nano-crystalline particles of magnetite, the charge transfer can still take place between these two different redox states within the nano crystals of magnetite. In such a situation, the activation energy is expected to show some changes with the progress of formation of nano crystals of magnetite, since the potential energy situation which has to be overcome is different in a glass (disordered state) and in a nano crystal (ordered state), even if the size of the nano particles are small and the distribution is quite narrow (see the **section 5.4**)

Therefore, the study of the electronic conduction behaviour of such a mixed system containing nano particles of magnetite assumes special significance, as also found out by Chakravorty on a system containing other TM ions [24], as shown later. The different sensitive magnetic studies like magnetization, Mössbauer and ESR spectra measurements at different temperatures and at different magnetic fields in case of Mössbauer, and the crystallization studies by SANS, basically compliment the study of electronic conduction behaviour of such a system with nano particles of magnetite within a basalt glass.

6.2.2. Preparation of Nano Particles and Conductivity Measurements

The creation of nano particles within a glass matrix is made in a much easier manner than that described in the previous **section 6.1** for switching glasses, wherein there are two different processes involved, viz. ion-exchange and reduction treatments. Here, in this case, the creation of nano particles is simply done within a basalt glass matrix by a heat-treatment process with the progress of nucleation and crystallization of the magnetite phase. The details are given in the **section 5.3.** Hence, only some details on the electrical measurement will be given here.

The three-terminal electrical measurements were made on disc-shaped samples of 1 cm in diameter and 2 to 3 mm in thickness, with gold electrodes evaporated on to the flat surfaces, which was made by polishing through a normal procedure. The DC conductivity measurements were made between 373 and 1000°K, by using a standard 'Electrometer'. The AC conductivity measurements were made only at room temperature as a function of frequency between 10 KHz and 200 MHz in a Q-meter with variable frequency range.

The conductivity measurements were also made down to liquid nitrogen temperature of 77°K for the samples heat-treated at 700°C and 800°C respectively. The DC conductivity measurement was started from the low temperature end at 77°K, by first cooling the sample with an app d electric field, and then taking the readings as the temperature was raised at a constant rates of 0.35°K min and 1°K/min respectively for 700°C sample. For the 800°C sample, the rates of heating were 1°K/min and 2°K/min respectively. This experiment is known as 'Thermally Stimulated Polarization Current' (TSPC). In the 'Thermally Stimulated Depolarization Current' (TSDC) measurements, after cooling the sample and doing the TSPC measurements, the sample is again cooled without any applied electric field and the readings are taken on reheating at the rate 1°K. These TSDC data were only recorded for 800°C sample.

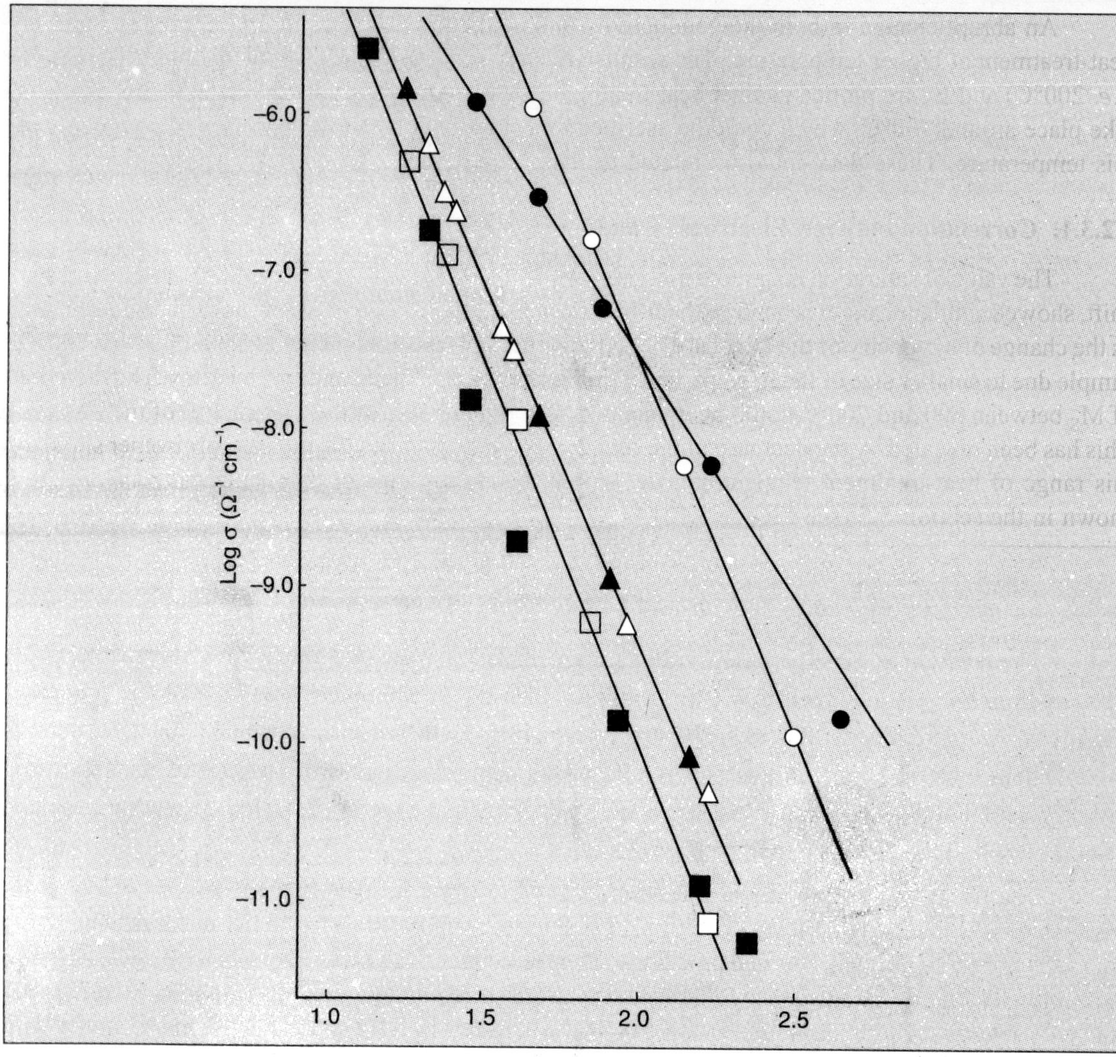

Figure 6.12 : The DC conductivity against 1/T : conductivity increases with heat-treatment temperature, for 900 glass (filled circle) - the intercept is lowered (see text).

6.2.3. DC Conduction Data of Nano Particles

Although log σT against 1/T plots are normally shown to elucidate the 'small polaron' hopping conduction [20], log σ against 1/T plots are presented here, as has been done by various other workers [17]. Such plots are shown in **Figure 6.12** in the temperature range 373 and 1000°K. There is some scatter in the experimental results, but they generally give straight lines. No appreciable time-dependent effects were observed during the conductivity measurements, thereby establishing the fat that the conduction is mainly electronic. The activation energy (E_a), calculated from the slopes of these straight lines, varies from 0.88 ev for the as-annealed glass to 0.60 ev for the basalt glass heat-treated at 900°C with a particle size of **7.0 nm.** For a $2BaO$-$3B_2O_3$ glass containing about 10% Fe_2O_3, the activation energy was found to be 0.93 ev [18, 19]. Therefore, these data could be considered to be consistent with the 'small polaron' hopping between isolated Fe^{2+} and Fe^{3+} ions within the basalt glass for the as-annealed glass.

An abrupt change in both the conductivity and activation energy is observed as a function of heat-treatment at higher temperature. The results are shown in **Figure 6.13,** in which log σ at 473°K (*i.e.* 200°C) and E_a are plotted against heat-treatment temperature. It is noted that the abrupt changes take place around 700°C, which could be ascribed to the formation of nano particles of magnetite at this temperature. These data could be related to those described in the **sections 5.4** and **5.5.**

6.2.3.1. Correlation between Electronic Conduction and Magnetic Data

The value of saturation magnetization (M_S) at 270°K, and the quadrupole splitting and the isomer shift, showed similar changes around the heat-treatment around 700°C. These data have been interpreted as the change of symmetry of the Fe ions. The remarkable 'superparamagnetic' behaviour of this 700°C sample due to smaller size of the nano particles **(5.5 nm)** of magnetite is quite noteworthy. The increase of M_S between 600 and 700°C could be associated with the increase of the 'symmetry' of the Fe ions. This has been revealed by the decrease of the 'quadrupole splitting' and the increase of 'isomer shift' in this range of heat-treatment temperature, *i.e.* within this nano range of magnetite particles, as also shown in the **section 5.5** (see **Figure 5.12**).

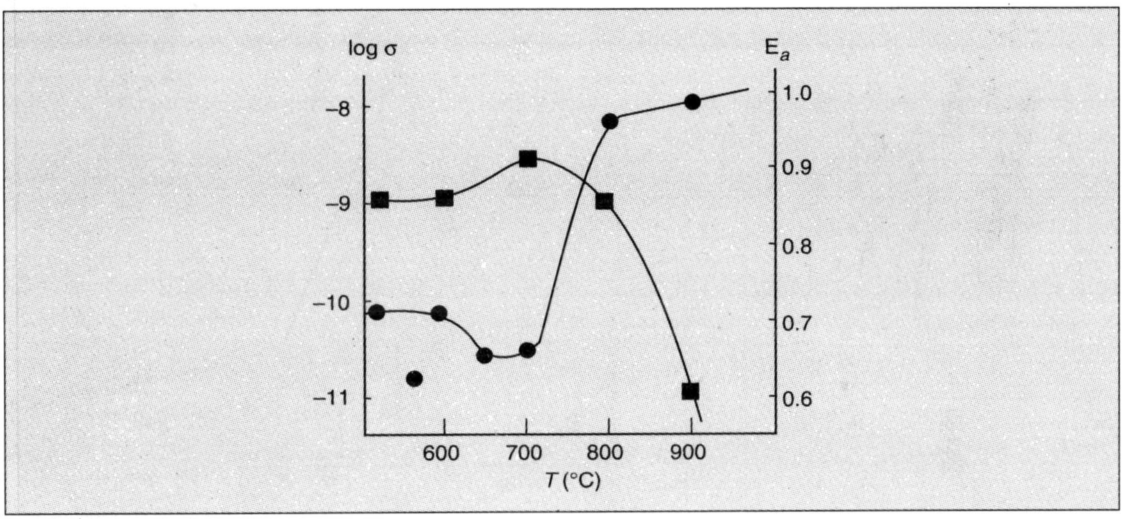

Figure 6.13 : The DC conductivity at 473°K (filled circle) and activation energy (filled square) against heat-treatment temperature.

This effect is not so pronounced between 700 and 900°C, as explained by the cation redistribution process of Neel [25], as the nano particles of magnetite grow in size from **5.5 nm** to **7.0 nm.** Moreover, the sharp decrease of activation energy after 700°C towards higher heat-treatment temperature might indicate that it is easier for the small polaron hopping to take place between two sites within the 'nano crystal' than within a disordered system like glass.

A structural definition can be given to the DC conductivity data in correlation with the magnetic data as follows : Below 700°C, the Fe^{2+} and Fe^{3+} ions are mainly in 'isolated octahedral sites' in the glass matrix, and the conduction is via electron or small polaron hopping between these ion sites. The initial nucleation of magnetite in the region of 600 and 650°C removes some of the Fe^{3+} ions from the glassy phase, and they are incorporated instead into the nano-crystalline particles of magnetite, where they are mainly in 'tetrahedral coordination'. Thus, some of the Fe^{3+} ions are in different structural coordination, and there are fewer number of "equivalent" Fe^{2+} and Fe^{3+} sites available for small polaron hopping conduction.

However, as the crystallization proceeds with increasing heat-treatment temperature, more and more of the Fe^{2+} and Fe^{3+} sites are created within the nano particles of magnetite, where the ratio of Fe^{2+} and Fe^{3+} ions in similar 'octahedral' coordination is optimized, and hence the conductivity will increase. The implication is that the 'easy conduction path' is via nano crystals of magnetite. Therefore, it should be noted that the maximum amount of magnetite available for precipitation from the basalt glass is about 2 mole%. However, the magnetite is a good *n*-type semiconductor with a very low activation energy ($\approx$ 0.0015 *ev*) and even a relatively small volume fraction of nano crystalline magnetite is just sufficient to cause a substantial change in the conductivity. Furthermore, the increase in conductivity is not simply due to a decrease in the activation energy. The 'intercept' of the straight lines of **Figure 6.12** also changes, which is particularly noticeable in the sample heat-treated at 900°C. This is an additional evidence of a change of the 'conduction path'.

6.2.4. AC Conduction Data of Nano Particles

The typical results of the dielectric measurements are illustrated in **Figures 6.14** and **6.15.** The AC conductivity (log $\sigma_{a.c.}$), the dielectric constant (ε') and the dielectric loss (log ε'') are plotted as a function of a wide range of frequency at only room temperature for the basalt glass heat-treated at 800°C for 8 hours. It is seen that while the AC conductivity increases with increasing frequency as expected, the dielectric constant shows a typical 'dispersion' with increasing frequency in the entire range. But the dielectric loss is an important parameter in any dielectric material, since it gives us a clue on the details of the conduction process.

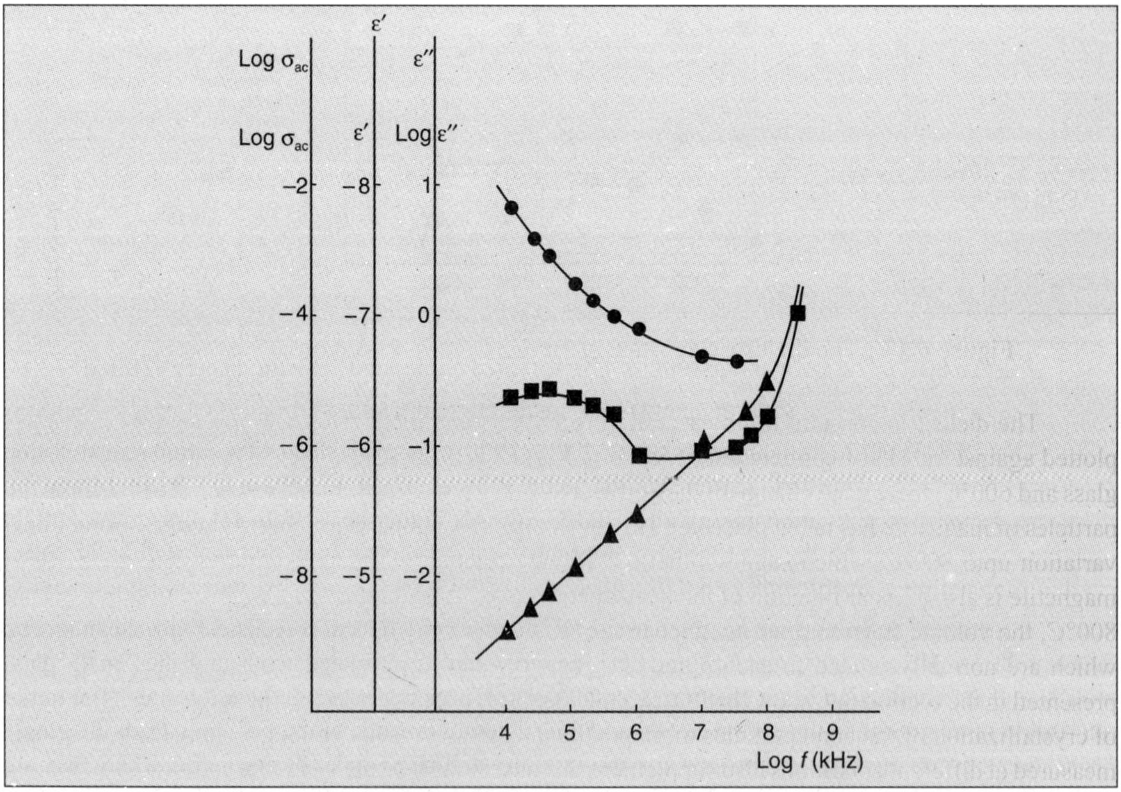

Figure 6.14 : The AC conductivity (filled triangle), dielectric constant (filled circle) and dielectric loss (filled square) against log (frequency) for the 800 sample.

It is interesting to note that there is a clear indication of a 'broad' relaxation peak maximum in the log ε″ plot that is typical of a 'hopping conduction' mechanism between two ion sites forming small polarons. This broad relaxation peak seems to be centred at about 50 KHz. But more complete data, particularly as a function of temperature, are necessary for a detailed interpretation since this relaxation of different small polaron sites participating in the hopping conduction process shows a strong temperature dependence giving rise to an activation energy. Unless this activated process matches with that of DC conductivity data, it is difficult to make any definitive conclusion on the mode of conduction process from the AC conductivity data.

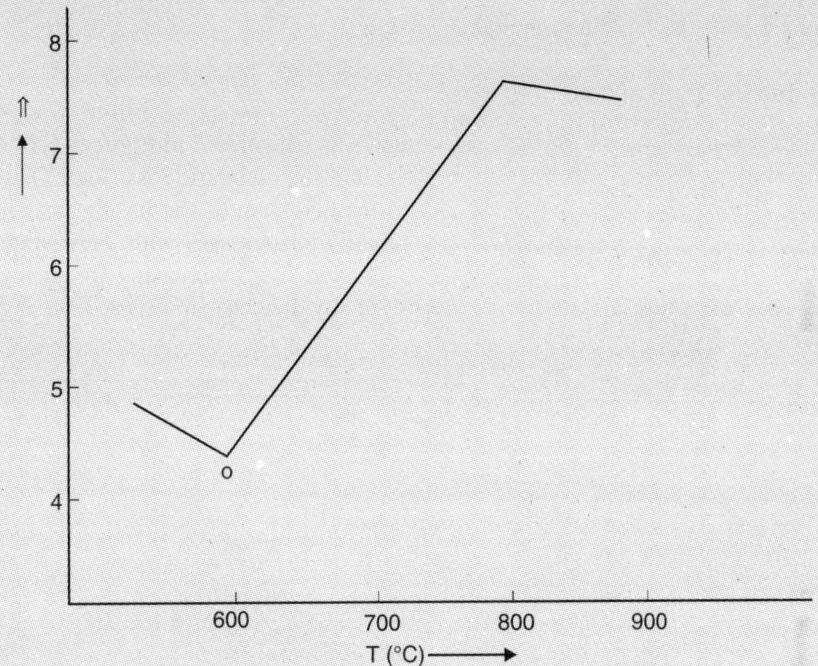

Figure 6.15 : The dielectric constant at 10^5 Hz against heat-treatment temperature.

The dielectric constant (ε′) measured at room temperature at a fixed frequency of 100 KHz is plotted against the heat-treatment temperature in **Figure 6.15.** It is seen that between the as-annealed glass and 600°C, there is no change of dielectric constant, as no significant nucleation of the small nano particles of magnetite has taken place yet. However, after this temperature, there is quite a strong linear variation upto 800°C, which suggests that the volume fraction of crystallization of nano particles of magnetite is also a linear function of the heat-treatment over this temperature range 600 - 800°C. After 800°C, the volume fraction does not appear to change as per the indication of the dielectric constant, which are normally related to each other [24]. It is quite plausible to mention that the SANS data, presented in the **section 5.7,** show that 'Ostwald Ripening' process sets in only in the higher temperature of crystallization of the nano particles of magnetite in the basalt glass. Various sensitive magnetic data measured at different conditions also supplement this observation on the dielectric constant as a function of the heat-treatment temperature (see the **sections 5.4** and **5.5**).

6.2.5. Verwey Transition of Nano Particles

The DC conductivity data as log σ vs. 1/T plots in the TSPC mode from 77°K to room temperature is shown in **Figure 6.16** for the 700°C sample for two different heating rates of 0.35 K/min and 1°K/min respectively, and those for the 800°C sample are shown in **Figure 6.17** for two different heating rates of 1°K and 2°K respectively. It is seen from **Figures 6.16** and **6.17** that there is a broad maximum around and below room temperature for the basalt glass heat-treated at 700°C and 800°C respectively.

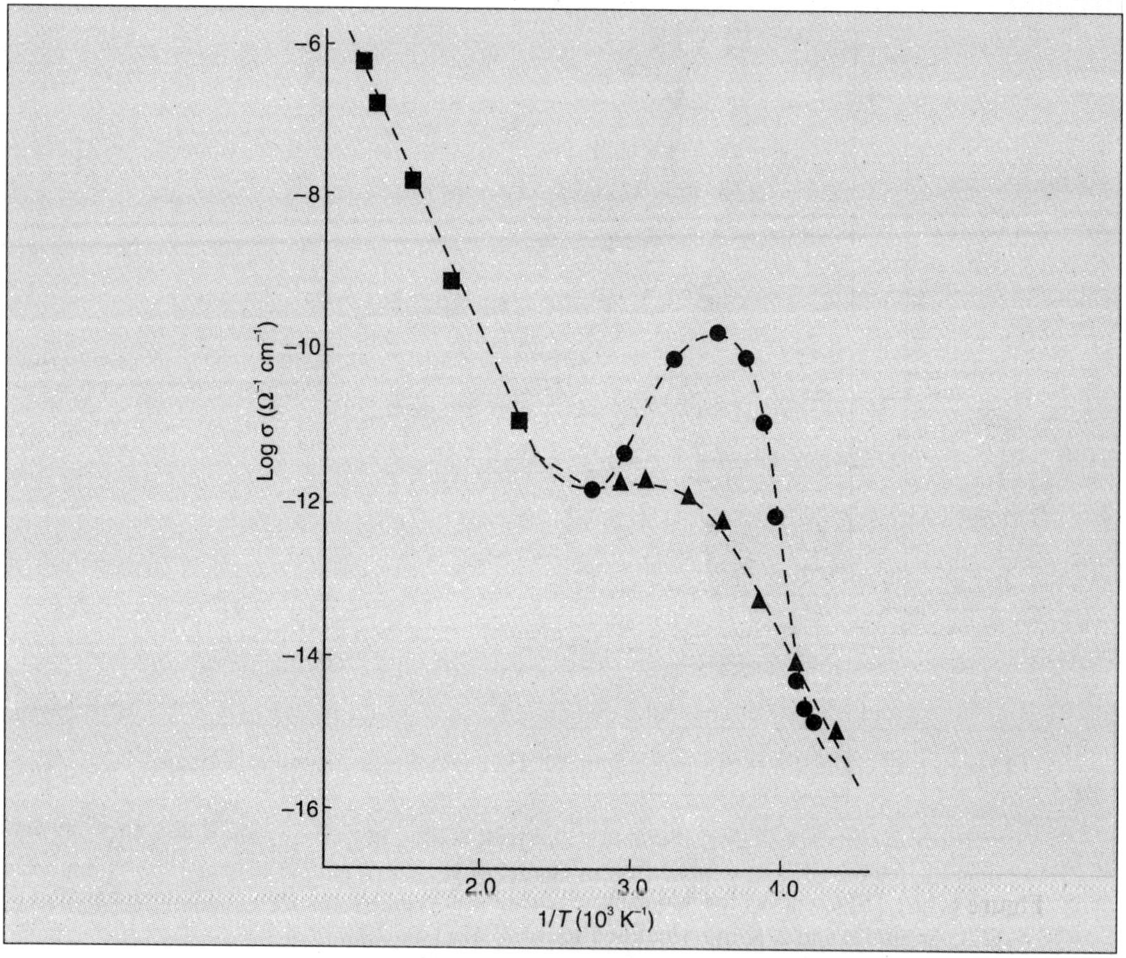

Figure 6.16 : TSPC curves for the 700 sample at the cooling rate of 0.35°K/min (filled triangle) and 1°K/min (filled circle)

The TSPC and TSDC data are similar to those reported in other glasses and dielectrics [26, 27], and they are normally related to the 'trapping and release' of charge carriers (*i.e.* ions or electrons). The trapping parameters may be derived from the TSPC and TSDC curves. However, in the present case, the situation may be quite complicated by the possible occurrence of Verwey transition in the nano-crystalline magnetite phase within the glassy matrix.

It is worth mentioning that M_S at ≈ 300°K increases by about 70% in the case of pure bulk magnetite in the particle size range of **20 nm-75 nm** [28], whereas in the present case it increases by about 50% between 650 - 900°C, *i.e.* within the nano particles size range of **4.5 nm - 7.0 nm.** This shows the 'remarkable' behaviour of ultra-fine nano particles of magnetite embedded within a glassy matrix. In the bulk crystalline magnetite phase, the Verwey transition temperature (T_V) is 119°K [29], *i.e.* well below the observed peaks shown in **Figures 6.16** and **6.17** respectively.

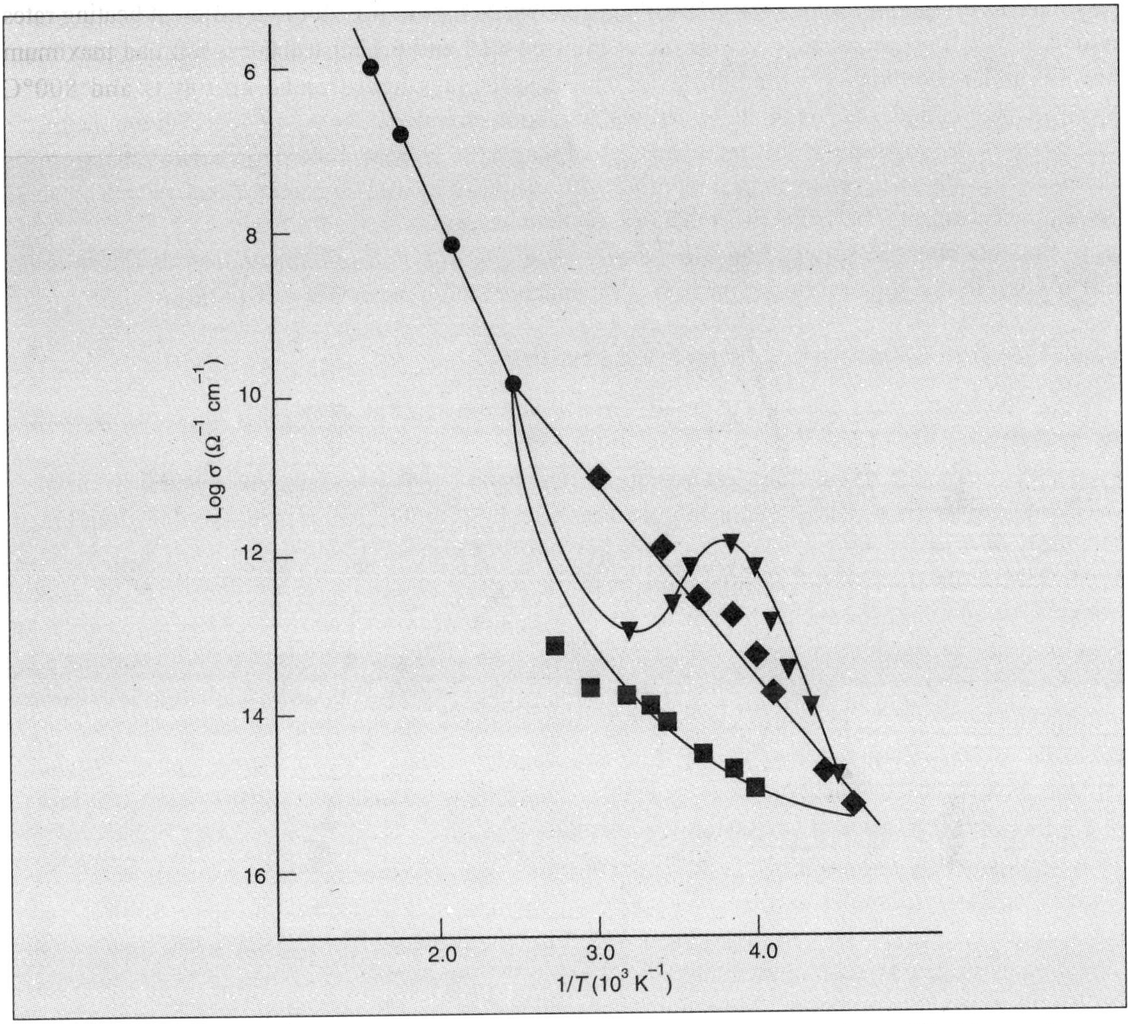

Figure 6.17 : TSPC curves for 800 sample at the cooling rate of 1°K/min (filled inverted triangle) and 2°K/min (inclined square), and TSDC curve at 1°K/min.

However, according to Krupyanski and Suzdalev [28], it should be noted that the Verwey transition temperature is a function of crystallite size of the nano particles of magnetite, and T_V is reported by these workers to be in the range 300°K to 350°K for nano crystals of magnetite of size **10 nm**. Rogwiller and Kundig have also shown that the Verwey transition temperature is spread over a range of temperature from 100°K and 300°K in the nano crystals of magnetite with a mean size of about **14 nm** [30]. The Verwey transition temperature in the present case (*i.e.* with a mean particle sizes between **5.5 nm** and **6.4 nm**) appears to be quite similar to the data of Krupyanski and Suzdalev [28] with a mean particle size of **10 nm.** However, further experimental work is needed to separate the possible contributions from the 'non-steady-state' thermally stimulated effects, and the Verwey transition temperature in the nano particles of magnetite. The tremendous effect of 'superparamagnetism' in this narrow range of sizes of nano particles in the basalt glass matrix may have some effect, which needs to be studied.

In summary, it can be said that the DC conductivity data indicates small polaron hopping between the isolated Fe^{2+} and Fe^{3+} ions in the as-annealed glass. With the increase of heat-treatment temperature, associated with the formation of nano particles of magnetite, the hopping can take place between these sites in similar coordination within the small nano crystals of magnetite in a basalt glass matrix. The abrupt change in both the conductivity and the activation energy can be correlated with the magnetic data in terms of formation of the nano particles of magnetite and the possible structural change with increasing heat-treatment temperature. Preliminary AC conductivity data support the small polaron hopping transport mechanism. The TSPC and TSDC data cannot be used to derive the 'trapping parameters' due to the possible effect of Verwey transition. However, the Verwey transition temperatures appear to be almost close to that found for nano particles of magnetite with a mean size of **10 nm.**

6.2.6. Electrical Conductivity of Other Nano Particles

In the recent years, the nano materials have been extensively studied due to the expectation that the new physics would enlighten us on the extra-ordinary properties exhibited by such systems of materials [31-34]. One of the important aspects of these nano-crystalline materials is their high surface-to-volume ratio and hence they contain larger volume fraction of grain-boundaries. The thin-films based on these nano materials have been exploited for the study of the structure and dynamics of the thin-films in confined geometries [35]. Recently, there has been a surge of activity in the domain of molecular dynamics simulations, and such type of work on silicon grain boundaries show that there is an existence of an 'amorphous equilibrium structure', as found out by an extensive work by the Phillpot group [36, 37]. Das et al [24] were able to create a metal core-metal oxide 'shell structure' with nano-scale dimensions within a silica gel matrix. The percolative configuration of the composite nano particle generates a large 'interface' in the disordered matrix [38]. A drastic change in electrical conductivity of the resultant structure as compared to that of the parent gel indicated that there is a marked influence of the 'interfaces' on the transport properties of such nano systems [24].

The target gel composition is chosen such that the copper phase on reduction formed a percolative chain of nano-sized copper particles. The gel composition in mole% was 60CuO - $40SiO_2$, with $CuCl_2.2H_2O$ and $Si(OC_2H_5)_4$ used as precursors. The composites of nanometer-sized copper core-copper oxide shell with sizes in the range of **6.1 nm to 7.3 nm,** which are dispersed in this silica gel, were synthesized by a technique consisting of reduction followed by oxidation of a suitably chosen precursor gel. The hot-pressed gel powders mixed with nano particles of copper dispersed in a silica gel showed electrical resistivity several orders of magnitude lower than that of the precursor gel. The electrical resistivity of different composites was measured over the temperature range of 30 to 300°C [24].

The activation energies for the core-shell nano-structured composites were found to be a fraction of that of the precursor gel. Such dramatic changes are attributed to the presence of an 'interfacial' amorphous phase. For these composites, the variation of resistivity as a function of temperature was analysed on the basis of Mott's small polaron hopping conduction model, as also done in case of basalt glass containing smaller nano particles of magnetite, as described in the previous section [23]. The effective dielectric constant of the interfacial phase as extracted from the analysis of the data was found to be much higher than that of the precursor gel. This has been explained by these workers as arising from the generation of a very high pressure at the interface due to the oxidation step to which the nano particles of copper are subjected [24].

In a related investigation, the conducting films consisting of silver particles of diameters ranging from **4 nm** to **12 nm** have been grown in glass-ceramic by subjecting the latter to a Li^+ to or from Ag^+ ion-exchange, which was followed by a suitable reduction treatment. The DC electrical resistance of these films has been measured over the temperature range 80-300°K. The resistivity data have been analysed in terms of the Ziman theory of electron-phonon scattering. The effective Debye temperature (θ_P) has been estimated by fitting the experimental data to Ziman's equation. The θ_P is seen to vary from 98 to 192°K for silver particle sizes ranging from **4.3 nm** to **11.0 nm.** The silver particle aggregates in this system have a fractal microstructure with fractal dimensions of around 1.6 and 1.9 respectively [39].

6.2.7. Impurity States in Electronic Conduction

Finally, in novel materials like 'quantum dots', the impurity states have to be theoretically mapped before embarking on their fabrication or their incorporation in the newer devices. The dopants are invariably added as a 'crucial element' in all the semiconductor devices for their proper functioning. The bound states are introduced in the forbidden gap due to the presence of such dopants in a semiconductor material. Hence, it influences both its optical and electronic transport properties (see the **section - 1.7.6** on optical transition in semiconductors). Therefore, we have to properly understand the 'impurity states' in the semiconductors nano-structures, which has attracted the attention of many workers in this new field, since the impurity states are strongly dependent on the following

1. Nature of the impurity states,
2. Strength of the impurity states, and
3. Width of the confining potential,

The spatial position of the impurity within the nano-structures is also important [40-44]. In order to optimise the performance of the devices based on such structures, all the above parameters obviously provide additional control. Although a large number of studies have been made during the past on impurity states in semiconductor nano-structures, there is still a 'gap' specially in the field of 'quantum dots', since it is an emerging field.

By using the variational technique, Bose [45] estimated the 'donor binding energy' for the lowest donor state in a 'spherical quantum dot' with 'finite potential barrier'. The impurity binding energy, expressed in terms of Rydberg's constant, was computed as a function of the composition (x) for GaAs-$Ga_{1-x}Al_xAs$ quantum dots. But the effective masses of electron have been taken as the same for both the well and the barrier material by various workers [43-46]. However, from the practical point of view, with varying compositions, both the barrier height as well as the effective mass of the carrier in the 'barrier material' gets affected. Such variation of effective masses was incorporated in the analysis by Bose[45].

The lowest electronic level and the associated binding energy is found to be more sensitive to Al concentration in the GaAs-$Ga_{1-x}Al_xAs$ quantum dot for low values of x, whereas both the potential barrier height and the carrier effective mass showed linear dependence on x. Moreover, the simplification usually made through the assumption of 'equal carrier effective masses' in both the dot and surrounding material is found to raise the binding energy. The consideration of the actual dependence of the electron effective mass in barrier material on the composition (x) lowers the ground electronic state and the associated impurity level, with respect to those where the effective mass is taken constant [45].

REFERENCES

1. D. Chakravorty, A. K. Bandyopadhyay and V. K. N agesh, 'Electrical properties of oxide-glasses containing bismuth and selenium granules', J. Phys. D Appl. Phys.), 10 (1977) 2077.
2. D. Chakravorty, Appl. Phys. Lett., 24 (1974) 62.
3. D. Chakravorty and C. S. Murthy, J. Phys. D, 8 (1975 L162.
4. A. Paul, J. Mater. Sci., 10 (1975) 415.
5. N. H. Furman, Standard Methods of Chemical Analysis, Van Nostrand, New York, (1962) pp.929.
6. R. J. Charles, J. Am. Ceram. Soc., 47 (1964) 559.
7. H. Rawson, Inorganic Glass Forming Systems, Academic Press, New York, (1967) pp. 205.
8. J. M. Stevels, Handbuch. Der Physik, Springler Verlag, Berlin, 20 (1975) pp. 350.
9. B. Abeles, Ping Sheng, M. D. Coutts and Y. Arie, Adv. Phys., 24 (1975) 407.
10. P. Sheng and B. Abeles, Phys. Rev. Lett., 28 (1975) 34.
11. N. E. Hill, E. Vaughan, A. H. Priceworth and M. Davis, Dielectric Properties and Molecular Behaviour, Van Rostrand-Reinhold, London, (1969).
12. C. H. Champness and D. Armitage, Proc. Intl. Conf. on Low Mobility Materials, Eilat, (1971).
13. S. Kirpatrick, Rev. Mod. Phys., 45 (1973) 574.
14. C. H. Seager and G. E. Pike, Phys. Rev., B10 (1974) 1435.
15. B. J. Last and D. J. Thouless, Phys. Rev. Lett., 27 (1971) 1719.
16. I. G. Austin and N. F. Mott, Adv. Phys., 18 (1969) 41.
17. G. S. Linsley, A. E. Owen and F. M. Hayatee, J. Non-Cryst. Solids, 4 (1970) 208.
18. A. K. Bandyopadhyay and J. O. Isard, 'Electrical conductivity of barium borate glasses containing mixed transition metal oxides', J. Phys. D (Appl. Phys.), 10 (1977) 99.
19. A. K. Bandyopadhyay and J. O. Isard, 'Electrical conductivity of borate glasses containing oxides of V, Fe and Cu singly and in pairs', "Amorphous and Liquid Semiconductors", Ed.: W. E. Spear, CICL, University of Edinburgh (U. K.), (1977) pp. 683.
20. A. K. Bandyopadhyay, J. O. Isard and S. Parke, 'Polaronic conduction & spectroscopy of semiconducting borate glasses containing vanadium', J. Phys. D (Appl. Physics), 11 (1978) 2559.
21. A. K. Bandyopadhyay, 'Thermodynamics of oxidation-reduction of vanadium in borate glasses', J. Mater. Sci., 15 (1980) 1478.
22. Y. Limb and R. F. Davis, J. Am Ceram. Soc., 62 (1979) 403.
23. J. Jurado Egea, A. E. Owen and A. K. Bandyopadhyay, 'Electrical conduction in basalt glass-ceramics- correlation with magnetite crystallization', J. Mater. Sci., 22 (1987) 3602.
24. D. Das, T. K. Kundu, M. K. Dey, S. Chakravorty and D. Chakravorty, 'Electrical conduction in composites containing copper core-copper oxide shell nanostructure in silica gel', Proc. Indian Acad. Sci. (Chem. Sci.), 115 (2003) 341.
25. L. Neel, Ann. Phys., 3 (1948) 137.

26. H. M. Gupta and R. J. Van Overstraeten, J. Phy. C (Solid State Phys.), 7 (1974) 407.
27. B. Dutta and D. E. Day, J. Non-Cryst. Solids, 48 (1982) 345.
28. Y. F. Krupyanski and I. P. Suzdalev, J. Phys., 35 (1974) 407.
29. E. J. W. Verwey, P. Haayman and F. C. Romeijin, J. Chem. Phys., 15 (1947) 181.
30. P. Roggwiller and W. Kundig, Solid State Commun., 12 (1973) 901.
31. Y. Imry, 'Nanostructures and Mesoscopic Systems', Ed. W. P. Kirk and M. A. Reed, Academic Press, New York, (1992) pp. 11.
32. D. Boese and H. Schoeller, Europhys. Lett., 54 (2001) 668.
33. Z. K. Tang, L. Zhang, N. Wang, X. X. Zhang, G. H. Wen, G. D. Li, J. N. Wang, C. T. Chan and Sheng Ping, Science, 292 (2001) 2462.
34. K. Jacobs, D. Zaziski. C. Scher, A. B. Herhold and A. B. Alivisatas, Science, 293 (2001) 1803.
35. H. Konrad, C. Karmonik, J. Weissmuller, H. Gleiter, R. Birringer and R. Hampelmann, Physica, B234-236 (1997) 173.
36. P. Keblinski, S. R. Phillpot, D. Wolf and H. Gleiter, Phys. Rev. Lett., 77 (1996) 2965.
37. P. Keblinski, D. Wolf, S. R. Phillpot and H. Gleiter, Philos. Mag. Lett., 76 (1997) 143.
38. S. Roy, A. Chatterjee and D. Chakravorty, J. Mater. Res., 8 (1993) 689.
39. B Roy and D Chakravorty, 'Electrical conductance of silver nanoparticles grown in glass-ceramic', J. Phys.: Condens. Matter, 2 (1990) 9323.
40. G. Bastard, Phys. Rev. B, 24 (1981) 4714.
41. G. Weber, P. A. Schulz and L. E. Oliveira, Phys. Rev. B, 38 (1988) 2179.
42. N. Porras-Montenegro, S. T. Perez-Merchancano and A. Latge, J. Appl. Phys., 74 (1993) 7624.
43. C. Bose, J. Appl. Phys., 83 (1998) 3089.
44. C. Bose and C. K. Sarkar, Solid State Electronics, 42 (1998) 1661.
45. C. Bose, 'Compositional dependence of binding energy of the lowest impurity state in spherical GaAs-$Ga_{1-x}Al_xAs$ quantum dots', Proc. Natl. Conf. on 'Nano-Science and Technology', J. U., Kolkata (India), January, (2005).
46. F. J. Ribeiro and A. Latge, Phys. Rev. B, 50 (1994) 4913.

Chapter 7

Optical Properties

PREAMBLE

The story of interaction of the electromagnetic waves, *i.e.* light, with matter dates back to Maxwell's time (see for details in the **section 1.7**). This interaction is very interesting for practical applications in glasses and a host of materials [1].

The optical properties of glasses are based on this "Interaction" of the medium with the energy of the electromagnetic waves. The interest in optical glasses is linked to their certain characteristics, which can be particularly written as

(*a*) Their Isotropy,

(*b*) High Level of Homogeneity,

(*c*) Large and Continuous Variation of Properties with Composition,

(*d*) To Obtain Samples without any Limitation on their Dimensions (contrary to the crystals).

In the majority of cases, the "short range order" only influences the optical phenomena, although the interesting optical properties of the crystals are observed without this inconvenience.

The optical glasses are used as prisms, filters, mirrors etc. for general purpose. For special purposes, they are also used as "Laser Glass" and recently as "Optical Fibres" for Communication of images and information (photonic guide).

Although no effort is made here to deal with the details of the optical properties of glasses or crystals, some preliminary definitions and explanations of some important optical properties are presented here to get some acquaintance with certain terminology that are definitely necessary. This has precisely been attempted in the following introduction. Moreover, this also has some relevance to the understanding of the optical properties of the nano materials.

7.1. INTRODUCTION

It is but natural that whenever we talk about 'optical properties' of materials, we immediately think of the 'glasses'. From the ancient times, the glasses are known as important and useful optical materials. The glasses show all types of usual optical properties, with which we are generally concerned in our daily life, *e.g.* refraction, reflection, transmission and absorption. However, there are many crystalline materials with and without centro-symmetric properties, which also show excellent optical properties due to their dielectric properties, *i.e.* the existence of the 'electric charges' and their spatial 'displacements'. While this so-called displacement gives rise to polarization, which in turn affects many important properties, that which will be discussed subsequently, like electro-optic and acousto-optic properties.

In this chapter, we are not discussing nano-optics, which is the hottest subject of the day, and which by no means surpasses all technological activities in the world today. This important topic like nano-optics will be dealt with in details in the **section 8.2.1.** Here, in this chapter, after a brief account of the so-called usual optical properties like refraction, reflection, absorption, etc., a description is given on coloured glasses and also about different ways, the colours can be developed in the materials, particularly with nano particles of gold, silver and copper. The description given here is not that vivid, but anyway it gives a sort of impression about the general behaviour of nano particles inside glasses and crystals, with many recent examples to acquaint the readers on the overall subject.

7.2. OPTICAL PROPERTIES

7.2.1. Some Definitions

The propagation of an electromagnetic wave in a material produces a displacement of electrical charge. For a sinusoidal wave, the change of speed and intensity are continuous in the complex refractive index (n^*), which is related to the complex permittivity

$$\varepsilon^* = \varepsilon' + i\varepsilon''$$

It is given by the relation as

$$n^{*2} = \varepsilon^*$$

By writing, $n^* = n + ik$, where n is the refractive index and k the absorption index , we get

$$n^2 - k^2 = \varepsilon'$$

and, $$2nk = \varepsilon''$$

the quantity $(\varepsilon' - 1) = \chi$ is the electrical susceptibility.

7.2.2. The Refractive Index and Dispersion

The refractive index is equal to the ratio of the speed of incident light in the vacuum (v_0), with respect to that in the material (v_m), $n = v_0 / v_m$. It depends on the wavelength and normally, it decreases as the wavelength λ increases. This variation carries the name of 'Dispersion' and can be defined by the relation

$$D = dn/d\lambda.$$

The variation of n and k are related to each other. The refractive index actually changes in an inverse sense compared to the dispersion in the region of strong absorption, which is called "Dispersive Anomaly".

7.2.3. The Non-Linear Refractive Index

The dependence of the refractive index on the intensity of light is governed by the electrical susceptibility of the third order. The refractive index is related to the average electric field of $\langle E^2 \rangle$ as follows :

$$n = n_0 + n_2 \langle E^2 \rangle$$

where n_0 is the usual linear index and n_2 is the coefficient of non-linear refractive index. The importance of non-linear index in the technology of making "High Power Lasers" has driven the scientists to search for glasses possessing a smaller value of n_2. It has been possible to show that the glasses with a smaller index and with a weakest dispersion behaviour like the 'fluoroberyllates' possess the smallest value of n_2. An excellent review exists in reference [2].

7.2.4. The Absorption Coefficient

From the absorption index (k) which is a function of the wavelength, λ, we can define the absorption coefficient, $\alpha = 4\pi k/\lambda$. For a homogenous material, the fraction of the intensity of light absorbed depends on α and on the thickness (dx) of the distance traversed

$$dI/I = -\alpha\, dx$$

The attenuation of the light traversing a material of thickness x is thus given by the formula

$$I = I_0 \exp(-\alpha x)$$

where I is the intensity of transmitted light, I_0 the intensity of the incident light, and α the absorption coefficient. In place of the transmission : $(I/I_0) \times 100$, we normally consider the "Optical Density", defined by the relation as

$$\text{O. D.} = \log_{10}(I_0/I)$$

For example, a 1% transmission corresponds to an O. D. = 2.

The technology of 'optical conductor' has popularized the concept of "Decibels" as a unit, which is related to the optical density as follows

$$1\text{ dB} = 1/10\ (\text{O. D.})$$

The losses in the fibre optics are indicated in dB/Km. It has to be noted that if we utilize the unit of O. D. or dB, the thickness traversed has to be specified as

$$\alpha = \ln(I_0/I)/x = 2.303\ (\text{O. D.})/x = 23.03\ \text{dB}/x$$

When the absorption is due to a particular ion (*i.e.* the chromophore), the absorption coefficient is then proportional to the concentration "c" of the ion. In this situation, we can write it as

$$\alpha = \varepsilon . c$$

where ε is called the "Extinction Coefficient". We can then write it as

$$I = I_0 \exp(-\varepsilon c x)$$

which is the "Beer-Lambert" Law.

7.2.5. The Reflection

The fraction R reflected under normal incidence is given by the formula of 'Fresnel' as

$$R = ((n-1)^2 + k^2)/((n+1)^2 + k^2)$$

which in the optical region of the spectra ($k \sim 0$), is reduced to

$$R = (n-1)^2/(n+1)^2.$$

Some Useful Units

The indices and the optical absorption are given as a function of the wavelength or the energy of the incident photon. The wavelength is generally expressed in microns (μ), milli microns ($m\mu$), angstroms (Å) or nano meters (nm) as follows

$$1\text{ nm} = 1\ m\mu = 10\ \text{Å} = 10^{-3}\ \mu$$

Between the energy (E) and the wavelength (λ), we have a relation

$$E\ (\text{eV}) = 1239.8/\lambda$$

and the frequency is written as

$$\nu\ (\text{cm}^{-1}) = 1/\lambda\ (\text{cm})$$

7.3. SPECIAL PROPERTIES

7.3.1. Accidental Anisotropy-Birefringence-Elasto-Optic Effect

In certain specific conditions, the glass can become anisotropic. The most frequent reason is the application of a mechanical stress, which induces a birefringence. The speed of propagation, *i.e.* the refractive index, thus depends on the orientation of the plane of polarization. Under the action of an axial stress (σ_z), the glass behaves like a uniaxial medium. It is diagrammatically shown in **Figure 7.1.** The speed of propagation of light parallel to σ_z is identical whatever be the orientation of the plane of polarization., whereas for a ray of light perpendicular to σ_z, the speed varies depending on whether the plane of polarization is perpendicular to σ_z (Ordinary Ray = OR) or parallel to σ_z (Extraordinary Ray = ER).

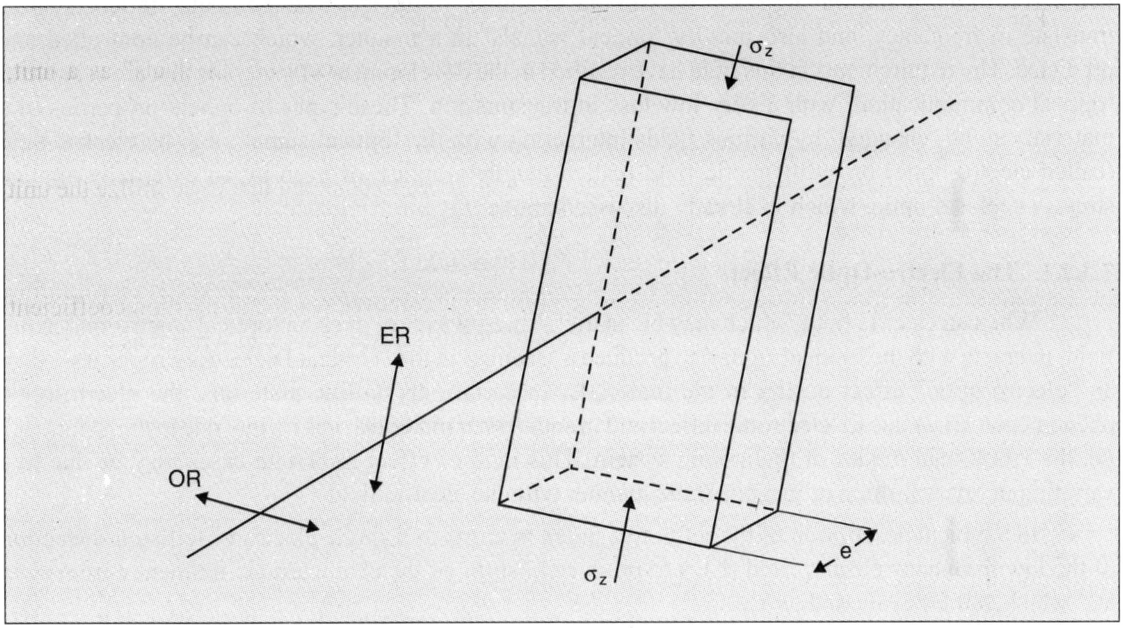

Figure 7.1 : Elasto - Optical Effect.

The glass will thus have two corresponding indices of refraction (n_0 and n_e) and the birefringence is then defined by the relation

$$\Delta n = n_e - n_0$$

which can be either positive or negative.

The "Lag", δ, for a thickness '*e*' of a glass is written as

$$\delta = e\,(n_e - n_0)$$

Usually, it is evaluated by the ratio *d*/*e* in **nm/cm.** The relative 'lag' is given by

$$r = (n_e - n_0)/n$$

where *n* is the index for the glass, which has not been subjected under any stress. It constitutes a measure of the "deformation". The relation between *r* and σ is as follows

$$r = \Delta n/n = B\sigma$$

which defines the elasto-optical constant **(B).**

If $\Delta n/n$ (a dimensionless quantity) is measured in mm/cm and σ in dynes/cm^2, B is measured in Brewsters. In this situation, $\Delta n/n = 10^{-6}\,B\sigma$. For a common glass, B = 2.6 Brewster.

It should be noted that there are also other causes of birefringence, which could happen like an elastic deformation, or due to a specific micro-structure inside the glass like phase separation, or the presence of oriented particles inside the glass.

7.3.2. Electro-Optic and Acousto-Optic Effects

There are systems, which are based on Laser technology that will need a set of extra technical items to the Lasers and wave guides. We can cite examples of devices that modulate, deflect, switch, translate in frequency, and also modify 'optical signals' in a manner, which can be controlled and predicted. The requirements in this field have resulted in the development of materials that are capable of 'optical communication' with a very low loss in transmission. These types of optical properties of a material can be 'changed' by various fields interacting with the 'optical signal', *e.g.* by electric field (called electro-optic) or by magnetic field (called magneto-optic), or even by an externally applied stress, *i.e.* elasto-optic, which is already discussed above, *i.e.* birefringence.

7.3.2.1. The Electro-Optic Effect

When an electric field, which may be 'static', 'microwave' or even an 'optical electro-magnetic' field, interacts with the 'optical signal' to produce a 'change' in the "Optical Dielectric Properties", then an "electro-optic" effect occurs in the materials. In certain crystalline materials, the electro-optic phenomenon arises due to 'electronic' effect and in some other materials, it is mainly due to the 'phonons', *i.e.* the vibrational modes of the atomic system. This kind of effect in certain cases may be due to a variation in linear fashion or in a quadratic manner with the electric field.

In a typical description of the refractive index in terms of a single electron-oscillator, the action of the low frequency electric field (E) is to produce a 'shift' of the characteristic frequency from n_1 to n_2, which can be expressed as

$$n_2 - n_1 = [2e\,v\,(\varepsilon + 2)E]/3mv_0^2$$

where, e and m are the electronic charge and mass, v is the inharmonic force constant, and ε is the low-frequency dielectric constant (since the field is of low-frequency). It is already known that the 'refractive index' n varies as $1/(\nu_2^2 - \nu_1^2)$, and hence the above equation directly expresses the linear variation of this refractive index with the electric field.

In the above description of a dipolar material, it was tacitly assumed that the dipole strength (μ) was a linear function of the externally applied electric field (E), which is expressed as

$$\mu = \alpha E$$

For a dipolar material with N number of dipoles, the polarization (P) is given as

$$P = N\mu = N\alpha E$$

So, the above description is for a linear case. However, for a crystal system lacking a 'centre of symmetry', *e.g.* a ferroelectric crystal like lithium niobate and lithium tantalate, the electric field can be also expressed in non-linear terms by involving the 'polartizability' of the atomic system [3, 4], since in this case, the polarization is not a linear function of the applied electric field (see also the **section 7.2.3**). Hence, it is expressed up to the third order as

$$P = \alpha_1 \langle E \rangle + \alpha_2 \langle E^2 \rangle + \alpha_3 \langle E^3 \rangle$$

where, $\alpha_1 = N\alpha$, and α_2 and α_3 are the coefficients with the non-linear terms, which are basically material constants.

One way to express the 'principal electro-optic' effect is to invoke the 'field-distance' product at half-wavelength ($\lambda/2$), *i.e.* $\langle E . L \rangle_{\lambda/2}$, where L is the 'optical path length'. This particular 'field-distance' product signifies the 'voltage' needed in order to produce half-wave retardation effect in a specific geometry of the sample or material, *i.e.* $L/d = 1$, where d is the thickness of the crystal through which the signal passes. The optical phase retardation (ϕ) can be expressed in radians as

$$\phi = (2\pi L/\lambda_0)[n_1(E) - n_2(E)]$$

where, λ_0 is the wavelength of the light in vacuum, and $n_1(E)$ and $n_2(E)$ are the electric field-dependent refractive indices. The form of their difference depends on the following :

(*a*) Crystal Symmetry,

(*b*) Direction of the Applied Electric Field, and also on

(*c*) Propagation and Polarization Direction of the Optical Beam.

As mentioned earlier, there are various important electro-optical materials like lithium niobate, lithium tantalate, potassium tantalate-niobate, calcium niobate, strontium-barium niobate, barium-sodium niobate, etc. In many of these crystals, Nb or Ta ion is octahedrally coordinated with six oxygen ions, which form the basic structural unit. The main property of the 'change' in refractive index with an applied electric field is exploited in the electro-optic materials in terms of a variety of applications such as

(*a*) Optical Oscillators,

(*b*) Frequency Doublers,

(*c*) Voltage-Controlled Switches in Laser Cavities, and of course

(*d*) Modulators for Optical Communication Systems.

Many of these devices or device accessories have been improved in recent years with novel nano materials. There is a constant search for newer nano-materials.

7.3.2.2. The Acousto-Optic Effect

In the above example of electro-optic materials, the refractive index changes with an applied electric field. Instead of electric field, if a crystal is strained, then also the refractive index can effect a change. This change of refractive index by strain is known as 'acousto-optic' effect. The crystal lattice has a potential, which can be changed by the action of strain that changes the shape and size of the molecular orbitals of the weakly-bound electrons. This causes a change in the polarizability and refractive index as well.

In a polarizable crystal, the strains have different values at different spatial directions, which are ultimately expressed as a strain tensor. Hence, the effect of the strain on the indices of refraction of a crystalline lattice depends on the direction of these 'strain axes' and also on the direction of the 'optical polarization'. These spatial dependence eventually guide the acousto-optic properties of the nano crystalline materials.

If a plane elastic wave is excited with a given crystal system, a periodic strain effect occurs with a spatial extent that is equal to the acoustic wavelength. Then, due to this strain effect, an acousto-optic variation of the refractive index occurs in the crystalline lattice, which is equivalent to a 'volumetric diffraction grating'. Based on this principle of partial diffraction of light incident on an acousto-optic grating at a proper angle, the acousto-optic devices are made. In an acousto-optic device, the use of a particular crystal depends on many factors, such as :

(*a*) The 'piezoelectric' coupling (produced by the strain in the crystal),

(*b*) The 'ultrasonic' attenuation, and also on

(*c*) The 'acousto-optical' coefficients.

The important acousto-optic crystals are lithium tantalate, lithium niobate and some other lead-based compounds. The refractive index of these material is about 2.2, and they are also transparent in the visible spectrum, *i.e.* **400 nm** to **700 nm** of wavelength. There are single domain and multi-domain materials, which have been superbly crafted by means of 'domain engineering' in the nano scale, which makes an 'integrated optical device' very efficient with precise control. On the application front, there are a variety of uses as components in the laser system, waveguides, couplers, modulators, diffractors and in optical detection (see the **section 8.2.1** on nano-optics) [2, 5].

7.4. THE COLOURED GLASSES

If a glass "selectively absorbs or scatters light" in a part of the visible spectra (VIBGYOR), it results in an irregular transmission of light and the glass appears to be coloured to the human eyes, which is sensitive only in the wavelength range of **400 nm** to **700 nm.** The impression of colour is actually a subjective "sensation" depending on the 'spectral sensitivity' of the human eye on the one hand and on the nature of the incident light on the other hand. By contrast, a spectral transmission curve is perfectly defined and is physically measurable. It can serve as a quantitative indication on the colouration of the glasses.

7.4.1. Absorption in Glasses

In the common glasses, the absorption in the visible is primarily due to the transition metal (TM) ions, which are characterized by their incomplete 3*d* shells, particularly, V, Cr, Mn, Fe, Co, Ni, Cu, etc. and to a lesser degree due to the presence of rare earth (RE) ions with incomplete 4*f* shells, and in some cases due to colour centres. This is guided by the 'Ligand Field Theory' in which it is postulated that the degeneracy of the 'electronic levels' will be lifted by the electrical field of the anions (*i.e.* oxygen ligands in the oxide glasses) surrounding the transition metal cations [6 - 7].

The colours produced depends on the oxidation state or redox state, and on the coordination number of the concerned ions. For example, Co^{2+} in a silicate glass is in a tetrahedral coordination with oxygen ions and produces a deep blue colour, whereas in a metaphosphate or borosilicate glass with coordination number of 6, it gives rise to rose colour.

The rare earth ions like Y, La, Gd, Yb and Lu give rise to colours due to the separation of levels in their 4*f* shells. These ions do not have any band in the visible region of the spectra. By contrast, Nd gives a strong red-violet colouration. The glasses doped with Nd^{3+} ions are the basis of "Laser Glasses". In the same way, the Pr ion gives rise to a green colour and the Er ion gives rise to a rose colour.

The above is an example where the 'ligand field' around a TM ion or RE ion will have a significant effect on its absorption characteristics and hence on the colour produced. Sometimes, a particular colour obtained, *e.g.* in organic dyestuffs, can be identified with a certain 'combination of absorbing ions' or 'chromophore'. In case of glassy materials, the 'chromophore' made up of the pairs Cd—S gives yellow colour. The ions Cd^{2+} and S^{2-} by themselves are colourless, since these ions have no absorption in the visible spectrum. However, when they are together associated, they produce lively yellow colour. An increase of temperature modifies the absorption and at a few hundred degree celsius, the glass becomes orange-red. The chromophores responsible for such beautiful colours is of 'nano-dimension', and it is of great scientific interest.

7.4.2. The Colour Centres : Photochromy

In the previous section, a brief description has been given about the basics of optics along with a general concept on how different colours is developed in different materials, like glasses, due to the absorption bands of TM ions or RE ions. Here, another brief description is necessary for 'photochromism' in coloured glasses, since in many of these materials, there is a fine distribution of 'nano particles' of 'colour-active' centres giving a particular colouration to the glass.

A prolonged exposure of a glass to the UV radiation of the Sun produces a colouration due to the change of valence of certain ions or combination of ions. This is called the phenomenon of "solarisation". If the glass contains Mn and Fe as impurities, we can write

$$Mn^{2+} + h\nu \rightarrow Mn^{3+} + e^-$$

where $h\nu$ is the photon energy of UV and e^- is the ejected electron, which is arrested somewhere in the glass structure, for example, on a site relative to Fe^{3+} as

$$Fe^{3+} + e^- \rightarrow Fe^{2+}$$

The "solarised colour centre" thus becomes stabilized and the glass takes a taint of violet colour, due to the presence of Mn^{3+}. This has been observed in the ancient glasses, which have been subjected to a prolonged exposure of light.

The highly reduced silicate glasses containing Eu^{2+} and Ti^{4+} under the action of photon energy develop the colour centres, which progressively disappear as the source of light is cut off. These glasses are called "photochromic" glasses. In this case, the reaction is as follows

$$Eu^{2+} + Ti^{4+} \Leftrightarrow Ti^{3+} + Eu^{3+}$$

and the colour centres responsible for giving colour is due to Ti^{3+} ions. The other example of 'photochromism' refers to the presence of nano-crystalline particles forming a well dispersed phase in the glass, which is ultimately responsible for colouration.

The colour can also result from the absorption of light by the interaction with the electrons, which are not associated with any specific ions, but arrested by the 'structural networking defect'. A typical example is 'colour centre' in many crystals. For glasses, the variety of sites to arrest them inside the glass makes the transition generally enough uniform across the spectra, which produces an uniform darkening (grey tinge) rather than a well defined colour.

The bombardment of a glass by the energetic particles or by irradiation by the X-rays or γ-rays produce a modification of the transmission, which can be suppressed by a thermal treatment of annealing, which is normally called thermal whitening or leaching. The atomic mobility thus becomes sufficient for the restoration of an 'unperturbed structure'.

This progressive modification of the transmission under irradiation can be troublesome for certain technical applications, *i.e.* nuclear reactors, space technology, etc., and we look for avoiding such consequences. In other cases by contrast, one looks for deliberately producing it, for example, in the case of variable transmission for lenses for spectacles and the glass for radiation dosimeter.

7.4.3. The Colour due to the Dispersed Particles

The glass consists of a medium where it is possible to produce a variable precipitation reaction under the influence of heat treatment or action of light, *i.e.* the photo-sensitive reaction. The ions of certain metals like Cu, Au, Ag, Pt dissolving in the glass can be reduced to a metallic state by incorporating reducing agents like tin oxide or antimony oxide in the composition.

7.4.3.1. The Gold Ruby Glass

The famous "Gold Ruby" glass is made by dissolving a small quantity of gold (0.01 to 0.02 wt%) in the glass. The glass containing Au^{3+} ions is then heated at a temperature higher than the annealing point and the gold ions are reduced as

$$Au^{3+} + 3e^- \rightarrow Au^0$$

and the necessary electrons are supplied by the reactions as

$$Sn^{2+} \rightarrow Sn^{4+} + 2e^-$$

or,

$$Sb^{3+} \rightarrow Sb^{5+} + 2e^-$$

In course of this heat treatment, first of all, it produces an agglomeration of Au atoms in the form of a "colloidal mass" and then in the form of small crystals, which are in the nano range. The glass, initially colourless, then takes a tinge of lively "ruby" during the final stage of this heat treatment, which is called the "striking" treatment.

The same phenomenon occurs from the interaction of light with the metallic particles : It does not refer to a phenomenon of "scattering", but an absorption by the gold sol. The corresponding theory is given by Mie [8] by using the Maxwell equation. Doremus [9] has shown that it refers to an effect of resonance of plasma, the electrons collectively oscillate at a characteristic frequency in the particles, whose dimensions are of the order of **20 nm.**

Very small gold "nano-particles" were studied by absorption spectroscopy by Lindfors et al [10]. Such particles are an interesting model system to study the transition from individual atoms to the optical properties of bulk matter. The nano-particles also have important applications in 'chemical sensing' and as 'labels' in biology. These workers used supercontinum light produced in a photonic crystal fibre to detect less than **10 nm** gold particles [10] by a detailed analysis of the 'confocal microscope images'. The cross section showed that the super-continuum light can be focused to almost a 'diffraction limited spot'. The scattering spectrum displays a clear resonance that can be used to distinguish 'gold nano-particles' from other 'scatterers' in a biological medium, which has a strong implication in the detection of some defects in the biological systems of many technical applications.

7.4.3.2. The Silver and Copper Rubies

Similar absorption is produced in the glass containing silver in solid solution. The dissolving Ag^+ ions that was originally colourless can be reduced to a metallic state Ag^0, the atoms are then fluorescent. The agglomeration of the Ag atoms to a colloidal state makes the fluorescence disappear, but it provokes a yellow colouration in its place, which can again be explained by the theory of Mie. The corresponding absorption around **396 nm** has been used in the study of diffusion of hydrogen in the glasses doped with silver, which serves as the "tracer".

The reduction of the ions like Cu^+, Ag^+, Au^{3+} can be effected by the photo-sensitive reaction by adding a small quantity (0.05%) of photo-reducing agents like CeO_2 in the glass. Under the action of UV irradiation at room temperature, there is an emission of an electron as

$$Ce^{3+} + h\nu \rightarrow Ce^{4+} + e^-$$

which serves to reduce the Cu^+ ions as

$$Cu^+ + e^- \rightarrow Cu^0$$

The atoms of Cu serve as the nucleation centres and a heat treatment permits the development of the colour in the irradiated part. Such glasses, called "photo sensitive" glasses, containing Cu, Ag or Au are usually produced for commercial applications. It allows the execution of real photography by using the sensitivity exposure to the UV light, which is followed by a development of heat treatment at a temperature of 500 - 600°C.

It is possible to precipitate small 'silver halide' crystals in a suitable glass and to obtain a transparent glass enjoying the property of 'photochromism'. The typical glasses are boro-alumino-silicates containing AgCl, AgBr or AgI in the form of small crystals of **8 nm** to **15 nm,** which is precipitated by a heat treatment between 400 and 800°C. The space between the particles is of the order of **100 nm.** The sensitivity and the kinetics of the darkening process of such glasses and their return to normalcy are influenced by the following :

(*a*) The Glass Composition ,

(*b*) The Nature of the Halogen Ions,

(*c*) The Particle Size (preferably in the nano range), and

(*d*) The Heat Treatment Schedule.

It is known that the addition of Cu increases the sensitivity to the light.

The system function as a reversible "photographic plate" : the absorption of a photon provokes a dissociation to Ag^0 and halogen. The metallic Ag^0 absorbs the light and colours the glass in grey. Contrary to the usual photo-sensitive layer in photography, the pair can recombine as the light is removed, which produces a "whitening" or "Colour Leaching".

Such systems are perfectly reversible and do not show the sign of "fatigue" up to about 300,000 cycles of 'darkening-whitening' cycles, which is contrary to the organic photochromic substances, which are found to be progressively degraded (*i.e.* ageing). These glasses find important applications in the glass lenses for spectacles.

7.4.4. The Luminescent Glasses

The colour of glasses just studied show the phenomenon of absorption in the visible spectra. The colours due to the "fluorescence" is shown by the electronic transition with an emission of a photon in the visible. One atom brought to an excited state by the absorption of a photon returns to its fundamental level with the emission of light → either immediately, *i.e.* fluorescence, or after a significant delay, *i.e.* phosphorescence.

The centres of fluorescence in the glasses can be either the metallic atoms in the "nano range" (*e.g.* Ag atom), or the crystalline phases (CdS), or certain ions → the most important of which are rare earth ions, which intervene in the amplification of coherent (stimulated) light, *i.e.* "The Lasers".

The synthesis of novel 'metal nano-particles' and their application to nano-optical materials were explored by Shiraishi et al [11-13]. The metal nano particles stabilized by organic molecules are now creating a new class of materials that are different from both conventional bulk materials and the atoms, giving one of the smallest building blocks of matter. The stabilizers play important roles is not only protecting the metal nano-particles, but also controlling the properties for optical functions.

The 'stimuli-responsive' colour change of colloidal dispersions of Au nano particles as an application as sensors : The gold nano particles protected by 3-mercaptopropionic acid were prepared by reducing tetrachloroauric acid in the presence of 3-mercaptopropionic acid. The colour of the dispersions changed from red to purple by adding hydrochloric acid, and changed back from purple to red with the addition of an aqueous sodium hydroxide solution. The change responsive to pH is reversible even after 5 repetitions. On the other hand, the colour of colloidal dispersion of Au nano particles stabilized by poly(beta-cyclodextrin) (PCyD) changed from red to purple with the addition of mercaptocarboxylic acid, suggesting the inclusion complex formation of the PCyD protecting Au nano particles with mercaptocarboxylic acid.

The electro-optic properties of 'liquid crystalline display' (LCD) system with the nano particles of palladium (Pd), which are protected by 'liquid crystalline molecules', was also explored by Shiraishi et al [13].

This study is aimed at synthesizing liquid-crystalline (LC) molecule-protected metal nano particles and developing novel liquid crystal display (LCD) materials. 4-Cyano-4′-pentylbiphenyl-covered palladium (5CB-Pd) nano-particles were prepared by UV irradiation of a tetrahydrofuran solution of palladium(II) acetate in the presence of 5CB. The nano particles of 5CB-Pd have higher solubility than fullerene in the liquid-crystalline medium. The twisted nematic LCD cell with Pd nano particles showed a frequency modulation (FM) response even under low voltages. The response of this FM-LCD is 10 times faster than that of the conventional device. This approach can be extended to other modes of LCDs.

7.4.4.1. The Laser Glasses

A solid Laser is a 'luminescent material' wherein the light emitted by the fluorescence from one of the "centres" stimulates the other 'centres' on its own in order to provoke the emission of light in phase with that of the first "centre" and in the same direction.

In order to obtain such a stimulated emission, it is necessary to provoke an "inversion of population", *i.e.* to create a situation wherein the species in the excited state are more in numbers than that in the fundamental or lowest energy state, so that the 'population inversion' can take place. By limiting to the case of only excitation by photon, which is called optical pumping, it can be shown that it is necessary that the "excitable" ions dispose at least "three" energy levels, as shown in **Figure 7.2.**

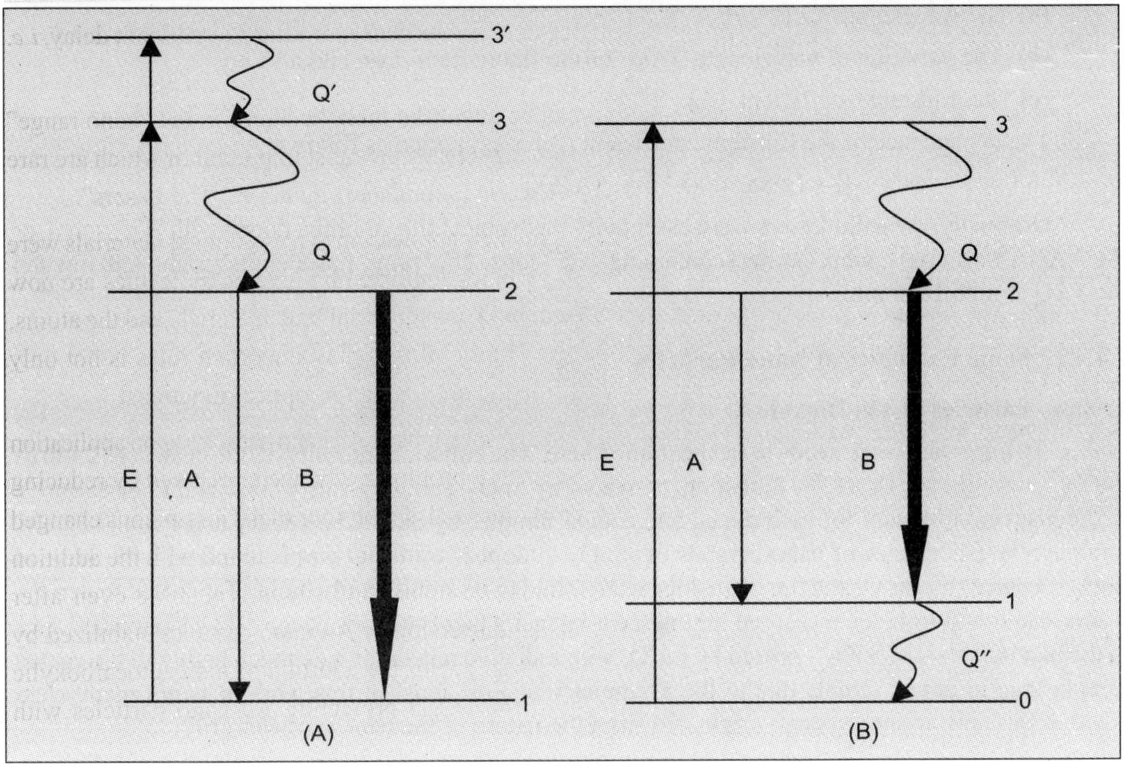

Figure 7.2 : Simplified Diagram of the Energy Levels for a Laser System (3 and 4 Levels).

This excitation, *i.e.* the optical pumping, elevates the atoms to the level **3** (or **3′**), wherein they have a chance to return either to the fundamental level with emission of a photon, or to move to an intermediate level **2**, by a non-radiative transition. This level **2** of fluorescence is of fundamental importance to understand the mechanism of Laser.

As the atoms return from the level **2** to the base level **1,** it emits a light of the same wavelength as the 'atom' which has originally stimulated this transition, and this 'atom' in turn stimulates another transition and so forth. Hence, the process continues. In the absence of the level **2** one could only have an equalisation of population density between the levels **3** and **1.**

Some systems have four levels. The effect of Laser is produced between the levels **2** and a level **1** above the fundamental level **0.** The excitation is generally produced by an 'external lamp', which emits a light, which is absorbed by the 'excitable' ions.

In actual situation, the active solid is placed between two 'mirrors' with the reflectivities as : $R_1 = 100\%$ and $R_2 < 100\%$. The light emitted between the mirrors provokes the 'stimulated emission' by the 'avalenche' effect. If N_2 and N_1 are the 'populations' in the higher and lower states respectively for a unit volume, for an inversion of population with $N = N_2 - N_1 > 0$ and a coefficient of gain/ion (β), it can be shown that the 'amplification' of light will be produced if the following condition is met as

$$R_1 R_2 . \exp((\beta N - \alpha)/2L) > 1$$

where, α is the normal absorption coefficient and L the length of the sample bar. The value of β depends on the indices of refraction *n*, which depends on the following :

(*a*) The wavelength (λ),

(*b*) The variation of wavelength ($\Delta\lambda$) of the fluorescent rays, and also on

(*c*) The Einstein Coefficient (A).

The dependence of the Einstein coefficient (A) is expressed as follows

$$\beta = (1/8\pi c) \cdot (\lambda^4/n^2) \cdot (A/\Delta\lambda)$$

Originally, the solid Lasers were essentially the 'rubies' (*i.e.* alumina doped with Cr^{3+} ions), or the YAG (Yttrium Alumina Garnet) containing Nd^{3+} ions. The 'ruby' Laser emits around **690 nm** and the YAG around **1060 nm.**

7.4.4.2. Some Examples of Nano Particles

1. Nano Particles of Tin Dioxide

An important work needs to be mentioned here. The emission intensity of the peak at **612 nm** of the Eu^{3+} ions due to $^5D_0 \rightarrow {}^7F_2$ transition, activated by SnO_2 nano-crystals were found to be sensitive to the nano environment for both doped and coated samples by Saha Chowdhury and Patra [14]. The luminescent efficiencies of nano-crystals of SnO_2 – 'doped' by Eu_2O were compared with those of SnO_2 - 'coated' by Eu_2O, and the intensities were found to be significantly higher for coated materials. It was also found from the measured luminescent intensity that Eu^{3+} ions occupy 'low symmetry' sites in the nano-crystals of SnO_2 – coated by Eu_2O, with radiative relaxation rate much higher in the coated sample than in doped sample due to the asymmetry of Eu^{3+} ions. In this work, it is not made clear whether this sort of 'asymmetry' originated from the nature of the nano-particles [14].

2. Nano Particles of Cuprous Oxide

The cuprous oxide (Cu_2O) particles coated with poly-acrylamide having diameters in the range from **4.8 nm** to **8.6 nm** were prepared by a chemical method. The optical absorption of these particles was found to be characterized by excitons with corresponding energies varying from 2.6 to 2.25 eV in the above particle size range. A second optical-absorption maximum signifying surface states within the band gap is exhibited by the specimens with relevant energies varying from 2.6 to 1.77 eV in this particle size range [15].

3. Nanoribbons of Zinc Sulphide

One-dimensional nano-structured materials set the pace of the recent trend in nano-materials research due to their wide range of potential applications in many nano-scale devices. It is known that ZnS is an important semiconductor used as 'phosphor materials'. Hence, some efforts have been made by Kar et al [16] into controlling the size, morphology and crystallinity of the ZnS crystals in order to fine-tune their physical properties.

The ZnS nano-ribbons were synthesized on a Si substrate by vapour liquid solid (VLS) process at 1100°C using gold as the catalyst, and the synthesized products were characterized by XRD, EDAX, SEM, TEM and photo-luminescence (PL) measurements. The ZnS nano-ribbons were found to be crystallized with the hexagonal wurtzite phase. The nano-ribbons were ultra long with width varying within **300-500 nm.** The nano-ribbons were also found to have good 'emissive property' with the blue emission centered at ~ **399 nm** [16].

4. Nano Particles of CdS

The optical and microstructural properties of nano-composite thin films based on the systems CdS-ZnO and CdS-Al_2O_3 were studied by Ray et al [17]. These nano materials were synthesized by sol-gel technique. The molar ratio of CdS with ZnO and Al_2O_3 were varied within the range of 20:80 to 50:50. The nanoparticles of CdS that are highly confined with radius **1.8 nm** to **4.7 nm,** which were estimated from the blue shift of the absorption edge, were obtained by using ZnO matrix, whereas for Al_2O_3 matrix the size varied from **2.8 nm** to **7.0 nm.** The microstructural characterization by high-resolution TEM (*i.e.* HRTEM) revealed well-resolved crystalline nano-particles in both cases.

The XRD studies showed reflections from the planes of CdS in the cubic phase for Al_2O_3 matrix, but for the case of ZnO matrix, the wurtzite phase of CdS was present. The band-gap of CdS-Al_2O_3 nano-composites was found to vary from 3.69 eV to 2.61 eV, whereas for CdS-ZnO nano-composites, it varied from 3.16 eV to 2.52 eV. The photo-luminescence (PL) studies indicated three 'surface-related' peaks at 2.33 eV, 2.19 eV and 1.89 eV for all the CdS-ZnO nano-composites. Only one prominent broad peak at 1.81 eV due to the 'surface defects' was observed in the PL spectra of CdS-Al_2O_3 nano-composite. This study revealed that the most effective capping of the CdS nano-particles was obtained by using Al_2O_3 matrix [17].

REFERENCES

1. R. S. Bera, A. K. Bandyopadhyay and P. C. Ray, 'Mathematical Physics for Engineers', New Age International, New Delhi, (2006) In Press.
2. M. Yamane and Y. Asahara, 'Glasses for Photonics', Cambridge University Press, Cambridge, (2000) pp. 159.
3. A. K. Bandyopadhyay and P. C. Ray, 'Perturbation analysis and memory in ferroelectric materials', J. Appl. Phys., 95 (2004) 226.
4. A. K. Bandyopadhyay, P.C. Ray and V. Gopalan, "An approach to the Klein-Gordon equation for a dynamic study in ferroelectric materials", J. Phys. — Cond. Matter, 18 (2006) 4093.
5. R. E. Newnham, 'Properties of Materials', Oxford Univ. Press, Oxford, (2002).
6. A. K. Bandyopadhyay, 'A study of interaction between copper & manganese in a soda -borate glass by ESR', J. Mater. Sci., 15 (1980) 1605.
7. A. K. Bandyopadhyay, 'Optical and E. S. R. investigation of borate glasses containing single and mixed transition metal oxides', J. Mater. Sci., 16 (1981) 189.
8. G. Mie, Ann. Phys., 25 (1908) 377.
9. R. H. Doremus, J. Chem. Phys., 40 (1964) 2389.
10. K. Lindfors, T. Kalkbrenner, P. Stoller and V. Sandoghdar, 'Detection of gold nanoparticles using supercontinuum whiter light confocal microscoy', Phys. Rev. Lett., 93 (2004) 37401.
11. Y. Shiraishi, D. Arakawa and N. Toshima, 'pH-dependent colour change of colloidal dispersions of gold nanoclusters : Effect of stabilizer', J. Eur. Phys., E 8 (2002) 377.

12. Y. Shiraishi, K. Maeda, H. Yoshikawa, J. Xu, N. Toshima and S. Kobayashi, 'Frequency modulation response of a liquid-crystal electro-optic device doped with nanoparticles', Appl. Phys. Lett., 1 (2002) 2845.

13. N. Toshima and Y. Shiraishi, 'Catalysis by metallic colloids : Encyclopedia of Surface and Colloid Science', Marcel Dekker Inc., New York, (2002) pp. 879.

14. P. Saha Chowdhury and A Patra, 'Understanding the influence of nano-environment on luminescence of Eu^{3+}-ions', Proc. Natl. Conf. on Nano-Science and Technology, J. U., Kolkata (India), January, (2005).

15. S. Banerjee and D. Chakravorty, 'Optical absorption by nanoparticles of Cu_20', Europhys. Lett., 52 (2000) 468

16. S. Kar, S. Biswas and S. Chaudhuri, 'Zinc sulphide nanoribbons', Proc. Natl. Conf. on Nano-Science and Technology, J. U., Kolkata (India), January, (2005).

17. S. Ray, S. K. Panda, S. Chakrabarti and S. Chaudhuri, 'Effective capping of CdS nano-particles embedded in different oxide host matrices', Proc. Natl. Conf. on Nano-Science and Technology, J. U., Kolkata (India), January, (2005).

Chapter 8

Other Methods and Other Nano Materials

PREAMBLE

In the description of different methods, which are used to prepare nano particles of alumina for various nano-composite materials, it is mentioned in **chapter 3** that there are some other methods, particularly the 'sol-gel' method, for preparing nano materials. Some of these methods will be described in this chapter. It has been mentioned that in high energy attrition milling, there is enough quantity of nano materials that can be produced for different purposes, since there are various types of attrition mill available in the market, as described in the **section 2.3.1.1.** The main drawback of other methods is that apart from some 'quality' problems, there is severe restriction on the 'quantity' of nano materials that can be produced. However, as claimed by various workers, as will be shown later, that there may not be any serious problem. In terms of 'purity', some of these methods are excellent and definitely deserve some mention in a book on nano materials. In this chapter, some attempts will be made towards this objective.

It should be clearly mentioned here that ceramic and glassy materials are not the only novel materials that can be prepared in the nano domain for some important applications in various fields. But, there are also a gamut of newer materials → specially called nano-optics, nano-magnetics, nano-electronics, etc. Moreover, there is a recent trend in a completely new domain of nano materials like quantum dots, nanowires, nanotubes for a variety of applications. By considering the technological importance and novel applications that are ever expanding, all these newer materials deserve some mention and definitely merit some space in this book. At the cost of doing injustice towards a proper discussion on these topics, some preliminary description is still attempted here for the sake of interested readers by encompassing most of these emerging areas, by keeping the well-known 'dictum' in mind that ***"it is always better to say something than nothing at all".***

8.1. PROCESS OF SYNTHESIS OF NANO POWDERS

There are various processes for the synthesis of 'nano powders' of different characteristics and purity for various purposes. Some of the most important processes are :

1. Sol-Gel Process
2. Electro-Deposition
3. Plasma Enhanced Vapour Decomposition
4. Gas Phase Condensation
5. Sputtering Technique

6. High Energy Attrition Milling Process.

(already explained in the **section 2.3.1**)

As said in the 'preface' of this book that there are many novel techniques for the preparation of nano powders, and their numbers are growing with the evolution of this subject of nano-materials. It is not possible to discuss all the processes within the scope of this book. Here, another point needs mention again that many of these processes are neither 'cost effective' nor the 'yield' of the "actual" materials is quite high enough for different purposes, as emphasized in the **section 3.1.** Moreover, many such processes were exclusively developed for a particular research programme for a given set of materials to yield certain desirable properties and many other plausible reasons, thereby limiting their applicability to some extent. Nevertheless, the novelty of these processes cannot be denied by any means, and some description is obviously necessary to highlight their 'principles' in order to get the readers acquainted with the latest trend in this fast emerging as well as fascinating fields of research. In this context, since 'Sol-Gel' method is the most talked-about in our common parlance, it will be mainly targeted for somewhat detailed description here in a relative sense.

8.1.1. General Principles of Sol-Gel Processing

The sol-gel process refers broadly to room temperature 'solution routes' for preparing mainly oxide materials. The process involves the hydrolysis and polymerization of metal alkoxide precursors of silica, titania, zirconia as well as other oxides. The solutions of precursors are reacted to form irreversible gels that dry shrink to rigid oxide forms [1]. The inter-disciplinary approach is followed in the sol-gel process (see the extensive work done by Klein on sol-gel in ref. [2]).

There are 'new materials', which are built up from the molecular level like [3] :

Use : (Environmentally Acceptable Precursors)

↓

Assemble to Size Approximating Device : (Chemical and Physical Transformations)

↓

Model Cooperative Properties

The concept of sol-gel processing is actually akin to the processing of the 'nano-structure'. First of all, this process starts with a nanometer-sized molecular unit, and it goes through the reactions, which are also on the nanometer scale. Since these 'molecular' sizes are susceptible to be scattered by light, the scale of the sol-gel process or rather the progression can be probed by the 'light scattering' (LS) technique. Obviously, some more sensitive techniques like small angle neutron scattering technique can also be employed for this purpose, as described for the nano particles of magnetite in the **section 5.7.**

The concept behind the sol-gel process is that a 'combination of chemical reactions' turns a 'homogeneous solution of reactants' into an → 'infinite molecular weight oxide polymer'. This polymeric unit is a 3-dimensional structural skeleton, which is surrounded by the 'inter-connected pores'. From an ideal point of view, this polymeric unit is isotropic, homogeneous, and obviously uniform in the nano-domain. Moreover, it can exactly replicate its mould and the miniaturization of all the features is possible without distortion. The gels contain 'pores', and the nano-phase porosity and the nano-structure of the gels are both scientifically and technologically important.

8.1.1.1. Precursor Alkoxides

Initially, the sol-gel process involves a 'homogeneous solution' as precursors. The alkoxides are the organo-metallic precursors for silica, alumina, zirconia, titania, etc. One of the most common systems consists of "Tetraethyl Orthosilicate" (TEOS), alcohol and water. This solution can react to an extent, where the molecular structure can no longer be reversed. This particular point can be considered as a 'critical point', which is known as "Sol-Gel Transition Point". In the whole structural constitution, the "Gel" is an elastic solid filling the same volume as the solution.

It is important to know about the "Common Alkoxides" for 'Sol-Gel Processing', which are :

1. TEOS
2. Trimethyl Borate
3. Aluminium Sec-Butoxide
4. Titanium Iso-Propoxide
5. Zirconium Iso-Propoxide

The last one is discussed in some details in the **section 3.7.2.** Now, the 'chemical reactions' should be described, which are the most fundamental in the entire process for the preparation of nano materials through 'Sol-Gel Route'.

8.1.1.2. Chemical Reactions in Solution

1. Non-Aqueous Process

The rate of chemical reaction depends on the following : (*a*) pH, (*b*) Concentration, and (c) Solution. In case of preparation of the 'alumina powder' from aluminium sec-butoxide, the following reactions occur

$$Al(OC_4H_9)_3 + H_2O \rightarrow Al(OC_4H_9)_2(OH) + C_4H_9OH \quad (1)$$

$$2Al(OC_4H_9)_2(OH) \rightarrow 2AlO(OH) + yC_4H_9OH \quad (2)$$

$$2Al(OC_4H_9)_2(OH) + 2H_2O \rightarrow Al(OH)_3 + 2C_4H_9OH \quad (3)$$

$$AlOOH \text{ or } Al(OH)_3 \rightarrow Al_2O_3 + zH_2O \quad (4)$$

Normally, a catalyst is used to start the reactions and control the pH of the solution. The series of reactions that occur are as follows :

1. The first reaction is the 'hydrolysis' to make the solution active,
2. The process (1) is followed by 'condensation-polymerization',
3. These reactions go along with further hydrolysis.

The 'molecular weight' of the oxide polymer is increased by these reactions, which result in either the mono-hydroxide AlOOH (boehmite) or the tri-hydroxide $Al(OH)_3$ (bayerite-4). When one desires to make the 'transparent activated alumina gels' from ASB, there is one well-studied composition for achieving the desired property, which involves an addition of little excess of water in the initial solution, *e.g.* 100 mL water/ASB. There is a need of the hydrolysis process to take place at slightly above 800°C in order to preferentially promote the formation of boehmite, rather than bayerite. The majority of the transition metal (TM) oxides have been prepared by non-aqueous sol-gel methods. Since these TM oxides can be fruitfully applied in certain optical devices, this technique assumes a special importance, and they also have a variety of applications.

2. Aqueous Process

The sol-gel processing route also involves the use of 'Aqueous Colloidal Sols'. This sol contains nanometer-sized particles, and hence it is quite natural to include such 'sols' as precursor materials, which admit that the mechanism for attaining the 'sol-gel transition' is quite different. The aggregation or agglomeration of the 'sol particles' is normally caused by changing the pH or the concentration in sols, like the well-known Ludox.

Then, there is a 'gelation' process. The sols can be gelled in such a way that the 'oxide skeleton structure' is a continuous linkages of the sol particles. But, there are some discrete features that make up this 'skeleton structure', which corresponds to the sol dimensions. The other features are obviously the "pores" within the 'secondary particles'. There is a difference between the chemical and structural aspects, *i.e.* between the non-aqueous alkoxide precursors and the aqueous sol precursors, which are usually blurred at a later stages of the sol-gel process.

8.1.1.3. The Process Details

1. Mixing

The 'mixing' is a first step for the single alkoxide, multiple alkoxide, and colloidal sol processes. Since the building blocks are nanometer in size, the particle size is smaller than the wavelength of the visible light.

2. Gelling

The 'gelling' is determined approximately as the time when the solution shows no flow. This is known as the 'gelation time', *i.e.* the 'time to gel'. Here, the viscosity plays an important role for the transition from a viscous liquid to a rigid structure.

3. Shape Forming

The sol-gel process can be used to make the 'bulk materials', which are done by casting and moulding . This process can also be used to make a micro porous 'preform', which is near 'net shape'. This preform is called 'Monolithic', which has its obvious reference to its continuity. The 'monolithic gels' can be formed from an alkoxide solution or from a colloidal sol. However, there is a difference between two types of monolithic gels, *i.e.* between the colloidal gels and the alkoxide gels. The main difference between them lies in their structures of the 'pores'. The colloidal gels have larger pores between the particles, whereas the alkoxide gels have smaller pores, which are less than **10 nm** in size. Obviously, by considering all the above points, the making of the 'monolithic gels' is quite a challenging problem in the sol-gel process. There are other areas of importance. These are "thin-films", "high aspect ratio fibre", etc. for optical transmission, and many more.

4. Drying

After having selected a particular geometry or shape and accordingly, designing the right chemical formulations, we have to take several operational steps further, which are generally common to the monolithic gels, the thin-films and the fibres. All these gels must be or have to be 'dried'. So, the 'drying' is an important step in the whole operation. The thin-films and the fibers can be dried quickly in air due to their smaller dimensions, since it has been found that all the observed 'shrinkages' are taken up along the thin dimension in the case of thin-films, and not in the 'plane of the substrate' if the thickness is below 1 micron.

However, for monolithic gels due to their thicker dimensions, the drying is more difficult. We would like to obtain a gel with a nano-structure and at the same time, it has also to be 'crack-free' gel. Accordingly, there are mainly two treatments of drying, which are in use. The first possibility consists of forming an 'aerogel', which is dried in an 'autoclave' by hypercritical technique. Then, the solvent is removed above its critical points. The result in aerogel is about 10% dense and shows no appreciable shrinkage. The second option is more simpler and it consists of a 'xerogel', which are dried by natural evaporation. The xerogels are 60% dense and undergoes a reduction of volume by about 35-65% during the drying process.

By going through the process of chemical reactions, gelation, and drying, the processed gel materials have several characteristics of the corresponding ceramic oxide, but they are more porous than the corresponding ceramic material, which are obtained by other processing routes. It is already known that both the water and the solvent escape through the "inter-connected" pores that remain open at the surface until the gel is fired at around 600°C or higher depending on the chemical composition.

5. Densification

The objective of any ceramic material is to obtain a dense or pore-free material by the various routes of sintering. The gel is obviously no exception in this regard. The ultimate aim of the 'sol-gel process' is to obtain a dense, *i.e.* pore-free, oxide material, and hence the final stage of processing is the 'sintering'. The smaller particles sizes in the nano range giving rise to a 'high surface area' of the gel is considered to be an advantage in terms of a 'higher driving force' for the sintering process. Therefore, the sintering process is expected to occur at lower temperatures than that required in the conventional powder compacts. This is the final step of the whole operation, which is independent of the fact that whether the compact was prepared by an aqueous or non-aqueous route. The main point lies in a 'driving force', *i.e.* the rate of diffusion of different atomic species, which makes the particles grow to a certain extent and which eventually removes the porosity in the final product, or a material is formed that is similar to conventionally processed material, which did not go through the sol-gel processing route [2] (See **section 2.2.3** for sintering of ceramics).

The presence of lower number of pores added with the high-purity uniform nano-structure become the 'hallmark' of the sol-gel process. However, there are still some challenges in each of the above mentioned processes. Apart from being cost-effective, very often the yield of the 'actual' material for further processing to different shapes is not too high. Moreover, in sol-gel route in particular, there are problems of segregation, contamination and unusual pore formation [3]. Still, the challenge of the sol-gel process is to exploit the "nano-structure" aspects of the process to derive the "real benefits" [1 - 3].

It should be pointed out here that even after proper drying and densification, the sintered gel still contains OH ions, which are considered impurities for optical fibre communication in terms of increasing the losses in dB/Km. Even for the preparation of nano-structured alumina, zirconia or any other material by the sol-gel route for some hi-tech applications, the presence of these OH ions is detrimental. In order to assess the OH content in the gel materials, it is better to do a study of the vibrational spectra of 'metal cation-OH' vibrations in the IR spectra [1]. But the analysis is complicated for determining the concentration of OH ions, since in the majority of cases, the deconvolution of the IR spectra were not done. In the gel glasses in the boro-silicate system, a detailed deconvolution of the IR spectra was done by computer in order to estimate even the concentrations of B-OH and Si-OH from the respective vibrational bands, which are distinctly different for each case of vibrations [4]. So, that sums up this sub-section on sol-gel process.

Does the inherent 'nano-structure' give us any new material, which can make wonders in the newer fields of electronics and communication ?

8.1.1.4. Behaviour of Some Gels

The literature on gels and their behaviour is so extensive that it is simply impossible to cover even the most important ones. So, only a few examples of the behaviour of some gels will be given here. Some of these gels are not really ordinary, since they have some extra-ordinary applications and a set of excellent properties. Some of these 'gels' will be described here to highlight the issue of the superiority of producing nano-materials through the sol-gel route.

1. Gallium based Nano-Materials

Some unique Ga-based nano-materials have been prepared by Sinha et al.[5]. The films were dip coated on the quartz substrates. The annealing of the films was carried out between 300°C to 1100°C in air atmosphere for one hour using a tubular furnace. The films were characterized by XRD, SEM, AFM, UV-VIS and PL techniques. The XRD spectra revealed that the films started to crystallize as α-GaO(OH) with orthorhombic structure, when annealed at 300°C. On increasing the annealing temperature, this structure collapsed and began to transform into α-Ga_2O_3. A pure α-Ga_2O_3 film having rhombohedral structure was obtained at 500°C [5].

A further increase of the annealing temperature indicated destruction of α-Ga_2O_3 crystal lattice, and the appearance of β-Ga_2O_3 structure. At 700°C and onwards, a pure β-phase with monoclinic structure was observed. It was claimed that the phase pure α-GaO(OH) and α-Ga_2O_3 thin films have been prepared for the first time to evaluate their optical properties. The SEM and AFM studies revealed that all the films were 'crack free' and very 'smooth'. The films with different structures showed both allowed and direct semiconducting transitions. The highest semiconducting band gap (5.27eV) was found for α-GaO(OH) among all the phases examined in this study whereas that for α-Ga_2O_3 was intermediate between α-GaO(OH) and β-Ga_2O_3 phases. Both strong green and blue emissions with low FWHM were observed in photo-luminescence study, and these gel-based materials have a great future [5].

2. Dye-Doped Gel Glasses

Many applications of the sol-gel process were focused on the optical properties. For example, the optical behaviour of organically dye-doped gel-glasses that were incorporated into the porosity of sol-gel glasses without deterioration of their photo-physical properties. For example, the spectral behaviour and chemical stability of dye-doped gel-glasses are quite interesting. The chosen molecular structure of these dyes, *i.e.* specific Dye Lasers as molecular probes, were used for the study of the surface properties of the porous cage where these molecules were entrapped, in terms of homogeneity, polarity and viscosity and dye stabilization. The feed-back from steady-state dynamics and polarization spectroscopy studies is crucial to establish the preparation of the optimal conditions. In particular, 'Organically Modified Ceramics' were successfully used for specific preparations [6-8].

3. Glass Dispersed Liquid Crystals

The 'glass dispersed liquid crystal' (GDLC) films, which are prepared by organic doping of Sol-Gel matrices, may be used as electro-optical devices. The films scatter light according to the number of droplets and the relative refractive indices of the LC and the silica matrix. The LCs are 'birefringent', *i.e.* they show elasto-optic effect. Therefore, their refractive index depends on the LC orientation and the optical angle of incidence. If the film is coated with transparent electrodes, and an electric field is applied, a reorientation of the LC director in the droplet occurs, producing a variation of the LC refractive index as 'seen' by the incoming light. If the refractive index of the sol-gel substrate matches the new LC index, the material changes from an 'opaque scattering state' to a 'transparent state'. This feature can be used for preparing devices for visual presentation, *i.e.,* the 'displays' [9].

The unaltered GDLCs switch from white opaque to colourless transparent states. If these materials are used for the 'displays', the colour needs to be incorporated for many applications. The direct-view, backlighted passive displays, usually include colour filters located between the backlight system and the electro-optical material. In GDLCs, the colour may be included in the sol-gel matrix or in the liquid crystal itself, allowing the preparation of GDLC colour displays [9].

4. Synthesis of Glass-Metal Nano-Composite

The glass-metal nano-composites incorporating ultrafine particles of iron, nickel, cobalt and manganese, respectively, in a silica glass matrix have been prepared by heat treatment of a gel derived from a sol containing silicon tetra-ethoxide and a suitable metal organic compound. The metal particles in all the nano-composites are isolated and spherical-shaped with diameters ranging from **3 nm** to **10 nm.** The films of these nano-composites with thickness of the order of a few micrometres have been prepared on glass slides by a simple dip-and-pull technique. The optical absorption spectra of the nano-composite films have been measured over the wavelength range **200 nm** to **2000 nm.** The 'effective medium' theories, due to Maxwell-Garnett and Bruggeman, respectively, have been used to calculate the optical absorption of these materials theoretically. The Maxwell-Garnett theory gives results, which are in better agreement with experimental data than those obtained from Bruggeman formalism. The filling factor '*f*' as estimated from the least-squares fit of the experimental results with the Maxwell-Garnett theory has a value in the range 1 to 4% [10].

5. Metal-Silica and Metal Oxide-Silica Nanocomposites

The nano-composites based on sol-gel processing of Me-SiO_2 and MeO-SiO_2 are quite interesting. Some modifications of the sol-gel process were investigated with the aim of modifying the characteristics of the nano-composites so that their properties might be modulated continuously. In absence of strong interactions among oxide nano particles and silica matrix, the number and therefore the size of the nano particles should not depend on the presence of links between the metal precursor and the silica matrix, which act as the nucleation sites. On the other hand, every cavity inside the matrix network constitutes a nucleation site, which affects the maximum size of the forming nano particles [11].

In order to improve the preparation process, one has to play on the factors determining the characteristics of the pores in the matrix. Therefore, attempts were mainly focused on varying those parameters, which affect the preliminary steps of the gelation process. The tests were performed in order to find the best preparation conditions, which allow the tailoring of the particle size in the final nano-composite.

In the case of Fe_2O_3-SiO_2 system, efforts have been also put into finding the conditions, which allow to obtain final nano-composites in which the iron oxide is in the form of pure maghemite The influence of performing mechanical treatments of the gels, of performing the drying step under supercritical conditions, of the gelation temperature and time, and of varying the amount and type of the solvent were tested [11].

The first attempt, *i.e.* mechanical treatment, dealt with the use of mechanical milling of the dried gels [12]. The ball milling was adopted in order to promote the fragmentation of the silica network that, in principle, should favour a better control of the microstructure of the composite allowing a redistribution of nano-sized particles [13]. A nano-composite with 28% iron oxide was treated at 300°C. The XRD spectra showed a sharpening of Fe_2O_3 reflections, which increases with milling times. The TEM observations indicated that the nano particle size distribution is more homogeneous after milling. These results suggested that the ball milling induces the crystallization of Fe_2O_3 nano particles by mechanical activation. The fine crushing of the particles results in an increasing amount of the number of small pores and in a decrease of particle size which is responsible for the better stability of Fe_2O_3 towards the transition.

8.1.2. Electro Deposition

First of all, it is better to highlight the advantages of electro-deposition for the synthesis of the nano scale materials, which are as follows :

1. Low processing temperatures minimize the 'inter-diffusion'.
2. Film thickness is controlled by monitoring coulombs.
3. Composition and defect chemistry are controlled potentio-statically.
4. Films can be deposited as complex shapes.
5. Non-equilibrium phases can be deposited.
6. Driving force is precisely controlled by the applied potential.
7. Technique is quite inexpensive.
8. Current time transient provides an in situ measure of the deposition process.

8.1.2.1. Electro-Deposition of Inorganic Materials

The chalcogenide materials have good semiconduction property. Hence, these materials assume significance in many electronics applications, particularly in the field of 'switching'. So, these materials need some mention here and we start this sub-section with the chalcogenides (see the excellent review by Switzer in ref. [2]).

1. Deposition of Metal Chalcogenides

It is quite interesting to mention about the deposition of the semiconductor materials of group II-VI compound semiconductors of metal chalcogenides, such as CdSe, CdS, CdTe. Both the anodic and cathodic processes are used to deposit metal chalcogenides. These two processes are described as follows :

The Cathodic Process

$$Cd^{2+} + 2e^- \rightarrow Cd \qquad (1)$$

$$HTeO_2^+ + 4e^- + 3H^+ \rightarrow Te + 2H_2O \qquad (2)$$

$$Cd + Te \rightarrow CdTe \qquad (3)$$

The Anodic Process

$$Cd \rightarrow Cd^{2+} + 2e^- \qquad (4)$$

$$Cd^{2+} + Te^{2-} \rightarrow CdTe \qquad (5)$$

In the 'cathodic process', the metal is electro-chemically oxidized in the presence of the chalcogenide ions. But, it does not allow the deposition of the metal chalcogenide onto the substrate other than metals. This is a true deposition process by which the component of the film is deposited from solution precursor. At the electrode substrate, Cd^{2+} and $HTeO^{2+}$ ions are electro chemically reduced to the elemental form, where they combine to form the metal chalcogenide. The reduction by the subsequent assimilation of Cd into the Te layer is a two-electron process. The Te electro-deposition is also very complex, which is the six-electron reductions product. H_2Te may also participate in the precipitation of CdTe as per equation (7), as shown below. The complication in the deposition process is that the homogeneous chemical reactions between electro-generated spaces can lead to impurities in the film.

$$Cd^{2+} + 2e^- + Te \rightarrow CdTe \qquad (6)$$

$$Cd^{2+} + H_2Te \rightarrow CdTe + 2H^+ \qquad (7)$$

$$2H_2Te + HTeO^{2+} \rightarrow 3Te + 2H_2O + H+ \qquad (8)$$

It has been found that the epitaxy shows higher perfection, when covered with a thin-film **(20 nm-30 nm)** of epitaxial CdS, which was grown by a 'chemical bath deposition'. The electrodeposited materials have been shown to be very successful with vapour-deposited materials for several applications, such as :

(*a*) Opto-electronic applications,

(*b*) Photovoltaic solar cells,

(*c*) Infrared detectors, and

(*d*) Smart goggles.

For the photovoltaic solar cells, the photovoltaic conversion efficiency is usually found to be in the 8-10% range [2].

8.1.2.2. Nano-Phase Deposition Methodology

1. The Growth in Nano Beaker

In this approach, our knowledge on 'electro-chemistry' is applied to produce materials in the nanometer range. This approach consists in growing nano-scale materials by the use of nano-beakers, which involves using the pores in nano-porous membrane as 'templates' in order to produce nano-sized particles of the desired material. The membranes, *i.e.* the anodized aluminium or track-etch polymers, have cylindrical pores of uniform diameter.

When a polymer, metal, semiconductor, or carbon is synthesized by the electro-chemical means, within one of these pores, a nano-cylinder of the desired material is obtained, which ultimately depends on the type of material and the chemistry of pores wall that decide whether this cylinder may be hollow, *i.e.* tubular, or solid, *i.e.* fibril. The metal nano-tube membrane can also serve as ion-selective membrane. The nano-tube diameter can be as small as **0.8 nm,** and the length of the nano-tube can span the complete thickness of the membrane. A metal nano-tube can act as "cation or anion selective", which actually depends on the applied potential.

2. The Scanning Probe Nano Lithography

For a surface with nanometer scale defects, STM is used to modify so as to induce the nucleation of the deposited material at these sites of the defects. The silver pillars **10 nm-30 nm** in diameter and **4 nm-10 nm** high on a STM-modified highly ordered pyrolytic graphite surface have already been reported in the USA, and thus a nanometer scale galvanic cell composed of copper and silver nano-pillars can also be fabricated.

3. The Epitaxial Growth of Quantum Dots

A short discussion on 'quantum dots' will be given later. Here, the epitaxial growth is mentioned. At Weizmann Institute, the Scientists have produced epitaxially oriented CdSe 'quantum dots' with diameters of about **5 nm** with a controllable spatial distribution and narrow size distribution by electro-deposition of the nano-crystals on the evaporated gold substrates. An interesting feature of this work is that the size of these 'quantum dots' is believed to be controlled by the strain that is induced by the mismatch between the CdSe and the Au lattice.

4. In-Situ Studies of Epitaxial Growth

When one of the precursors is in low concentration in the solution and the layer is grown at high over potential, the composition is graded throughout the layer with a $(\text{time})^{1/2}$ dependence. The superlattices grown at lower potential in which both the reactants are deposited under activation control have square profiles, *e.g.* for three Pb-Ti-O superlattices that were grown by pulsing between 70 MV and 150, 230, or 260 MV versus SCE.

The composition of the 150 MV layer is relatively constant at 64% Pb, whereas the Pb content of the 260 MV layer varies from 39 to 76% through the layer. The graded composition profile may be desirable for applications, *e.g.* by grading the composition, the lattice may inhibit 'misfit dislocation formation' in strained-layer superlattices. In semiconductor devices for optical or electronic applications, it is desirable to have square composition profiles. The electro-chemical method is ideal for both measuring and tailoring the interface symmetry, and the composition profile is in real time on a nano meter scale [2].

8.1.2.3. Electro-Deposition of Nano Composites

As explained in the **section 3.6,** the nano-phase materials are also of considerable interest because of their enhanced mechanical properties and abrasion-resistance relative to those of the bulk materials. One reason for the different properties of these materials is that an increasing fraction of the atoms occupy sites at interfaces, and hence the atomic packing at the interfacial positions is much better to resist any external mechanical force. It has been estimated that only about 3% of the atoms in a material are at the boundaries when the grain size is **100 nm,** but this increases to 25-50% when the grain size approaches **5 nm.** This is also evident in the small nano-sized magnetite grains, where some spins are disorganized, as the particle size is smaller (see the **section 5.5.2** for spin canting in the nano particles of magnetite).

In metals, there is typically an increase in the yield strength with decreased grain size as described by the well-known Hall-Petch relation, which describes the yield strength as a linear function of the inverse square root of the grain size, as explained in the **section 4.4.1,** for the fracture toughness of sintered nano-particles of SiC. This behaviour is due to the influence of the grain boundaries on the 'dislocation motion'. As the grain size approaches the nano scale, there is a large increases in strength to be obtained. In the nano regime, the crystallite size becomes smaller than the characteristic length scales associated with the generation of the dislocations and glide, which are the typical processes that determine mechanical behaviour in metals.

The deviations from the Hall-Petch relationship have been observed in nano-crystalline materials, even softening at the smallest grain sizes. This softening has been observed in nano-crystalline electro-deposits of Ni with grain sizes below approximately **12 nm.** The thick-films of metals with nano-crystalline or amorphous structures for mechanical applications can be produced by co-depositing a 'metalloid element', such as phosphorous boron with nickel and other iron group elements. Another approach is to take advantage of the high super-saturation that is achievable during the very high peak current densities possible in pulse plating. This is the electro-chemical version of 'splat cooling' of molten metal to make metallic glass for various magnetic applications.

The electro-chemical deposition is a very attractive processing route for the synthesis of composite materials. The low processing temperatures minimize the problem of chemical interaction and thermally induced stresses that are often serious problems in the conventional sintering, vapour-phase , or liquid metal processes, which are used to fabricate various composite materials. A simple approach is to suspend particulate material in the plating electrolyte and co-deposit this with the metallic matrix. This can be accomplished both by electrode-less deposition and by electro-plating. The commercial applications of this approach include co-deposition of alumina, silicon carbide , or diamond with a metal such as nickel. The challenge in this work is to prevent the 'agglomeration' of the particles prior to co-deposition.

The electro-chemical scheme for growing composites is electro-chemical infiltration. This approach of electro-chemical infiltration has been used to fill the **5 nm** pores of a silica 'xerogel' film with nickel. In this manner at room temperature, 3-dimensional nanoscale networks of metal and ceramic were synthesized. There will be certainly an increased emphasis on the electro-deposition of nano-phase materials for magnetic applications, as well as for optical and electrical applications. Another big area of application should be the co-deposition of very dissimilar materials. The low processing temperatures of electro-deposition allow the co-deposition of materials that would tolerate each other at high temperatures used for the traditional thermal processing.

8.1.3. Plasma-Enhanced Chemical Vapour Deposition

The Plasma-Enhanced Chemical Vapour Deposition (PECVD) is widely used to produce thin films. The volume of literature on the subject is quite large, since this technique is used to produce a variety of materials for diverse applications. Here, a brief description is given.

The PECVD is the family of deposition processes, which are broadly defined as 'chemical vapour deposition' (CVD). In conventional CVD, the 'precursor gases' are thermally decomposed by the temperature of 500-1000°C. However, in the PECVD processes, they proceed at temperatures that are much lower because the energetic electron gas of a plasma is capable of highly dissociating the feed gas. Even when the feed gas and the substrate are near room temperature, the PECVD is useful for deposition on sensitive substrates that are damaged by high temperature or, in the case of semiconductor production, where dopant redistribution is an important concern. So, the rise of temperature is not desirable for such a process. The concurrent 'ion bombardment' of the PECVD film by the plasma may also modify the properties of the film during the deposition process (see the excellent review by Hopwood in ref. [2]).

There are two classes of PECVD process, which are commonly practiced :

(*a*) Direct method, and

(*b*) Remote method.

In the 'direct' method PCVED, the precursor gases, inert gas dilutants, and the substrates are directly in the region. The formation of reactive species and deposition precursors occurs in the active plasma region by many possible 'reaction pathways'. These pathways include electron impact dissociation, dissociative collisions with electronically excited atoms, and the reactions involving the products of recombination, as well as the by-products of the film formation.

In the 'remote' PECVD techniques, which generate a plasma separately from the region of the deposition. The inert or non-depositing gases are introduced into the plasma region. The excited species and radicals diffuse 'downstream' to the substrate area, where additional reactants are supplied. The long-lived metastable excited states of He supply the energy required for the dissociation of SiH_4 and subsequent deposition of amorphous silicon. The number of reaction path is greatly reduced compared with direct PECVD, making control of the deposition a simpler task. However, the deposition rates for the remote PECVD are generally lower than that of the direct method.

One of the greatest successes of PECVD is the deposition of polycrystalline films of diamond and cubic boron nitride. Many methods have been used for the decomposition of hydrocarbon precursors, including **dc** hot filament, oxygen-acetylene flame, and the microwave direct and remote diamond deposition. Typically, highly diluted mixtures of methane (~ 1%) in hydrogen are used to create PECVD diamond films. The substrate temperatures for the production of PECVD diamond fall in the range of 400-900°C, with the lower temperature of depositions requiring the addition of oxygen.

At lower temperature (< 400°C), the PECVD of methane and other hydrocarbons produces amorphous films of hydrogenated carbon (C:H). The film exhibits a high degree of four-fold bonding, similar to that of diamond. Typical hydrogen concentrations are 20-40%. Many of the properties of the films are similar to those of the 'actual diamond', and therefore the material is referred to as diamond like carbon (DLC). The hardnesses are normally in the range of 20-50 GPa, compared to greater than 100 GPa for crystalline diamond. Although plasma decomposition of methane results in the deposition of thin films, DLC hardnesses are quite low (~ 5 GPa) without energetic (> 50 ev) 'ion bombardment' by substrate biasing. However, the excessively energetic ion bombardment of the film degrades the diamond like properties. It is also observed that 'ion bombardment' reduces the atomic hydrogen concentration in the film.

The silicon dioxide and silicon nitride films have long been mainstay insulators in the electronics industry. The PECVD of SiO_2 at 300°C was accomplished by introducing oxygen radicals from a weak, inductively coupled, plasma source in a tube furnace. SiH_4 dilute in N_2 was introduced down stream from the discharge at the inlet of the furnace. An advantage of a remote PECVD process in the deposition of thin-films is a reduction in the damage caused by the energetic electrons and ions found in the plasma.

The electron cyclotron resonance (ECR) plasma sources are frequently used in the deposition of silicon oxide. ECR plasma produce highly dissociated radical fluxes at very low pressure (~ 10^{-4} torr), thus proving usable deposition rate **(20 nm-40nm/min)** at sufficiently low pressure in order to reduce gas-phase nucleation of particles. Since SiO_2 is used as an inter-metal dielectric insulating layer in the fabrication of integrated circuits, the planarisation of the film that is deposited over thin pattern "wires" or an inter-connection is needed. The application of a bias to the substrate during the deposition of SiO_2 results in the re-sputtering and planarisation of the film.

However, there are also disadvantages of the PECVD processes, such as incorporation of impurities, *i.e.* the particles of silicon are found to contain large amounts of hydrogen from the silane precursor. These impurities can be detrimental for many applications [2].

8.1.4. Gas Phase Condensation of Nano Particles

8.1.4.1. *Gas-Phase Condensation Methods*

The particles smaller than **10 nm** can be produced by condensation of evaporated metals [2]. The metal is introduced into a low pressure (0.5-4 torr) inert gas environment by evaporation from a temperature controlled alumina crucible. The metal vapour cools rapidly through collisions with the inert gas, and then it becomes a super-saturated vapour, and undergoes homogeneous nucleation. The particles are expected to both nucleate and grow in a narrow region immediately adjacent to the evaporation source [14].

The particle size is increased by : (*a*) Higher pressure of the inert gas, and (*b*) Higher atomic weight. These factors increase the confinement of the metal to the growth region. The critical diameter for the spontaneous formation of 'particles' is in the order of **0.2 nm-1 nm** for evaporated alumina. Experimental evidence indicates that the solid phase growth of a particle after nucleation is by the coalescence of the small-nucleated particles rather than by impinging adatoms.

The particles formed by gas-phase condensation may be collected on liquid nitrogen filled cold finger. By using a Teflon scraper, the particles are periodically removed from the cold finger and compacted by pressures 1-2 GPa into pellets of synthesized nano materials [2, 14].

8.1.5. Sputtering of Nano Crystalline Powders

The 'sputtering' at high pressure is an effective method of producing nano particles [2]. Normally, the system uses a standard dc/rf magnetron to sputter target metal into a high-pressure inert gas ($0.23 < P < 1.5$ torr). The gas-phase nucleation of the particles occurs at these pressures, when there is the formation of the 'film' for conventional magnetron sputtering [15].

The particles are collected on an LN_2 'cold finger' positioned 10 cm from the magnetron. The thermal gradient provides 'convective circulation' of gas between the target and the collection finger. A scraper is used to transfer particles from the cold finger to compaction device. The production rate is ~ 0.05 mg/s using a 75 mm target and 1.2 kW of power. The advantages of the sputter method over the evaporation methods are :

1. No chemical reaction with evaporator crucibles (giving improved purity),
2. Improved control of the process,
3. Production of stoichiometric alloys and inter-metallic compounds, and
4. Production of oxide and ceramics.

The nano-composite in which the particles are incorporated into a matrix produced by 'low pressure' conventional sputtering have also been produced. Thus, the particles of Mo at 0.2 to 0.6 torr was produced, while alternately sputtering aluminium films at 0.002 torr.

The formation of Mo particles by sputtering has demonstrated three pressure regimes :

1. $P < 150$ mtorr

No particle formation occurs,

2. $200 < P < 400$ mtorr

Particle formation is observed, with smaller particles forming at higher pressure,

3. $P > 600$ mtorr

Agglomeration attributed to decreased target sputtering rate, hence lower densities of Mo. The decrease in rate is due to the decreased ion bombardment of the target caused by the collisions in the plasma sheath. For this type of experiment, the target-substrate distance can be 10 cm [2].

In a typical sputtering experiment, it is possible to vary the target-to-substrate distance, e.g. the nucleation distance of Mo particles are found to be 8 cm at 150 mtorr. The nucleation distance decreases to 4.5 cm, when the background pressure is doubled to 300 mtorr. The reduction of the mean free path at higher-pressure result in an increased density of vaporized Mo available for the particle nucleation. The sputtered 'particle size distributions' are found to follow a log-normal distribution. This provides evidence that the particles grow by the coalescence of the 'particle nuclei' in the sputtering process [2, 15].

Finally, it should be mentioned that for the preparation of the 'preforms' for optical fibres, many similar processes are utilized, such as MCVD, PMCVD, PCVD, OVD and AVD. Although the fibre optics technology is the latest in the field of research on communication, and these processes are on the constant evolutionary path, this particular aspect has not much relevance for nano materials, except that some of these techniques are quite popular for making different nano materials for a specific purpose. Hence, due to the shortage of space, they are not discussed here.

8.2. IMPORTANT NANO MATERIALS

As said earlier, a book on nano materials cannot be concluded without any mention on the newer materials, or rather on newer 'frontiers' of technology of materials being opened up 'by the day', or shall we be little bit conservative and say 'by the month' → well, almost so ! Here, in this section, some newer materials will be described neither in terms of any specific importance or order nor in terms of any superiority of their applications. Since there is literally an explosion in terms of the amount of "nano technology" work being done in the areas of bio-science, this subject is totally excluded here, in order to avoid doing injustice to this new and emerging subject. Instead, some aspects of nano-optics, nano-magnetics and nano-electronics will be described within the limited space in this last chapter. The readers are referred to a number of good books on the subject being published regularly, mainly in the USA, which contain the latest technological development [16].

It should be clearly mentioned that there is a very active research branch in almost every University in the USA and some of them even have a special Institute for dedicated research on nano science and technology, which are simply excellent in their noble activities. There are some small to medium sized hi-tech minded companies, who are also actively involved in the evolution of different techniques for the preparation of newer nano materials. It is simply not possible to mention them by name. Since it is a short section of the last chapter, this unintentional omission might be excused.

8.2.1. Nano-Optics

Preamble

How to make an "Integrated Optical Circuits"

In order to create high performance optical materials 'integrated optical circuits' with hi-tech properties for a large number of innovative applications, we strongly need a new kind of technology. The "Nano-Optics" is invariably a novel class of technology that takes full advantage of the following :

(*a*) Unique interaction of electromagnetic radiation or light with sub-wavelength,

(*b*) Nano-scale patterned materials, and

(*c*) Nanotechnology-enabled fabrication methods to create :

1. A broadly applicable 'optical device', and
2. A 'manufacturing platform' to achieve this goal.

It is well-known that the 'optical circuits' are used in different areas as :

A. Consumer Electronics,

B. Communications,

C. Industrial and Lab Electronics Instruments,

D. Medical Diagnostic Electronics, and

E. Defense and Security (both military and non-military) Applications.

These applications are meant for harsh environments and several demanding requirements or specifications for really high performance and obviously high reliability. The nano-optic devices offer significant benefits as :

(*a*) Smaller in size and lighter in weight,

(*b*) Ease of integration, and finally

(*c*) Lower overall costs.

In an electro-optic circuit or interface, there is 'optical portion' where we have to face the biggest challenge for optimum design for high performance, efficiency, robustness, and cost. The reason is that the 'photons' carry information in a different way than 'electrons' do, and both these ways are not manipulated or guided in the same manner.

Thus, it is through the above qualities, the nano-optical materials deliver 'challenging alternatives' to conventional bulk 'optical solutions', and obviously, it also offers the necessary 'functionality and form factors' that conventional optics cannot really match or deliver.

Materially speaking, a nano-optic device first of all consists of a "nano-patterned layer" between two thin-film layers on a passive substrate. It is by going to the 'roots' of the materials problems, we can better understand the difference between nano-optics and conventional optics, since a 'change of the material' changes the function, often resulting in unwanted compromise between material properties and optical functionality. So, a short description is given on the structure and function of nano-optical materials.

8.2.1.1. Structure and Function

A properly patterned material on the nano-scale with structures, whose critical dimensions are several times less than the wavelength of light at which they operate, gives rise to the unique optical properties of nano-optic devices, thereby creating 'optical materials' with highly 'useful' as well as 'modified' optical functions.

The nano-optic devices can perform many passive optical functions as : A. Polarization filtering, B. Phase retardation, C. Spectral filtering, and D. Management of propagation, *e.g.,* lenses and beam splitters. These functions are achievable with a proper combination of materials and structures in the field of nano-optics. For both free-space and waveguide-based applications, we can design various functions.

The idea is to design nano-optic devices for an operation over any wavelength range. The fundamental technique is applicable to UV, Visible, and IR wavelengths, but there are proper materials and their structural dimensions, forming the core of an 'optical system'. In practice, the applications of nano-optics need a complete 'optical system', which consists of the nano-patterned material and other related materials that include the optical substrate and thin-film coatings. A dramatic reductions in overall size and weight can be achieved by such nano-structured layers made on a broad range of substrates so that we can customize the optics to the application, or make it on the surface of another optical component in an optical circuit. The nano-optic structures are defined as a combination of material and physical attributes, which are divided into different types as follows :

1. Structural

A. Pattern, *i.e.* Linear, Mesh, and Circular,

B. Dimensions, *i.e.* Period, Thickness, and Duty Cycle, and most importantly

C. Spatial Variations, *i.e.* Chirping, Arrays, and Multiple Layers.

2. Material

A. Nano-Structured Material, *i.e.* Dielectrics, Metals, and Polymers,

B. Inter-Structure Filling Material,

C. Adjacent Thin Film Materials, and

D. Substrate Material, *i.e.* Glass, Plastics, Dielectrics, and Crystalline Material.

3. System

A. Relationship of the Nano-Optic Structures with the Optical-Beam Path, and

B. Actuator Materials and Structures, *i.e.* Liquid Crystal, Solid-State Materials, Electro-Optic Polymers, and MEMS.

4. Functions

A. Switching,

B. Spectral Filtering,

C. Polarization Filtering, and finally

D. Phase Modification.

The fundamental aspect of nano-optics material is a nano-structure of sub-wavelength size, which has both optical and physical advantages and whose elements have microstructures one or more orders of magnitude smaller than the wavelengths of the incident light (see later in the **section 8.2.1.6**). In this dimensional situation, they can interact with light both according to the principles of classical optics of Maxwell and also in the quantum level, giving rise to 'quantum optics'. Therefore, many nano-optical devices show unique optical attributes as

A. High polarization discrimination in a micron-thin space,

B. Achromatic phase retardation, and

C. Single layer combinations of optical functions, *e.g.* polarization and spectral filtering.

Moreover, the above typical behaviours give rise to devices in order to achieve their optical effects over relatively short distances. It is known that the nano-optic devices are usually very thin, *i.e.* one micron or less in thickness, and similar to thin films, often the nano-optic structures can be applied as a coating layer during a complex processing. As said earlier, this smaller dimension has a strong positive effect for integrated optics. Also, the quantum optical behaviour make the 'optical performance' to be readily customized.

In addition, the modification of the material properties by adding nano-structures gives rise to a continuous spectrum of optical functionality. In practice, the optical wavelengths can be incrementally and accurately varied to meet 'specific application requirements' through a combination creating 'hybrid optical' materials.

8.2.1.2. Preparation of Nano-Optics

There are a variety of preparation methods for nano-optic structures. In order these nano-structures to be commercially useful, it is necessary to produce them in large volumes. Otherwise, it is not practically optimum optical device, since there is a multiple 'application specific' variations, which should be readily integrated with other relevant technologies. There are various criteria to be satisfied as :

A. Achieve 'economies of scale',

B. Ensure 'target market opportunity' to be broad and interesting enough; and finally

C. Facilitate 'easy integration' of nano-optical elements into more complex systems.

In order to achieve the above objectives to some extent, there is a 'self-assembly' process for forming regular nano-scale structures with properly controlled conditions, which has a lack of flexibility as a major difficulty in terms of useful structures and materials that limit its functioning. Hence, it is not a very popular technique.

The other available 'fabrication approach' at present is 'nano-lithography', which is similar to standard preparation technique of 'wafer-based process' for semi-conductors, giving a lot of benefits for commercial production. It consists of creating an 'image in a polymer resist layer'; then this 'image' is used as an etching mask to transfer the nano-scale pattern into the target material. There are various steps involved in making the optimum methods for certain set of nano-structure patterns, materials, and volume requirements. The process steps are as follows :

1. Interference Lithography

By using the UV light sources, it readily creates 'interference patterns' with useful dimensions. The benefit of this method is its 'simplicity'. However, there is a drawback in creating complex shapes and arrays.

2. Electron-Beam Lithography

By using a focused beam of electrons, it writes arbitrary and complex patterns. The benefit of this method is that almost any pattern can be created. But, there is a difficulty for commercial production in that the individual 'wafer processing times' can be somewhat longer.

3. Nano-Patterned Replication

By using the above two methods, it is possible to create a 'master plate' capturing the inverse of the desired pattern. A printing-like transfer method then patterns the polymer resist. The benefit of this method is high-fidelity reproduction allowing the following :

(*a*) Extremely small feature size,

(*b*) Complex patterns,

(*c*) Effective material independence (during repetitive production), and

(*d*) Different structures in the same production line.

The difficulty with this method is that it is not applicable when only a 'fixed interference pattern' is required, or for 'limited runs'.

8.2.1.3. Integration Modes

In a mixture of individual optical component technologies, which are often used and which gives rise to a minimum integration, so the component capabilities seriously hamper the circuit designs. Hence, there is a need of proper integration. The above methods of nano-lithography have been successfully developed in some companies in the USA that support a range of 'integration modes' for making the optical circuits as follows :

1. By exploiting the advantage of the 'unique optical behaviour of nano-optics', as mentioned above, an optimization process can be used, as in the case of traditional 'multi-element', 'multi-technology' optical circuit architectures.
2. By using the 'layering technique' of applying one layer on the top of the other layer to create 'aggregate optical effects', we can make 'monolithic' self-integration.
3. By organizing various 'optical functions' into an 'array structure' via nano-pattern replication, 'spatial integration' is established.
4. By adding a nano-optic layer or layers to functional optical materials, the 'hybrid integration' is possible to be achieved.
5. By electronically controlled optical devices, an integration of nano-optics is feasible with optically active layers, such as MEMS Structures or Liquid Crystals.

It should be pointed out that when combined with other nano-optics or other materials to make 'monolithic integrated optics', the benefits of the 'alignment in assembly' eliminates the complication and cost of multi-device lamination, thereby improving the reliability of the system. Moreover, in order to avoid an individually aligned discrete optical system, an 'arrayed' nano-optical device can be used in 'multi-path' or 'multi-beam' optical circuits.

8.2.1.4. Applications of Nano-Optics

There are many applications of nano-optic devices, but here we will briefly talk about only three of these applications, which are very important. Some more applications like SOEs will be described in the **sections 8.2.1.5** and **8.2.1.6.**

1. Optical Disk Drives, Communications, and Projection Displays

A nano-optic polarization beam splitter/combiner (PBS/C) provides 180 degrees of effective separation. It means that one polarization is transmitted and the other is reflected via a submicron-thick nano-structured layer. This device's beam separation and thin form allow adjacent components to be closely coupled compared with conventional optics, reducing existing optical path lengths by up to one-half. In turn, shorter optical path lengths reduce beam divergence, improving coupling efficiency and reducing power loss. Many optical circuit designs also incorporate a wave plate adjacent to the PBS/C.

2. Digital Imaging, Communications, and Sensors

Physically, a nano-optic enabled variable optical attenuator (VOA) consists of a crossed pair of polarizing nano-structures sandwiching a liquid crystal cell. This is done by applying an electrical field across the liquid crystal that will rotate the liquid crystal molecules and control the fraction of light that is passed through the structure, *i.e.* tuning. This 'tunable optic' addresses the frequent need to control and block the output power of polarized light sources, such as optical transceivers, in order to dynamically optimize their signals and to facilitate maintenance. Since the nano-optic polarizer is directly integrated into the liquid crystal cell construction, this allows the entire component to be less than 1 mm thick.

3. Near Field Optical Microscopy

The future applications opened by the photonic local probe based methods overpass the simple concept of near-field optical (NFO) microscopy for the detection of local signals emitted by individual molecules, *i.e.* single molecular detection (SMD). The NFO is entering a new era in which the concept of imaging a sample with a single molecule has become a reality. This is a new class of nano-optics experiment in which it is intended to use the 'optical tunnel effect' to control the 'optical energy transfer' between several delocalized detection or injection, *i.e.* optical addressing of individual molecules.

8.2.1.5. Photonic Band Gap

In the computer industry, there is a strong demand for a very fast processor in the range of ~ 1000 GHz. In order to address this problem of such an incredibly faster processors, the researchers have thought about a different route of processing information, *i.e.* the optical route, by means of different "optical components" made from so called 'Photonic Crystals', which are also known as 'quasi-crystals'.

By comparing the movement of the 'electrons and holes' in a typical semiconductor, we have an easier way to understand the behaviour of 'light' in a photonic crystal. For example, in a silicon crystal, the atoms are arranged in a diamond lattice structure, and the electrons moving through this lattice experience a periodic Bloch potential (as described in the **section 1.5.1**) as they interact with the silicon nuclei via the Coulomb force. This type of interaction of the electron in a periodic lattice gives rise to the formation of 'allowed and forbidden' energy bands, as described in the **section 1.5.2.**

In the actual materials, the situation is little different. Here, the electrons can have an energy within the band gap if the 'periodicity' of the lattice is "broken" by an impurity/defect or a missing atom. If we consider a block of 'transparent dielectric material' containing a number of 'air holes' which are all arranged in an ordered lattice pattern, then a moving 'photon' will pass through the dielectric medium, which has a high refractive index, *i.e.* the 'photon' knows its route. This contrast in the refractive indices experienced by a 'photon' can be used as an 'analogy' → to the 'periodic potential' experienced by an electron that moves through a crystal lattice. This is a kind of 'energy confinement', or we can call it the 'broken symmetry', which gives rise to a formation of the 'allowed' energy bands, which are clearly separated by the 'forbidden' energy bands. This is something that is called as a 'Photonic Band Gap' (PBG).

In the recent years, there has been a considerable amount of activity in the field of fabrication of **2D** and **3D** 'periodic arrays', which consists of mono-dispersed magnetic nano-particles. There are simple and easy methods to prepare a **3D** magnetic nano-particle arrays. Firstly, this is done by making a **3D** 'opal matrix' from uniform silica spheres, and secondly, by infiltrating it with various magnetic materials such as Fe, Ni, Co. This second process is done by Electro-Deposition method, or Chemical Vapor Deposition (CVD) techniques, as described in the **section 8.1.2.** For other non-magnetic materials, the infiltration under high pressure method can also be used.

Recently, some important work has been done in the fabrication of the silica opals, with variable sphere sizes by a sedimentation process and then infiltrate them by the above magnetic materials through the use of the electro-deposition technique.[17]. The 'optical characteristics' of these opals has also been done and such 'infiltrated opal systems' are found to possess a variety of unique structural, optical, electrical and magnetic properties [18-20].

When the nano-particles of a magnetic material infiltrates an 'opal system', it is important to characterize the 'spatial arrangement' of these magnetic nano-particles and then correlate it with the overall magnetic properties. For example, the 'dipolar interaction' between the neighbouring particles will play an important role in these materials, and an interesting magnetic domain pattern could develop during the reversal process of the magnetization. These issues are addressed with the help of an heightened approach of 'Small Angle Neutron Scattering' (SANS), as has been done in a simpler case of nucleation of nano particles of magnetite, as described in the **section 5.7.** The characterization of the magnetic and non-magnetic infiltrated materials by the 'electronic transport' measurement has also been done by these workers [20].

8.2.1.6. Optical Chips > Semiconductor to MEMS

There was a time when researchers wondered on the future of 'optical chip technology', which gives us the 'speed' of communication. In the recent years, a dramatic change is taking place with a focus shifted from 'semiconductor technology' to "MEMS" and "Nano-Optics" including "Photonic Crystals". Due to the needs of network specialists and the advancement of the technology, such changes can be well appreciated. With the massive expansion of the telecommunication networks, there has to be a tremendous progress in the 'optical networking equipment' in order to meet the ever increasing challenges of 'speed', 'capacity', and finally the 'reliability'. However, the integration of the 'circuits' is of fundamental necessity at the heart of all 'optical networking equipment'.

It has to be understood that the 'common semiconductors' have dominated the communication scenario as the mainstream 'technology' from the early days of optical networking almost till to-day. However, the advancement of the cutting edge 'integrated circuit solutions', which are particularly engineered for 'photonic networks', assumed significance in the recent years. Now, they are competing hard in order to remove 'semiconductors' from their dominant position, in terms of new 'optical chip enabling technology'.

At this stage, we have to understand and at the same time ask ourselves as regards the 'time frame' as well as to 'what extent' will MEMS and Nano-Optics supplant the so-called common semiconductors? However, we need to ask ourselves some more important questions as

1. How are the 'Optical Chip' markets evolving ?
2. What roles are 'Nano-Optics' and 'Photonic Crystals' playing ?
3. How reliable will be MEMS and Nanotechnology ICs ?
4. What will be the 'Cost Implications' of these new 'Technologies' ?

Our future opportunity for the fabrication of the 'optical chips', from Semiconductors → MEMS → Nano-Optics (including 'photonic crystals') will be known by answering some or all the above questions. ***For all of us, that's what is important to know in order to redefine the rules of the game for 'optical processing'.*** **[21]**

8.2.1.7. Subwavelength Optical Elements (SOEs)

In the earlier days, the 'optical elements' were plagued with the following problems :

(*a*) Reduction of Size,

(*b*) Manufacturing Cost, and

(*c*) Reliability Problems.

These problems are tackled by a new generation of 'circuit materials', like wafer-based nano-fabrication technique, as briefly mentioned above. This new category of 'optical components' are known as 'sub-wavelength optical elements' or SOEs, which have physical structure much smaller than optical wavelength. New arrangements of 'optical-processing functions' have greater 'density', more 'robust performance' and greater 'degree of integration', which have to be understood in the realm of 'physics' involved in the 'interaction' of these 'fine-scale surface structures' with 'light' that could possibly alter approaches to 'optical system design'.

However, there are technology limits too, which have to be also understood in terms of increasing the number of components and module density. Also, the 'flexibility' in the design of the optical components is rather limited, the 'transferability' from design to design is reduced, and the properties of many discrete optical elements are relatively 'fixed'. However, the realization of the SOEs are based on nano-technology in terms of optical elements with the following advantages :

(*a*) Excellent optical properties,

(*b*) Easy to integrate with other optical materials in a broad range of configurations,

(*c*) Self-integration by allowing a combination of 'optical functions',

(*d*) Increase reliability, and an increased flexibility in optical component design, and finally

(*e*) A broad range of optical effects can be achieved by the manipulation of nano-structures.

In conventional optics, we describe the behaviour of 'optical elements' by the principles of reflection, refraction, diffraction and interference effects. However, since SOEs are very small in the nano level, the mathematical description of traditional optical behaviour is not quite sufficient, since the 'quantum-mechanical' effects are operative at this stage. Let us consider the 'refraction', which is an important property used in many optical components. Our conventional wisdom tells us that different materials should be used to get different refractive indices. However, with the same SOE material, it is possible to get different refractive indices by simply adjusting its physical structure.

The 'birefringence' or 'elasto-optic' effect has been described in the **section 7.3.1.** Here, let us take an example of a SOE structures, which can be used to create an 'artificial birefringence' effect. For a SOE, the grating design is very important in terms of dimensions. If d is the grating width and A is the grating period, then the refractive index of the TE wave is written as n_{TE}, *i.e.* the electric vector parallel to the grating grooves, and that of the TM wave as n_{TM}, *i.e.* the electric vector normal to the grating groove. These refractive indices are expressed as :

$$n_{TE}^2 = F\, n_1^2 + (1 - F)\, n_2^2$$

$$n_{TM}^2 = n_1^2\, n_2^2/[F\, n_2^2 + (1 - F)\, n_1^2]$$

where, n_1 and n_2 are the dielectric constants of the 'grating material' and that of the 'fill material' respectively, and F is the 'filling factor' of the grating and is defined as F = d/A. By choosing the SOE material and adjusting the grating 'filling factor', a 'birefringence effect' can be achieved, which is much larger that those achieved with standard components (see the extensive and excellent work done by Park & Kustal [22]).

There are many SOEs, which have periodic patterns, and hence they could be viewed as 'optical gratings'. When incident light is perpendicular to a grating surface, the conventional grating equation can be expressed as

$$A \sin Q_m = m\,\lambda$$

where, Q_m is the diffraction angle, m the grating order and λ is the wavelength. Depending on the 'operating wavelength, when the 'grating period' is less than λ, as in the case of SOEs, the incident light is still subject to 'grating diffraction'. However, all the diffracted optical energy of the incident light will enter zero order, with no high-order light existing in physical space. Therefore, the SOEs have relatively uniform performance across a broad range of wavelengths and wide acceptance angles.

Despite all the good things about SOEs, it has entered the 'market' little late simply due to the fabrication capability in order to serve the telecommunication industry in terms of cost-effectiveness and the requirements of high energy facilities for making sub-wavelength grating structures. These are made by 'electron-beam lithography'. Although there are difficulties in the creation of nano-structures, recently in order to broad-base the pattern design, a technique called "nano-print lithography" has been successfully developed. This process consists of four steps as follows [22] :

(*a*) Making a mould inscribed with the complement of the desired nano-structure,

(*b*) Impressing this mould into a resist-coated wafer,

(*c*) Separating the mould, and

(*d*) Selectively removing the resist with reactive ion etching to transfer the nano-pattern to the target material [22].

The physics of the SOE depends on rigorous application of the boundary conditions of Maxwell's equations to describe the interaction of light with the structures. At the wavelengths of light used in telecommunications, *i.e.* **980 nm** through **1,800 nm**, the structures required to achieve those effects have some dimensions on the order of tens to a few hundred nanometers. At the lower end of the scale, obviously single-electron or quantum effects may also be observed.

Finally, due to the diameter of the incident light beam is often much larger than the structures of the grating, the effect on the transmitted light is really the statistical sum of numerous 'local interactions' between the light and the grating. For example, if a light beam measuring 300 microns in diameter is incident on a two-dimensional grating with a size of the structure as roughly **100 × 100 nm**, more than a million nano-structures will be illuminated. Therefore, varying the grating dimensions spatially across the incident beam front allows additional control over optical-processing effects. This can be considered a real achievement [22].

8.2.1.8. Novel Properties of Nano Vanadium Dioxide

The ultrafast insulator-to-metal transition in nanoparticles of VO_2, is obtained by ion implantation and self-assembly in silica. The non-magnetic and strongly correlated compound VO_2 undergoes a reversible phase transition, which can be photo-induced on an ultrafast time scale. The transition of vanadium dioxide nano particles takes place from a transparent to a reflective, mirror-like state, at an 'ultrafast' speed less than 100 femtoseconds (a tenth of a trillionth of a second). According to this work, mainly carried out at the Vanderbilt University (USA), this effect has a size limit in the sense that 'it does not occur in particles that are smaller than about 20 atoms across or **10 nm**'[23]. This opens the door that are transparent at low temperatures and block out sunlight when the temperature rises. However, other applications are possible, such as 'nano-sensors', which could measure the temperature at different locations within human cells, or as 'ultrafast optical switches', which could be used in communications and optical computing.

How vanadium dioxide or VO_2 can turn from a transparent insulator into a reflective metal so rapidly has wondered many physicists. The change from insulator to metal is called a phase transition, *i.e.* a type of 'Mott Transition'. The phase transitions in solids generally occur at the speed of sound in the material, but vanadium dioxide makes the switch 10 times faster. Hence, a definite explanation is needed for such a rapid change.

In the nano particles, a prompt formation of the metallic state results in the appearance of 'surface-plasmon resonance'. This is a form of 'electron wave' that only occurs on the surfaces of metals and is responsible for the glowing colours of 'stained glass'. The detection of this effect confirmed that vanadium dioxide can switch all the way from transparent to reflective in less than 100 femtoseconds (a tenth of a trillionth of a second).

As said earlier, this switch effect had some size limits, which shows that the 'nano particles' undergo the same phase transition as 'thin films'. It does not occur in particles that are smaller than about **10 nm**. The researchers have established that it is possible to raise and lower the temperature at which the 'insulator ↔ metal' transition takes place by as much as 35°C by adding small amounts of impurities, and this temperature effect leads to newer applications.

It is relatively easy to change the material's transition temperature to body temperature, *i.e.* 37°C by adding precise amounts of impurities. Such doped nano particles would be small enough to measure the temperature at different locations within an individual cell, and when injected into the body, it could pinpoint 'hot spots' by turning into microscopic mirrors.

Obviously, there will be other applications for such a fast 'phase transition' effect. For example, the creation of an 'ultrafast' optical switch is being explored by putting a layer of nano particles of vanadium dioxide on the end of an optical fiber. A large ultrafast enhancement of optical absorption in the near-infrared spectral region that encompasses the wavelength range for optical-fiber communications has been observed. It is possible to further tailor the response of the nano particles by controlling their shape. Such a switch could be useful in communications and optical computing in future [23].

8.2.2. Nano-Magnetics

The research in the field of nano-magnetics is directed towards a range of 'topics' related to the development and applications of novel magnetic materials and devices at nano-scale dimensions. These include nano-magnetic materials and devices directly related to the current and future magnetic storage technologies such as disk drive storage and probe storage based on MEMS, magnetic random access memory (MRAM), and magnetic quantum cellular automata (MQCA). Obviously, there are more to add to this list of newer materials for advanced applications.

The main issues involve the physics of devices and the fabrication techniques of 'magnetic probe heads' at nano-scale dimensions, *i.e.* the 'recording heads' with dimensions down to a few tens of nanometers. These materials have been fabricated by using 'focused ion-beam' (FIB) nano-fabrication techniques with the following list of goals :

1. Development and Characterization of Nano-Crystalline Materials (Application : Advanced Recording Media),
2. Micro-Magnetic behaviour of Soft Magnetic Materials,
3. Recording Properties of Nano-Crystalline Alloy and Superlattice-based Media Materials,
4. Recording Processes at Nano-scale Dimensions, etc.

The researchers achieved 'record track densities' in excess of 400 ktpi (~ 60 nm track width) by using above-mentioned nano-probe recording heads and specially prepared media. The micro-magnetic behaviour of magnetic '*nanotubes*' has also been studied.

The researchers are also focusing their attention on the applications of nano-crystalline materials and nano-scale devices for achieving extremely 'high density recording', *i.e.* above 1 Terabit/in^2. At present, the ability of the nano-materials could give a density of 1/10 - 1/5th of that number, as recorded into a magnetic recording medium. The individual magnetic grains are of ~ **9 nm** in diameter, which form the recording medium. At these dimensions, the conventional recording schemes employed to-day are rapidly approaching the fundamental "super-paramagnetic" limit in areal bit density, above which the recording data become unstable. It is widely believed that 'longitudinal recording' will almost become obsolete at approximately 200 Gb/in^2.

The perpendicular magnetic recording will enable us to sustain the current great strides in technological advances for the next several generations of mass storage solutions. The technology is technically the 'closest alternative' to 'conventional longitudinal recording', while it is capable of extending the 'super-paramagnetic density' limit beyond what is achievable with 'longitudinal recording'. The recording densities above 1 Terabit/in^2 are conceivable utilizing perpendicular recording. In order to support such a nano-scale technology, the major innovations in both magnetic recording heads and media are necessary.

8.2.2.1. Magnetic Semiconductors

In the **sections 5.4** and **5.5,** the details of magnetization and Mössbauer experimental data are given for very small nano particles of magnetite within a glass matrix, which show several interesting magnetic phenomena even within this small nano range. These interesting aspects include most importantly 'anisotropy', 'magnons' and 'spin canting', whose theoretical developments are neatly presented in the **section 1.6.1**. These properties may be very important also for the study of newer brand of nano-magnetic materials for information storage and processing. Thus, it opens up the door for the future possibilities in the field of nano-technology, which is presently directed towards the understanding of 'hetero-structures' of silicon, germanium and carbon. These are not new materials, but their 'heterostructures' are new to the world of nano-magnetics [24, 25], which are proving to be the key materials in today's field of information technology. One of the important category of such materials is III-V compounds.

The new class of materials such as 'Magnetic Semiconductors' are definitely fascinating, since they couple the complementary functionalities of the 'spin properties' of ferromagnets and the 'electronic properties' of semiconducting materials. This brings us to a new category of materials, which are based on GaAs. The ferromagnetism in these materials is evident due to the presence of randomly substituted magnetic Mn ions.

The recent efforts have been on GaMnAs based materials such as $Ga_xMn_{1-x}As$ alloys, which are grown by 'Molecular Beam Epitaxy' or MBE. In these materials, the Mn ions provides the charge carriers, *i.e.* the holes and carries spin (S = 5/2) making both the electrical and magnetic properties tunable. But, another important aspect is the ferromagnetic transition temperature, *i.e.* Curie Temperature or $T_c > 100$ K in (Ga,Mn)As materials. It also shows the 'long spin coherence times' [26-28], which is a necessary ingredient in the rapid development of information storage technology based on magneto transport effects.

But in the above important investigations, the solubility of manganese ions in the GaAs matrix plays a crucial role. The equilibrium solubility of Mn in GaAs limits the T_C. It should be mentioned that in many magnetic materials, T_C is either below or near room temperature. This severely limits their technological applicability in devices at or above room temperature, and hence we require high T_C materials. Therefore, both these limitations are to be tackled together.

One clever way is to choose a 'ferromagnetic semiconductor system' with a type of 'digital superlattice structure'. By creating a thinner non-magnetic 'spacer layer' in the superlattice structure, the effective Mn concentration is increased and thus overcoming the solubility limit on Mn. At the same time, it is possible to have a higher Tc material [26-28]. In this type of system, the ferromagnetic ordering of Mn layers and their magnetic coupling is unknown. Moreover, very little is known about the magnetic anisotropy of these systems, and the type of the interlayer coupling and its strength. Various newer type of experiments are to be performed on such materials, like it has been done on nano particles of magnetite, as described in the **sections 5.4** and **5.5**.

8.2.2.2. Spin Electronics (Spintronics)

This subsection is a continuation of the previous subsection on magnetic semiconductors. During the last decade or so, one of the new and highly potential inter-disciplinary fields that has emerged in the horizon of nano science is the 'Spin Electronics', which is popularly known as "Spintronics". Apart from various electronic effects in materials in the quantum or nano level, the spin dynamics assumes a great importance, as written above. The main purpose of 'spintronics' is to understand this spin dependent property and its implications in the domain of newer potential applications, *e.g.* in 'high density information storage media' and the non-volatile 'memory devices', which is now known as "Magnetic Random Access Memories" or in short as MRAM, which would combine the advantages of magnetic memories and those of DRAM. It should be pointed out here that in the field of 'ferroelectricity', there is also a new class of materials like lithium niobate and lithium tantalate, which show high non-volatile memory called FRAM, and there is a possibility of 'infinite memory' in such materials [29].

This exciting new field of 'Electronics and Magnetics', namely 'Spintronics', has attracted great attention recently. This is based on the very basic fact that electrons have spin as well as charge. Within the context of spin-electronics, the electron spin, as well as charge, is manipulated for the operation of information processing circuits. The advantages of 'spintronics devices' over traditional 'semiconductor devices' include the following :

(*a*) Much faster switching times,

(*b*) Reduced power consumption,

(*c*) Non-volatile memory, and

(*d*) Increasing levels of miniaturisation.

The research into 'spintronics devices' has mainly focused on methods of switching the magnetic configuration and there is a growing interest in the use of spin-polarised current to switch magnetic devices. There are several advantages associated with the use of spin-polarised current rather than external magnetic fields, *i.e.* the most important being no more (or much reduced) cross-talk and low power consumption. The mechanism involved in the current-induced (*i.e.* spin-polarised current) switching of magnetisation is not clear yet, and some new insights of the switching process need to be given in this area.

It is important to make devices, but the characterization of some properties need to be also done, particularly by novel techniques. With magnetic recording data rates approaching GHz range and the data density exceeding 150 Mbit/mm^2 (100Gbit/in^2), the 'dynamic magnetization mapping' at 'nanometer-length' scale and 'sub-nanosecond' temporal resolutions is demanded for material analysis and device characterization. This is beyond the limits of traditional imaging techniques. The 'Time-Resolved Magneto-Optical Scanning Kerr Effect Microscopy', *i.e.* in short called TR-SKEM, has recently emerged and demonstrated its power in imaging fast dynamics of the sample magnetization directly, but it does not image the strain field around it. This is done at femtosecond and picosecond temporal resolution.

With the advent of TR-SKEM, the study of 'spin dynamics' in nano-scale contact 'spintronic devices', by polarisation current-induced switching at picosecond temporal resolution, the activity in this field is now taking shape. In TR-SKEM, the transient spin-polarized current pulses are generated by a 'photo-conductive switch', which is triggered by 'picosecond laser pulses'. The magnetization of the devices can be manipulated directly by feeding the 'spin-polarized current' into the devices themselves. The device materials for 'spintronics' are prepared by different methods of nano-fabrication, involving : (*a*) Optical Lithography, (*b*) Electron-Beam Lithography, and finally by (*c*) Focussed Ion-Beam (FIB) Lithography.

In the domain of 'spintronics', in the 'advanced magnetic recording materials', the study of 'spin dynamics' at 'submicron spatial resolution' and 'picosecond temporal resolution' is also done by using a high-field TR-SKEM. The magnetization is pumped by sub-nanosecond magnetic field pulses with high amplitude, which are generated by feeding high voltage pulses with picosecond rise time to a microstrip line structure. The 'magnetic recording materials' that are normally studied include the following :

(*a*) Thin film with perpendicular anisotropy,

(*b*) Patterned thin films with in-plane or perpendicular anisotropy,

(*c*) Nano-scaled single crystals of Fe, Co and Ni elements.

The 'spintronics' is quite a new area. With the emergence of newer materials and devices in the horizon of 'spintronics' along with some powerful techniques of chatracterization of the devices through the study of 'spin dynamics', there is definitely a great future for 'nano-magnetics'.

8.2.3. Nano-Electronics

Preamble

A brief overview of quantum mechanics is given in the **section 1.2** and a detailed analysis of the band structure is given in the **section 1.5.** Here at the cost of some repetitions, something needs to be said about quantum mechanics before embarking on the nano electronics front It is often told to us about quantum mechanics that the particles also behave like waves. In fact, the waves and the particles are just two complementary ways of describing a quantum system. The wave or particle are basically 'terms' that are descriptive for objects. Actually, the wave function is what the object is all about. Something that looks like a particle can be described as a 'superposition of waves', and something that looks like a wave that can be described as a 'superposition of localized particles'. Usually, we are used to the simplest description. A superposition of waves means that it is some type of combination of two or more waves. The quantum computers are potentially so powerful due to the principle of superpositions. At the end of the day, the uncertainty principle is one of the consequences of quantum theory in that there is always some kind of uncertainty in the computing world.

There are objects which are described by quantum mechanical principles, but they are not freely moving. They are either 'attracted to' or 'repelled by' other objects, which is called "quantum confinement". Such quantum confinement makes things behave in a significantly different way.

The way quantum objects behaviour is defined by their wave function, which is a solution of 'Schödingers' wave equation. The way different objects interact can be described using "matrix mechanics", which is devised by Heisenberg and Born. However, these interactions can equally well be described using the wave equation.

The quantum mechanics has been generalised into quantum field theory, which includes the effects of special relativity. The quantum field theory explains why fermions have half-integral quantum spin, and also predicts the existence of anti-particle counterparts to normal particles. Finally, by taking care of the non-linearity of a given system, the quantum systems can be chaotic, which is a situation that is described by the 'quantum chaos'.

8.2.3.1. The Semiconductors

The semiconductors are materials that nearly conduct electricity, and lie between conductors and insulators. The semiconductors are useful because by changing their structure or composition, we can make them either conduct electricity, or not conduct electricity. In addition, they interact significantly with light, which strongly depends on their band structure that in turn depends on the composition, as explained in the **section 1.7.6.** Due to this control on composition/structure, we can design semiconductor devices that have a variety of useful properties → amplifying or detecting electrical current (*i.e.* the electrons), as well as detecting, amplifying, or emitting light (*i.e.* the photons).

The semiconductors are materials where the highest occupied band is completely full of electrons, but where the next band up is close by. Although there is no space for the electrons to move around on the band, it is not too hard for an electron to gain the extra energy required to leap up into the nearby higher band, where it can move freely carrying an electric current. This leaves a gap, or 'hole' in the lower band, which can also move. Since we can easily supply the extra energy from outside, or by intuitive design, we can control the way the semiconductor conducts electric current.

The electric current is carried by both 'electrons' and 'holes' in semiconductors. These electric charge carriers move differently through semiconductors with different compositions, so we can make different sorts of semiconductor-structure. Since the electrons and holes have opposite electric charges, they can attract each other and form 'excitons'. These carriers do not move freely through the semiconductor. The crystal lattice of atoms does affect their motion by making them more likely to travel at certain speeds, but a perfect crystal will not cause them to be deflected. However, their motion is affected by imperfections in the lattice, as well as by the phonons.

8.2.3.2. The Semiconductor Structures

If we put two different semiconductor materials side by side, it is found that the electrons and holes will have preference of one material over the other. If we visualize the electrons to be residing on a 'cage', then we will find that electrons from one cage in one material can easily fall into an adjacent partially filled cage on a lower band from another material, so that we can say that the electrons "prefer" some materials over others. It is a kind of 'friendship' of the electrons in a cage of a given material with the cage of a second material. The holes also can do this, but since they are voids or gaps, they simply fall upwards.

By putting a piece of preferred semiconductor next to another, we can make a device where the electrons will only want to travel in the direction of the preferred material thereby creating a 'diode'. By surrounding some preferred semiconductor by a different semiconductor, the 'traps' can be created for the electrons and holes that lead us to a 'quantum well' , a 'quantum wire', or a 'quantum dot'.

8.2.3.3. The Quantum Wells

A 'quantum well' is a potential well that is obviously very small. This 'well' is like a cage in which the carrier particles (*i.e.* the excitons) can be trapped, in much the same way that light can be trapped between the mirrors. By creating 'layers of different semiconductors', it is possible to make particular 'layers' to act as 'traps' for excitons. These 'trapped particles' can be considered to be in 'quantum confinement'.

Due to this particle trapping, the 'quantum confinement effects' naturally restrict the motion of the carriers in some way, *i.e.* restrictions on what the particles want to do. In a 'quantum well', the excitons can move freely sideways in the plane of a 'thin layer', but they might like to move in the forward and/or backward directions as well. These particles are constrained or confined in the forward-backward directions, *i.e.* a **2D**-confinement. For any particular sideways motion, there could now be various ways in which the carrier is confined in the layer → giving rise to quantum wells. These 'quantum wells' have many useful properties, since it is possible to exactly engineer the forward-backward confinements, and since they are relatively easy to make. As a result they are now widely used to make 'Semiconductor Lasers' and other useful devices. This 'quantum confinement effect' can be taken further and make semiconductor structures → that brings us to the 'quantum wires' and 'quantum dots'.

8.2.3.4. The Quantum Wires

From 'quantum well', we can go ahead one step further in the semiconductor structure, *i.e.* by making not a layer but only a thin 'wire' of the preferred semiconductors (*i.e.* a chain like structure). These trapped excitons can now be considered to be in **1D** 'quantum confinement'. In this case, an exciton is only free to choose its trajectory along the wire. However, for each motion of its movement, the exciton could have various ways of being confined. This is a little bit tedious for the exciton as well as for us to explain, which creates a great technological challenge to make a defect-free good 'quantum wire'. This has naturally triggered a flameboyant activity into this exciting field of research. That takes us to the another interesting area like 'quantum dots'.

8.2.3.5. The Quantum Dots

In the above descriptions, for the charge carriers like excitons, from a 'semiconductor structure' that constrains them in a thin layer **2D**-quantum confinement, we get a 'quantum well', and by going one step ahead, *i.e.* by not making a layer, we can create only a thin **1D**-quantum wire. Now, we can go another step ahead, *i.e.* the final step, and put the excitons in zero dimension-quantum confinement in just a tiny dot of a given semiconductor material. Hence, in this situation, the exciton only has 'confined states', *i.e.* there are no freely moving excitons. This implies that it behaves in a very different and newer ways. Although this quantum dot consists of many thousands of atoms, this would have a number of allowed states that can be counted. Each state would be separated from the others. This indicates that the 'quantum dot' is more like a 'single atom', rather than 'many atoms'. Due to this peculiarity, a 'quantum dot system' may make the ideal 'Laser Material', which is very much focused. Moreover, it may be an ideal 'building block' with which to make a quantum computer.

8.2.3.6. Quantum Computers

Quantum-computers offer the possibility of vastly increased computing power. Conventional computers operate using classical logic, even though much of the electronics they are made with relies on quantum mechanics. In a conventional computer, each logic-gate and memory-cell operate by using "bits" that can be in one of two states, 0 and 1. In a quantum-computer we can have superpositions → so a quantum-logic-gate can be in two states at once. In order to store a quantum-bit (or Q-bit) we could use the spin of an electron, or, more practically, the particular quantum states of an atom, or perhaps a quantum dot made from semiconductor materials.

The superpositions in a Q-bit means a quantum-computer can run many calculations in parallel using the same hardware. At the moment, however, quantum-computing is in its infancy, and while theoretical work continues at an incredible speed, so far only small collections of quantum-logic gates have been experimentally tested. The making of quantum-computers is hard because the quantum-logic elements need to be super-isolated from the rest of the universe. A single atom floating in an atomic trap can be used to make Q-bits, and there is potential for using quantum dots to make Q-bits that are embedded in a solid, and therefore easier to integrate into a quantum-computer.

8.3. OTHER IMPORTANT NANO MATERIALS

8.3.1. Microelectronics for High Density Integrated Circuits

Due to size limitation even with scaling down, MOSFET technology cannot continue forever. It will hardly go beyond a few nanometer, even if adequate lithographical technology is available. As a result, the search for the new principles of operation of the small-size devices is becoming more and more important. They possess radically different properties from those of bulk semiconductors. This change in the effective dimensionality offers fascinating changes in electric, magnetic, optical and vibrational properties. The electron mobility is high in those devices. The researches on the nano devices or quantum devices continue to be both challenging and exciting as novel structures with different materials having different properties are developed. They are useful for millimeter and submillimeter wave applications. They have potential advantages that make them attractive for nonlinear functions. It is possible to realize high frequency, low power consumption and low dimensional devices [30].

The application of soft computing tools on nano device modeling can help in the optimization of system parameter of the quantum (nano) devices to get the desired characteristics. The on-line optimization during fabrication is also possible with the application of softcomputing tools like 'Genetic Algorithm' (GA) [31] or 'Artificial Neural Networks' (ANN) [32].

However, the quantum devices have inherent limitations like material and process related limitations, power limitation, wiring limitation, quantum mechanical limitation and system architecture limitation. The most likely candidate for future ultra-dense digital circuits is 'single electronics', which has links with the motion of a single electron. Single electron device provides a complete and concise representation for most digital functions encountered in logic-design applications.

The generality and robustness of the effect and the relative simplicity of the device structures make the single-electronics the most likely candidate for future ultra dense digital circuits. As a single electron is sufficient to store an information which is not in the case of 'Transistor or CMOS' circuits, the circuit has advantages of reducing the power consumption. In single electron devices in which one bit of information is represented by a few electrons, the power consumption is drastically reduced. The single electron devices that allow manipulation of individual electron are ultimate forms of the electron devices. Their potential integration level is obviously extremely high due to its small size. An extremely low 'operation power' solves some of the instability and reliability problems. The 'speed-power' product is predicted to lie close to the quantum limit set by the Heisenberg's uncertainty principle. The processing speed of such device will be nearly equal to electronic speed. The exquisite sensitivity is about five orders of magnitude better than conventional solid-state MOSFET transistors. The integration density is much higher than that available in the existing VLSI / ULSI circuits.

There is, at the current time, a growing interest in the possibilities of designing electronic circuits using evolutionary techniques. During the decade since their discovery, single-electronic tunneling devices have received a great deal of attention, both in terms of the physics of the coulomb blockade and for potential devices applications. The single-electron tunneling (SET) devices exploit effects that arise due to the quantized nature of charge. These effects have been observed in systems of small metal structures, in semiconductor structures, and in structures made from conducting polymers. Since these effects are ubiquitous in small structures, they are likely to have an impact on any future nano-scale electronic circuits. These devices can be useful in low power circuits since only a few electrons are needed to be transported. The dense memories where bits are represented by the presence or absence of only a few electrons can also be realized with the use of single electron devices.

Biswas et al [32] investigated the properties and applicability of circuits based on single electron devices. An approach to such new devices was built on the basis of the concept of binary decision diagrams. The unit function of this device was a simple two way switching. A binary decision diagram represented a digital function as a 'directed cyclic graph' with each node labeled by a variable. It provided a complete and concise representation for most digital functions encountered in logic-design applications in order to realize some simple and also complicated digital circuits. All logical circuits starting from NOT gate to CPU of a digital computer can be realized with the help of single electron circuits. The 'timing analysis' together with 'propagation delay' in case of single electron device based digital circuits can also be estimated.

The single electron circuit based multiplication scheme was presented as an efficient application of single electron devices. A comparison-based table was generated to establish that the use of single electron devices cannot only increases the density of integration, but also make the execution much faster. Hence, it is seen that the time taken by normal conventional circuit is approximately three times of the 'single electron' circuit based multiplication circuit. Hence, the efficiency of 'single electron' circuits based multiplication scheme is better than that of conventional multiplication circuit [32].

8.3.2. Si/SiGe Heterostructures for Nano-Electronic Devices

The semiconductor nano-structures are attractive candidates due to their potential in making substantial improvements in optical and electrical properties over those of conventional two-dimensional structures. The modern epitaxial techniques have made the growth of the 'atomic scale features' possible in the 'vertical direction'. The advent of nano-lithography has made the fabrication of 'sub-100 nm' feasible in the 'lateral dimensions'. With suitable combinations of such characteristic 'length-scales' in the 'vertical' and 'lateral' directions, the 'quantum-mechanical confinement' in one, two or three dimensions becomes possible in a wide array of structures so that the conduction electrons have quantized energy levels. A lot of interesting results have been reported on the growth, properties and device fabrication using semiconductor nano-structures over the last decade [33-37].

The epitaxial growth of SiGe/Si films can be used for the self-assembled growth of a three-dimensional islands. The synthesis of Si and Ge nano crystals embedded in a high bandgap insulator silica matrix could be utilized to achieve nanometer range structures without sophisticated nano-lithography techniques. Both the nano crystals of silicon and germanium are promising candidates for 'light emission' in the visible wavelength range due to 'quantum confinement' of carriers, and for flash electrically erasable and programmable read-only memory devices, which is due to the 'Coulomb Blockade' phenomena.

The growth of self-assembled Ge nano-structures, the light emission in the visible range due to 'quantum confinement' and the 'charge storage' characteristics of Ge nano crystals embedded in SiO_2 matrix have been studied by Ray [38]. The 'multi-quantum well' semiconductor heterostructures based on SiGe/Si are attractive for their applications in TeraHz devices due to the compatibility with planar integration technology and the absence of the reststrahlen absorption band that is present in III-V and II-VI compound semiconductors. The details of the characteristics of the TeraHz emitter devices fabricated using molecular beam epitaxially (MBE) grown Si-Ge quantum well structures have also been reviewed by Ray [38].

In particular, Si based nano-structures received a lot of attention for their possible use in opto-electronic applications. The 'silicon microphotonics', a technology that merges photonics and silicon microelectronic components, is rapidly evolving.

8.3.3. Piezoresistance of Nano-Crystalline Porous Silicon

Since the pioneering work of Smith, the 'piezo resistance' of silicon has been used as a sensing principle for mechanical sensors such as 'diaphragm' type pressure sensors, accelerometers, gyroscopes, etc. [39]. As the 'miniaturization' advances towards the molecular level with the ultimate goal to mimic 'biological systems', "nano-electro-mechanical systems" or NEMS have gained tremendous importance [40]. However, as the dimensions of the systems are shrunk, the bulk properties get modified like the mobility, concentration, effective mass and so on. Hence, there is a shift from MEMS to NEMS through the replacement of silicon by nano-crystalline silicon, and the piezo resistive coefficient that determines the 'sensitivity' of the sensor also changes.

It has been observed that the piezoresistive coefficient of nano-crystalline silicon is 1.5 times greater than that of mono-crystalline silicon [41]. This enhanced sensitivity of the nanometric piezoresistors has found applications in high resolution measurements like AFM cantilevers, data storage cantilevers [42]. But the fabrication of such 'nanowire piezoresistors' is complex and requires '*e*-beam lithography' and 'reactive ion etching' techniques, which are also expensive. The porous silicon, being a natural nano-crystalline silicon material, has been used as an active pressure sensing medium [43].

It has been shown by Pramanik and Saha [44] that the piezo-resistivitive coefficient of nano-crystalline silicon is 55% more than that of mono-crystalline silicon. The porous silicon being a natural nano-material with silicon nano-crystallites confined between silicon dioxide and voids has the potential for highly sensitive piezo resistive pressure sensors. It can be easily fabricated by 'anodic etching' of silicon and it has been observed experimentally that an optimized porous silicon membrane yields 3 times higher sensitivity than that of conventional bulk silicon membrane. The enhancement behaviour in the piezoresistive coefficient has been explained on the basis of hole energy consideration and finite element stress analysis [44].

8.3.4. QMPS Layer with Nano Voids

In thin silicon solar cell technology, the trapping of light by texturisation of the silicon surface and multiple reflections are usually involved. The 'nano voids' in thin silicon layer aids the scattering process significantly. It is known that silicon as a material has low absorption coefficient at visible and infrared wavelength of light. This problem can be overcome by incorporating nano-sized voids in the silicon body. The light scattering by nano-voids and related enhancement of light path lengths due to diffraction phenomena by the nano-crystallites between the voids open up the phenomena of significantly increased light absorption by the thin silicon layer [45]. The 'Quasi Monocrystalline Porous Silicon' or QMPS layer is basically a mono-crystalline silicon layer containing voids in its body, which is formed by annealing of porous silicon of low porosity at high temperature [46]. Thus, it is prudent to calculate the light absorption in such layers, by considering the scattering and diffraction effects occurring in such thin structures. The theory will not only be limited to QMPS layers, with uncontrolled void distribution, but it will be valid for every 'thin silicon layer' containing uniform void distribution in its body. Recently, Brendel et al. [47] has studied the absorption in QMPS layer experimentally and has advanced a model for explaining their experimental results. However, their model involves complicated calculations.

Banerjee et al [48] modeled the absorption in QMPS layer by considering the diffraction of light by the nano-crystallites between the voids as well as the scattering of light by the voids. The QMPS layer was formed by high temperature annealing of a double layer porous silicon, which was modeled as a mono-crystalline silicon layer containing nano voids of radii in the range of **5 nm - 50 nm** in its body. This structure helps in light trapping due to scattering of light by nano-voids that enhances the light absorption almost ten times in the whole solar spectrum due to enhancement of optical path in the layer. Thus, the material become cost-effective by reducing the active layer thickness ten times. A model was developed where absorption, transmission and scattering, including multiple scattering, have been considered to explain the enhanced absorption due to nano-voids in the layer. This theory is applicable to lower radii ($a < \lambda/10$) as well as to higher radii ($a > \lambda/10$) particle. For higher radii particles, instead of Mie, diffraction theory of light is used in this case to account for scattering, which is much simpler to manipulate. In this work, the simulated values of total absorption in a QMPS layer has also been compared with Brendel's experimental findings and excellent matching has been observed [48].

8.3.5. MEMS based Gas Sensor

The semiconducting metal oxides like SnO_2, ZnO, TiO_2 have long been used for detecting poisonous (CO) and inflammable gases (CH_4) by their change in conductivity. For sensing gases, the temperature of the sensing oxide layer should be raised to a particular value requiring a large amount of power. In order to reduce the operating temperature, three approaches are taken. Firstly scaling down the sensing layer materials to nano-dimension, secondly by micro-machining the silicon substrate to reduce thermal mass, and finally by the combination of the above two, *i.e.* nano-structured metal oxide MEMS approach.

There are various parameters that characterize the performance of the gas sensors, but the three most important parameters are :

(*a*) Sensitivity,

(*b*) Response Time, and finally

(*c*) Selectivity.

Amongst these parameters, the selectivity tuning of chemical gas sensors poses a real challenge. The selectivity and also better sensitivity can be achieved by judicious choice of operating temperature, modification of material structure, and by the use of dopants and catalysts [49-51]. Out of these various options available, the modifications of the material architecture present a more acceptable idea. A number of material properties are modified as the crystallite size is lowered to the nano dimension and it can easily be concluded that the gas sensing properties of the gas detecting materials like MOX will also be affected by reduction in the crystallite size down to a few nanometers. An important property of nano materials is their enormously increased surface to volume ratio compared to their bulk counterpart [52], providing ample opportunities to lower the operating or sensing temperature of the metal oxide semiconductor (MOS) gas sensors.

The rapid development of integrated-circuit (IC) technology during the past decades has stimulated many initiatives to fabricate chemical sensors on complementary metal oxide semiconductor (CMOS) substrates [53, 54]. A variety of micro-mechanical structures, including cantilever beams, suspended membrane and free-standing bridges have been produced using micro-machining or micro-electromechanical system (MEMS) technology, combination of special etchants, etch stops and sacrificial layers [55-58]. The CMOS or CMOS-MEMS technology provides excellent means to meet some of the key criteria of chemical sensors as :

(*a*) Miniaturization of the devices,

(*b*) Low power consumption,

(*c*) Faster sensor response,

(*d*) Batch fabrication at industrial standards,

(*e*) Low cost factor, and

(*f*) Greater sensitivity [55-58].

There is an impact of nano-structures on the physical principles of the semiconductor gas sensors. Out of the several challenges that MOX gas sensor research is facing, two most important ones are selectivity of the sensors and their operating temperature. Now it is widely recognized that the surface plays a key role in metal oxide based gas sensors [59]. This domination of surface atoms provides adequate 'dangling bonds' that increase 'adsorption sites' for chemisorptions of ambient reducing gas. Thus, the answer to the problems of selectivity or lower operating temperature possibly lies in the arena of nanoscience and nanotechnology, as pointed out by Bhattacharya et al [60].

The optimum sensitivity is obtained for small grain size, large Debye length, and relative low carrier concentration. In addition, the values of carrier concentration and mobility are strongly dependent on grain size as well as the electronic structure of the gas-oxide and oxide-oxide interfaces. Nanostructured materials do not necessarily confine themselves to nanocrystals or nanopowders. The material architecture has been manipulated to the range of nanopins, nanorods, nanobelts, nanotubes, nanopores and so many novel architectures. SnO_2 nanobelts have been studied as stable and highly sensitive gas sensors for CO and NO_2 detection [60].

Finally, the role of MEMS in gas sensing needs little elaboration. Before the discovery of micro-electromechanical structure or MEMS, the gas sensors were mainly ceramic based materials, which have very high power consumption of the order of 200mW-1W due to their excessive thermal mass. Moreover, the response time was also very high. However, with MEMS and proper thermal isolation between the sensor element and the substrate, this power consumption has been scaled down to about only 30-150 mW and the response is also faster in case of MEMS based gas sensors.

In summary, it has been found that there are a variety of possibilities to fabricate micro-machined substrates compatible with thin as well thick film metal oxide layers [60]. The advantages compared with the well-established ceramic sensors are : a reduction of power consumption, a faster thermal time constants, which allow faster temperature modulations, the possibility to integrate the sensitive layer with control and signal evaluation electronic on one chip and easy integration of sensor arrays.

Moreover, with the incorporation of the nano-structured metal oxides, the operating temperature get reduced substantially with respect to microstructured materials. So, it was concluded that MEMS with nano-materials (oxide) on them as sensing media might therefore play an important role to fulfill the demands of the market for low power, cheap and high performance devices.

REFERENCES

1. M. Prassas, J. Phalippou, L. L. Hench and J. Zarzycki, J. Non-Cryst. Solids, 48 (1982) 79.
2. A. N. Goldstein (Ed.), 'Handbook of Nanophase Materials', Marcel Dekker Inc., New York, USA, (1997), (excellent coverage on nano materials).
3. D. W. Hoffman, R. Roy and S. Komarneni, Ceram. Bull., 62 (1983) 375.
4. A. K. Bandyopadhyay, R. Jabra and J. Phalippou, 'Association of OH Groups with Boron and Silicon Atoms in SiO_2—B_2O_3 Glasses by Infra-red Spectroscopy', J. Mater. Sci., 8 (1989) 1464.
5. G.Sinha, K.Adhikary and S.Chaudhuri, 'Variation of optical properties with structure in gallium based nano-crystalline oxide thin films', Proc. Natl. Conf. on Nano-Science and Technology, J. U., Kolkata (India), January, (2005).
6. M. L. Ferrer, F. del Monte and D. Levy., J. Phys. Chem., B 105 (2000) 11076.
7. F. del Monte, M. L. Ferrer and D. Levy, Langmuir, 17 (2001) 4812.
8. F. del Monte, M. L. Ferrer and D. Levy, J. Mater. Chem., 11 (2001) 1745.
9. D. Levy, 'The Encyclopedia of Materials : Science and Technology', Vol. 7, The Optical and Dielectric Properties of Materials : 'Optical materials based on sol-gel technology', Pergamon, Elsevier Science, (2001) pp. 6449.
10. S. Roy, D. Chakravorty and D. L. Agarwal, J. Appl. Phys., 74 (1993) 4746.
11. C. Cannas, D. Gatteschi, A. Musinu, G. Piccaluga and C. Sangregorio, J. Phys. Chem., B 102 (1998) 7721.
12. C. Cannas, A. Musinu and G. Piccaluga, 'Synthesis and Methodologies in Inorganic Chemistry', Eds. S. Daolio, E. Tondello and P.Vigato, (1997) pp. 7.
13. S. Takehuci, T. Isobe and M. Senna, J. Non Cryst. Solids, 194 (1996) 58.
14. H. Gleiter, Prog. Mater. Sci., 33 (1989) 223
15. K. Baba, N. Shohata and M. Yonezawa, Appl. Phys. Lett., 54 (1989) 2309.

16. H. Nalwa (Ed.), 'Handbook of Nanostructured Materials and Nanotechnology', Academic Press, San Diego, USA, (2000), (the most exhaustive book on nano materials).

17. Xu Lianbin, Chem. Commun., 43 (2000) 997.

18. E. V. Charnaya, J. Phys. Cond. Matt., 10 (1998) 7273, and A.G. Sergey, J. Apple., 12 (1994) 6671.

19. G. Byron, Adv. Mater., 12 (2001) 1605.

20. V. Kozhevnikov, ETOPIM-6 (2002) Conference Proceedings.

21. A.K. Bandyopadhyay, P.C. Ray and V. Gopalan, J. Phys. — Cond. Matter, 18 (2006) 4093.

22. Y. K. Park and H. Kustal, Military and Aerospace Electronics, July (2005), and Comms Design.com in the Internet.

23. R. Lopez, 'Photoinduced phase transition in VO_2 nano crystals : Ultrafast control of surface-plasmon resonance', Optics Letters, 30 (2005) 558.

24. H. Ohno, Science, 281 (1998) 951.

25. F. Matsukara, H. Ohno, A. Shen and Y. Sugawara, Phys. Rev. B, 57 (1998) R2037.

26. J. M. Kikkawa and D. D. Awashlom, Nature, 397 (1999) 139.

27. D. D. Awashlom and N. Samarth, J. Magn. Magn. Mater., 200 (1999) 130.

28. I. Malajovich, J. J. Berry, N. Samarth and D. D. Awashlom, Nature, 411 (2001) 770.

29. A. K. Bandyopadhyay and P. C. Ray, J. Appl. Phys., 95 (2004) 226.

30. W. W. Keller et al, Appl. Phys. Lett., 60 (1996) 1808.

31. Makoto Saen et al, I E I C E Trans. Electron, E 81 - C (1998) 1.

32. A. K. Biswas, A. A. Sheikh, S. K. Sarkar, M. K. Naskar, S. K. Sarkar and S. S. Chowdhury, 'Nano and giga challenges in microelectronics for high density integrated circuits', Proc. Natl. Conf. on Nano-Science and Technology, J. U., Kolkata (India), January, (2005).

33. K. Das, S. Maikap , Je-Hun Lee, M. NandaGoswami, R. Mahapatra, G. S. Kar, A. Dhar, H.N. Acharya and S. K. Ray, Appl. Phys. Lettr., 84 (2004) 1386.

34. K. Das, S. Maikap, A. Dhar, B. K. Mathur and S. K. Ray, Electronics Lettr., (2003) 1865.

35. S. K. Ray, T. N. Adam, R. T. Troeger, J. Kolodzey, G. Looney and A. Rosen, J. Appl. Phys., 95 (2004) 5301.

36. S. K. Ray et al., Proc. EMRS 2004 Spring Meeting, Strasbourg (France), May (2004).

37 T N Adam, R. T. Troeger, S. K. Ray, P. C. Live and J. Kolodzey, Appl. Phys. Lettr., 83 (2003) 1713.

38. S. K. Ray, 'Si/SiGe heterostructures for nanoelectronic devices', Proc. Natl. Conf. on Nano-Science and Technology, J. U., Kolkata (India), January, (2005).

39. S. E. Lyshevski, 'Micro and nano-Electromechanical systems: Fundamentals of Micro and nano-engineering', CRC Press, Boca Raton, Fl, (2000).

40. S. A. Campbell, 'The Science and Engineering of Microelectronic Fabrication', Oxford University Press, New York, (2001).

41. T. Toriyama, Y. Tanomoto and S.Sugiyama, 'Single crystal silicon nanowires piezoresistors for mechanical sensors', I E E E Journal of MEMS, 11 (2002) 605.

42. P. Vettiger, G. Cross, M. Despont, U. Drechsler, U. Drurig, W. Haberle, M. I. Lutwyche, H.E. Rothuizen, R. Stytz, R. Widmer and G. K. Binnig, 'The millipede more than 1000 tips for parallel and dense data storage', Dig. Tech. Papers, 11th International Conference on Solid State Sensors and Actuators (Transducers), (2001) pp.1054.

43. P. M. Fauchet, J. von Behren, K. D. Hirschman, L. Tsybeskov, S. P. Duttagupta, 'Porous silicon physics and device applications : A Status Report', Phys..Stat.Sol (a), 165 (1998) 3.

44. C.Pramanik, H.Saha, 'Enhanced pressure sensitivity of nanocrystalline porous silicon', Proc. Natl. Conf. on Nano-Science and Technology, J. U., Kolkata (India), January, (2005).

45. D. Majumdar, S. Chatterjee, M. Dhar, S. K. Dutta and H. Saha , 'Light trapping in layer-transferred quasi-monocrystalline porous silicon solar cell', Solar Energy Materials & Solar Cells, 77 (2003) 51.

46. M. Banerjee, D. Majumdar, S. K. Dutta, S. M. Hossain and H. Saha, 'Quasi-monocrystalline porous silicon : a promising active material for solar cell', 12th IWPSD, IIT Chennai, Dec., (2003).

47. R. Brendel and H. Seel, 'Optical absorption in crystalline Si films containing spherical voids for internal light scattering', Thin Solid Films, 451-452 (2004) 608.

48. M. Banerjee, S. K. Datta and H. Saha, 'Enhanced optical absorption in QMPS layer with nano voids', Proc. Natl. Conf. on Nano-Science and Technology, J. U., Kolkata (India), January, (2005).

49. S. M. Sze, 'Semiconductor Sensors', John Wiley and Sons Inc, New York, (1994).

50. G. Sberveglieri, Sensors and Actuators B, 23 (1995) 103.

51. N. Yamazoe, Sensors and Actuators B, 5 (1991) 7.

52. V. Lantto and P. Romppainen, Surf. Sci., 192 (1987) 243.

53. J. M. Bustillo, R. T. Howe, and R. S. Muller, "Surface micro-machining for micro-electro-mechanical systems, 'Proc. I E E E', 86 (1998) 1552.

54. W. Menz, J. Mohr, and O. Paul, 'Microsystems Technology', Wiley-VCH, Berlin, Germany, (2001).

55. G. T. A. Kovacs, 'Micromachined Transducers', MCB McGraw Hill, New York, (1998).

56. M. Madou, 'Fundamentals of Microfabrication', CRC Press, Boca Raton, CRC, Fl., (1997).

57. J. W. Gardner, V. K. Varadan, and O. O. Awadelkin, 'Microsensors, MEMS and Smart Devices', Wiley, New York, (2001).

58. M. Gad-el-Hak, 'The MEMS Handbook, CRC Press, Boca Raton, Fl., (2002).

59. See other references in (51).

60. P. Bhattacharya, R. Adhikari, H. Saha and S. Basu, 'Role of nano-structured materials for MEMS based gas sensor', Proc. Natl. Conf. on Nano-Science and Technology, J. U., Kolkata (India), January, (2005).

Index

NEW AGE

BEST SELLING BOOKS ON MECHANICAL ENGINEERING

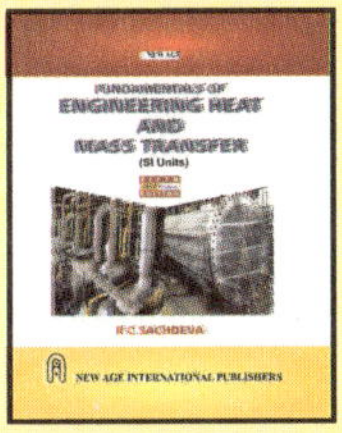

ISBN: 978-93-86070-96-8
PB Price: ₹499.00

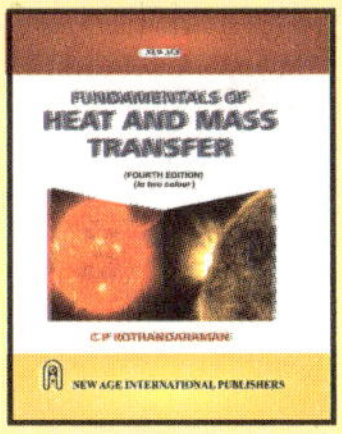

ISBN: 978-81-224-3364-7
PB Price: ₹499.00

ISBN: 978-93-86649-30-0
PB Price: ₹225.00

ISBN: 978-81-224-3889-5
PB Price: ₹99.00

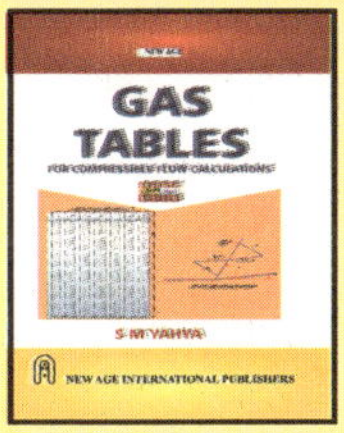

ISBN: 978-81-224-3952-6
PB Price: ₹199.00

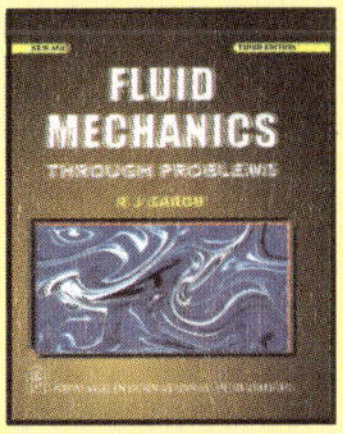

ISBN: 978-81-224-3016-5
PB Price: ₹399.00

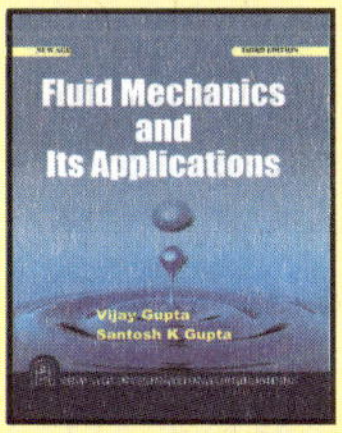

ISBN: 978-81-224-3997-7
PB Price: ₹350.00

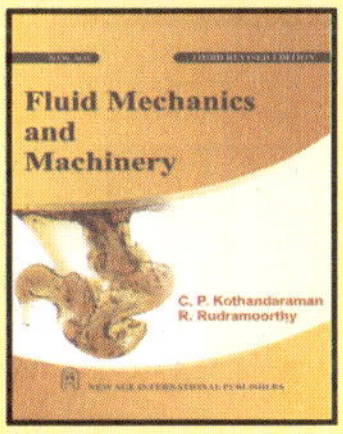

ISBN: 978-81-224-3398-2
PB Price: ₹399.00

ISBN: 978-81-224-2662-5
HB Price: ₹899.00

ISBN: 978-93-86649-91-1
PB Price: ₹499.00

ISBN: 978-93-86418-84-5
PB Price: ₹499.00

ISBN: 978-81-224-2761-5
PB Price: ₹299.00

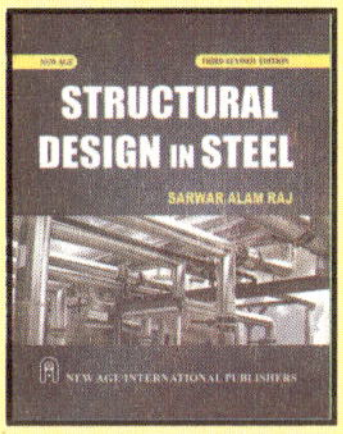

ISBN: 978-93-86418-21-0
PB Price: ₹499.00

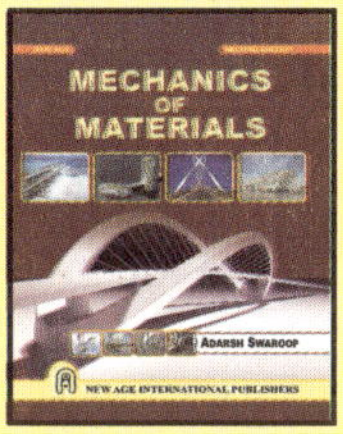

ISBN: 978-81-224-3605-1
PB Price: ₹499.00

ISBN: 978-81-224-2809-4
HB Price: ₹1250.00

ISBN: 978-81-224-2289-4
PB Price: ₹350.00

ISBN: 978-93-86286-76-5
HB Price: ₹499.00

ISBN: 978-93-86286-02-4
PB Price: ₹450.00

ISBN: 978-81-224-3968-7
PB Price: ₹180.00

ISBN: 978-81-224-3015-8
PB Price: ₹475.00

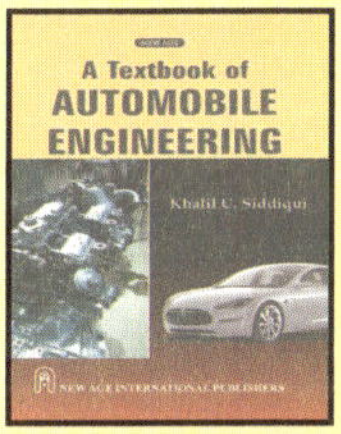

ISBN: 978-81-224-3072-1
PB Price: ₹250.00

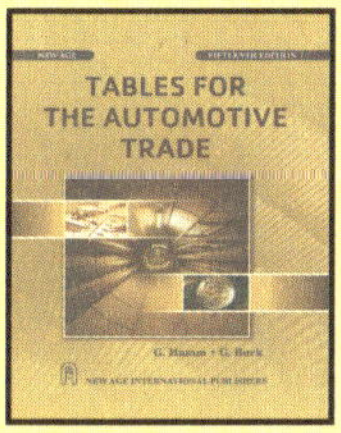

ISBN: 978-81-224-3412-5
PB Price: ₹199.00

ISBN: 978-81-224-1782-1
PB Price: ₹299.00

ISBN: 978-81-224-2666-3
HB Price: ₹795.00

ISBN: 978-81-224-2230-6
HB Price: ₹975.00

Publishing Globally

NEW AGE INTERNATIONAL (P) LIMITED, PUBLISHERS

22, Golden House, Ansari Road, Daryaganj, New Delhi-110 002 (India) • Tel.: +91 (11) 23262370, 23262368 • Fax : 43551305
E.mail: sales@newagepublishers.com Visit us at www.newagepublishers.com • Bangalore (080) 26756823 • Chennai (044) 24353401
Cochin (0484) 2377303 • Guwahati (0361) 2513881 • Hyderabad (040) 24652456 • Kolkata (033) 22273773 • Lucknow (0522) 2209578 • Mumbai (022) 24927869

NEW AGE
Publishing Globally
BEST SELLING BOOKS ON MECHANICAL ENGINEERING
Engineering Thermodynamics
ISBN: 978-81-224-1750-0
PB Price: ₹299.00
APPLIED THERMODYNAMICS
ONKAR SINGH
ISBN: 978-81-224-3686-0
PB Price: ₹495.00
REFRIGERANT TABLES AND CHARTS including Air Conditioning Data
ISBN: 978-81-224-3873-4
PB Price: ₹125.00
REFRIGERATION AND AIRCONDITIONING DATA BOOK
ISBN: 978-81-224-3503-0
PB Price: ₹150.00
REFRIGERATION AND AIR CONDITIONING
Manohar Prasad
ISBN: 978-81-224-3694-5
PB Price: ₹399.00
Fundamentals of METAL FORMING PROCESSES
B L Juneja
ISBN: 978-81-224-3089-9
PB Price: ₹375.00
MANUFACTURING SCIENCE AND TECHNOLOGY
ISBN: 978-81-224-2675-5
PB Price: ₹325.00
Principles of ENGINEERING METALLURGY
ISBN: 978-81-224-2202-3
PB Price: ₹195.00
ADVANCES IN MANUFACTURING ENGINEERING AND TECHNOLOGY
M ADITHAN
ISBN: 978-81-224-2674-8
PB Price: ₹695.00
NON-DESTRUCTIVE TESTING TECHNIQUES
FIRST REVISED EDITION
RAVI PRAKASH
ISBN: 978-81-224-2588-8
HB Price: ₹499.00
OPTIMIZATION TECHNIQUES
Chander Mohan • Kusum Deep
ISBN: 978-81-224-2385-3
PB Price: ₹499.00
FUNDAMENTALS OF OPTIMUM DESIGN IN ENGINEERING
(SECOND EDITION)
S S BHAVIKATTI
ISBN: 978-93-86286-10-9
HB Price: ₹699.00
STRATEGIC COST ANALYSIS FOR PROJECT MANAGERS AND ENGINEERS
ISBN: 978-81-224-2837-7
HB Price: ₹395.00
ALL ABOUT MACHINE TOOLS
HEINRICH GERLING
ISBN: 978-81-224-1826-2
PB Price: ₹199.00
WESTERMANN TABLES
For the Metal Trade
ISBN: 978-81-224-1730-2
PB Price: ₹199.00
ENGINEERING MECHANICS
S S BHAVIKATTI
ISBN: 978-93-86418-00-5
PB Price: ₹450.00
ENGINEERING MECHANICS
H M RAGHUNATH
ISBN: 978-81-224-4037-9
PB Price: ₹299.00
ENGINEERING MECHANICS
ISBN: 978-81-224-2396-9
PB Price: ₹295.00
INTRODUCTION TO MECHANICAL ENGINEERING
ISBN: 978-81-224-2178-1
PB Price: ₹295.00
Problems and Solutions in Mechanical Engineering
ISBN: 978-81-224-2100-2
PB Price: ₹250.00
MATLAB FOR MECHANICAL ENGINEERS
RAO V DUKKIPATI
ISBN: 978-81-224-2270-2
HB Price: ₹899.00
ENGINEERING MECHANICS PROBLEMS with MATLAB
ISBN: 978-81-224-2601-4
HB Price: ₹695.00
MACHINE DRAWING
ISBN: 978-81-224-4054-6
PB Price: ₹399.00
CNC MACHINES
ISBN: 978-81-224-3426-2
PB Price: ₹275.00
CAD/CAM/CIM
P RADHAKRISHNAN
S SUBRAMANYAN
V RAJU
ISBN: 978-81-224-3980-9
PB Price: ₹475.00
Contact your nearest bookseller to buy these books. If you are unable to get it, you could send us Demand Draft in favour of NEW AGE INTERNATIONAL (P) LTD., payable at New Delhi with the total amount of book value at the address below:
FREE SHIPPING IN INDIA
NEW AGE INTERNATIONAL (P) LIMITED, PUBLISHERS
7/30 A, Daryaganj, New Delhi-110002
Visit us at www.newagepublishers.com
(CIN: U74899DL1966PTC004618)